The Circular Economy
Meeting Sustainable Development Goals

ISSUES IN ENVIRONMENTAL SCIENCE AND TECHNOLOGY

How to obtain future titles on publication:

A standing order plan is available for this series. A standing order will bring delivery of each new volume immediately on publication.

For further information please contact:
Sales and Customer Care, Royal Society of Chemistry, Thomas Graham House, Science Park, Milton Road, Cambridge, CB4 0WF, UK
Telephone: +44 (0)1223 432360, Fax: +44 (0)1223 426017, Email: booksales@rsc.org
Visit our website at books.rsc.org

ISSUES IN ENVIRONMENTAL SCIENCE AND TECHNOLOGY

EDITORS: SADHAN KUMAR GHOSH AND GEV EDULJEE

51

The Circular Economy
Meeting Sustainable Development Goals

ROYAL SOCIETY
OF **CHEMISTRY**

Issues in Environmental Science and Technology No. 51

Print ISBN: 978-1-83767-069-7
PDF ISBN: 978-1-83767-198-4
EPUB ISBN: 978-1-83767-199-1
Print ISSN: 1350-7583
Electronic ISSN: 1465-1874

A catalogue record for this book is available from the British Library

The Royal Society of Chemistry is a charity, registered in England and Wales, Number 207890, and a company incorporated in England by Royal Charter (Registered No. RC000524), registered office: Burlington House, Piccadilly, London W1J 0BA, UK, Telephone: +44 (0) 20 7437 8656.

For further information see our website at www.rsc.org

Preface

The concept of the circular economy as a system that retains value by deploying regenerative processes such as reuse, recycling, recovery and remanufacturing is now well entrenched in both policy and practical terms. Its initial focus on mimicking nature's biological cycle by closing the waste loop has broadened, incorporating other resource efficiency measures within the paradigm such as minimising the use of primary raw materials, designing products that are more readily reused, recycled or remanufactured, and moving from the outright ownership of products to acquiring its functionality as a service – from car rentals to the management of office space. Circular economy principles, or more generally the concept of "circularity", has now been applied to a wide range of activities and sectors: cities and the built environment, manufacturing processes and the agricultural sector being some examples.

In practical terms implementation of circular economy precepts has been patchy. The linear take–make–dispose model of production and consumption still dominates the global economy, despite many examples of good practice: the high rate of reuse and recycling of consumer products in Low- and Middle-income Countries (LMICs) and the adoption of new business models by the corporate sector being but two. Nevertheless, as a general rule, increasing global wealth has gone hand in hand with increasing consumption, profligacy and wastefulness, putting greater pressure on natural resources and despoiling the environment. The challenge is to reverse this trend by decoupling economic growth from its downsides – in essence, to achieve sustainable growth, wealth creation and improvements in the lives and livelihoods of citizens by (among other measures) transitioning to a circular economy.

This leads to the related theme of sustainable development. Although not originally conceived as formally connected as a theoretical construct, the two

Issues in Environmental Science and Technology No. 51
The Circular Economy: Meeting Sustainable Development Goals
Edited by Sadhan Kumar Ghosh and Gev Eduljee
© The Royal Society of Chemistry 2024
Published by the Royal Society of Chemistry, www.rsc.org

concepts overlap in respect of their overall aspirations and goals. The circular economy most obviously engages directly with the economic pillar of sustainable development, and although it seemingly engages only indirectly with the latter's social and environmental pillars, it is now the dominant view of policymakers and of practitioners that the economy, and implementing circular economy-informed projects in particular, can be legitimised only if all three pillars of sustainable development are respected. In other words, transitioning to a circular economy is expected to, and should be designed to, deliver economic as well as environmental and social benefits – examples being tackling marine pollution and creating jobs. Under this interpretation the circular economy contributes to several of the United Nations' sustainable development goals and their related targets, Goal 11 (Sustainable cities and communities) and Goal 12 (Responsible production and consumption) being direct touchpoints, with the social and environmental aspects of others such as Goals 2 (Zero hunger), 3 (Good health and wellbeing), 4 (Quality Education), 5 (Gender equality), 6 (Clean water and sanitation), 7 (Affordable and clean energy), 8 (Decent work and economic growth), 9 (Industry, innovation and infrastructure), 13 (Climate action), 14 (Life below water) and 15 (Life on land) being indirect examples.

The purpose of this book is to explore in eleven chapters how circular economy principles and their applications relate to the sustainable development goals under scenarios covering a range of activities and sectors, and the extent to which the circular economy can contribute towards achieving these goals. Chapters 1 and 2 (by Sadhan Kumar Ghosh of the Sustainable Development & Circular Economy Research Centre and Gev Eduljee of Resource Futures, respectively) present an overview of the concepts, their adoption in policy, legislation and international standards, and the current state of play with respect to their implementation. In Chapter 3 Mike Webster of Systemiq examines the efficacy of the circular economy in LMICs, emphasising the need for a pragmatic approach that recognises the particularities of their developmental needs.

Chapter 4, by Ann Stevenson of Resource Futures, discusses the challenges faced by Small and Medium-sized Enterprises (SMEs) in manufacturing. Social, cultural and political dynamics define how they perceive risks in engaging or not in actions to transition to a circular economy. In Chapter 5 Deborah Andrews of London South Bank University and Beth Whitehead of Operational Intelligence explore the role of design in the value chain focusing on the electronics, textiles, furniture and construction sectors to identify the potential for circularity and the extent to which design can influence and contribute to circular practice and sustainable development goals. In Chapter 6 Hyeong-Jin Choi and Seung-Whee Rhee of Kyonggi University study the role and responsibility of consumers in a circular economy society in conserving resources and consuming environmentally friendly products, identifying six success factors: clear policy targets, allocation of responsibilities by stakeholders, awareness, transparency of information, communication and building infrastructure. Chapter 7 by Fabiula D. B. de Sousa of Universidade Federal de Pelotas examines the literature on the role of plastics in achieving the UN sustainable development goals. Although contributing to the achievement of the majority of the goals, marine pollution caused by plastics outweighs the totality of their positive contribution. Applying circular economy strategies such as reduction, recycling and service life extension was fundamental to helping solve the socio-environmental problems that plastics may cause.

Chapters 8 (Peter Vangsbo of Arup Denmark) and 9 (Purva Mhatre-Shah and Amos Ncube of EarthShift Global) are complementary. Chapter 8 on circularity and sustainable cities emphasises the need for the entirety of the city-wide supply chain to work together to implement circular economy actions, highlighting trends that should be embedded in future city planning. Chapter 9 on construction and the built environment identifies the circular economy as having a direct positive impact on ten of the seventeen sustainable development goals. Supporting initiatives include stakeholder coordination, capacity building and knowledge sharing, using tools such as material flow analysis or material stock assessment for traceability information on resources and deploying AI or blockchain technology for resource modelling.

Chapter 10 by Zoë Lenkiewicz of The Global Waste Lab Circular focuses on the management of biowaste and its contribution to the UN sustainable development goals. Following circular economy priorities, the chapter discusses the reduction of food waste, recovery processes to produce beneficial materials such as compost, biochar and biogas, and problems such as groundwater pollution and disease vectors associated with the mismanagement of biowaste. Finally, Chapter 11 by Dominika Ptach, Deborah Andrews and Simon P. Philbin of London South Bank University explore sustainability and circularity in the data centre industry, a sector that has seen recent rapid growth. The authors identify impacts that include high energy use with concomitant emissions, resource depletion, critical raw materials extraction and unethical labour practices, and review opportunities for the sector to contribute to the sustainable development goals.

The chapter contributors are drawn from a range of backgrounds and disciplines, reflecting the broad sectoral and international applicability of the circular economy and of sustainable development, and the challenges and opportunities for their enhancement. Our intention is for the book to provide a source of information for the benefit of researchers, managers in the corporate and public sectors, policymakers, aid agencies and practitioners implementing strategies and projects on the ground, as well as contributing to study material and supplementary reading for environmental science, sustainable development, waste management and circular economy courses.

Sadhan Kumar Ghosh and Gev Eduljee

Contents

Issues in Environmental Science and Technology No. 51
The Circular Economy: Meeting Sustainable Development Goals
Edited by Sadhan Kumar Ghosh and Gev Eduljee
© The Royal Society of Chemistry 2024
Published by the Royal Society of Chemistry, www.rsc.org

Chapter 2   Sustainable Development and the Circular Economy: Concepts, Progress and Prospects   29
Gev Eduljee

Chapter 3   The Circular Economy in Low- and Middle-income Countries – A Tool for Sustainable Development?   65
Mike Webster

Circular Economy and Sustainable Development Goals: Policy, Legislation and ISO Standards

SADHAN KUMAR GHOSH*

Sustainable Development & Circular Economy Research Centre,
International Society of Waste Management Air and Water,
29/6 J. N. Ukil Road, Kolkata 700041, India
*E-mail: sadhankghosh9@gmail.com

1.1   Introduction

Depletion of natural resources and nature loss are perceived not only as a moral or an ecological issue, but also as economic, developmental, human health and justice issues, as the most vulnerable populations are the most affected. It is also an intergenerational justice issue, as we are leaving a complex legacy to our children, their children and future generations to come. The rate of extraction of natural resources has been increasing day by day, creating problems for all living creatures, including humans. In the words of the Secretary-General of the Organisation for Economic Co-operation and Development (OECD):[1]

Growth in materials use, coupled with the environmental consequences of material extraction, processing and waste, is likely to increase the pressure

Issues in Environmental Science and Technology No. 51
The Circular Economy: Meeting Sustainable Development Goals
Edited by Sadhan Kumar Ghosh and Gev Eduljee
© The Royal Society of Chemistry 2024
Published by the Royal Society of Chemistry, www.rsc.org

on the resource bases of our economies and jeopardise future gains in well-being. This Outlook can help decision makers understand the direction in which we are heading and help to assess which policies can support a more circular economy.

The tension between the state of the planet, ever-growing human and consumption pressures, ever-increasing rates of extraction of natural resources, and our awareness and readiness to respond, are a sign of a society in transition, a society at its greatest fork in the road, a turning point, and facing its deepest system change challenge around what is perhaps the most existential of all our relationships: the one with nature. The nexus between water–energy–food has become the focus area globally with huge wastage of these resources accompanied by scarcity in several parts of the world. Global population and economic growth have increased the demand for resources, energy and food, resulting in mass production and mass consumption. This generates vast amounts of waste, exacerbating global environmental challenges, in particular climate change, resource depletion and marine pollution.

Chapter 1 discusses two key concepts expressed in the title of this book: sustainable development and the circular economy. Complementary to Chapter 2, this chapter focuses on issues relating to implementation: policy, legislation and standards developed by the International Organization for Standardization (ISO). Following a discussion on some of the pressures on the environment, in particular the consequences of resource depletion and resource wastage, policy implementation supported by legislation is discussed as it has been applied in a selection of countries, and in the European Union. Finally, standards developed by the ISO and relating to the circular economy are presented.

1.2 Resource Depletion, Waste Generation and Environmental and Developmental Pressures

According to the latest United Nations Population Fund (UNFPA) white paper, the world's population is now 7.954 billion and it is expected to surpass 9.7 billion by 2050.[2] The World Bank estimates that this will be accompanied by a significant increase in the world's annual waste volume, from 2.01 billion tonnes today to 3.4 billion tonnes in 2050.[3] Because of disparities in economic growth, daily *per capita* waste generation in high-income countries is projected to increase by 19% by 2050, compared to >40% in Low- and Middle-income Countries (LMICs), with greenhouse gas emissions increasing from 1.6 BT CO_2eq in 2016 (5% of global emissions) 2.6 BT CO_2eq by 2050 if current waste management practices remain unchanged.

Waste generation is a consequence of increased raw materials' extraction, transformation into goods and products, their use by consumers and their subsequent discard, activities that rise with population growth coupled with economic growth, increased prosperity and increased purchasing power. Global material use has tripled over the past four decades, with annual global extraction of materials growing from 22 billion tonnes (1970) to 70 billion tonnes (2010).[4] Global material use has been accelerating. Material extraction *per capita* increased from 7 to 10 tonnes from 1970 to 2010, indicating improvements in the material standard of living in many parts of the world.

The report identified the large gaps in material standards of living that exist between North America and Europe and all other world regions. Annual *per capita* material footprint for the Asia Pacific region, Latin America, the Caribbean and West Asia is between 9 and 10 tonnes, or half that of Europe and North America, which is about 20 to 25 tonnes per person. In contrast, Africa has an average material footprint of below 3 tonnes *per capita*.[4]

Climate change, resource depletion, rapid and unplanned urban growth and delivering adequate and sustainable housing are amongst the greatest challenges of the 21st century. To illustrate the latter challenge, countries such as India instituted policies and schemes to build houses for its homeless citizens through the Pradhan Mantri (Prime Minister) Awas Yojana (Urban) Mission launched in 2015 (also see Section 1.4.3). The mission provided central assistance to stateside implementing agencies for providing houses to all eligible families/beneficiaries from the urban poor with a target of building 20 million affordable houses by 31 March 2022. Clearly the call on building and other construction materials was considerable, leading to the need for circularity in construction industries by using recycled, repurposed products and demolition materials.

1.3 Sustainable Development Goals and the Circular Economy

1.3.1 Sustainable Development

In 1962 American biologist Rachel Carson, with her book *Silent Spring*, alerted the world to the environmental impact of chemical pesticides, inspiring the global environmental movement. *Silent Spring* is widely considered as one of the most important environmental books of the 20th century. In January 1969 the Santa Barbara oil spill released over 15 million litres of oil onto the California coast, killing thousands of animals. This disaster, which was the worst oil spill until that time, captured global attention. It became another alarm for the world to consider humans' responsibility for conserving the environment. This led to the first Earth Day (22 April 1970), when 20 million people got together to celebrate our planet. Calling attention to a model of unsustainable economic growth, the UN established the Brundtland Commission in 1983, from which the seminal report of 1987 resulted (see Chapter 2), defining the framework for sustainable development.

Approved by the United Nations General Assembly (UNGA) as part of the September 2015 Resolution A/RES/70/1 *Transforming Our World: The 2030*

Agenda for Sustainable Development,[5] SDGs are an outcome of the intergovernmental process initiated at the United Nations Conference on Sustainable Development in Rio de Janeiro in June 2012.[6] In September 2015 all member states of the United Nations agreed to adopt the 2030 Agenda. It is a universal action plan for global cooperation on sustainable development for the period 2015 to 2030. Under the commitment "leave no one behind", the 2030 Agenda defined 17 Sustainable Development Goals (SDGs) with 169 targets. These are discussed in Chapter 2; the SDGs are summarised here in Table 1.1 for convenience.

SDGs have a strong focus on the means of implementation that include finance, capacity building, trade, policy, institutional coherence, multistakeholder partnerships, data, monitoring and accountability, public governance and technology.[7] A number of targets under SDG 16 and SDG 17 also focus directly on the means of implementation, which "are key to realizing our Agenda and are of equal importance with the other Goals and targets".[5] The SDG framework emphasises national planning and regular progress reviews on the national level, complemented by voluntary reviews through the High-level

Table 1.1   UN General assembly sustainable development goals.

No	Sustainable development goal
1	End poverty in all its forms everywhere
2	End hunger, achieve food security and improved nutrition and promote sustainable agriculture
3	Ensure healthy lives and promote well-being for all at all ages
4	Ensure inclusive and equitable quality education and promote lifelong learning opportunities for all
5	Achieve gender equality and empower all women and girls
6	Ensure availability and sustainable management of water and sanitation for all
7	Ensure access to affordable, reliable, sustainable and modern energy for all
8	Promote sustained, inclusive and sustainable economic growth, full and productive employment and decent work for all
9	Build resilient infrastructure, promote inclusive and sustainable industrialization and foster innovation
10	Reduce inequality within and among countries
11	Make cities and human settlements inclusive, safe, resilient and sustainable
12	Ensure sustainable consumption and production patterns
13	Take urgent action to combat climate change and its impacts
14	Conserve and sustainably use the oceans, seas and marine resources for sustainable development
15	Protect, restore and promote sustainable use of terrestrial ecosystems, sustainably manage forests, combat desertification, and halt and reverse land degradation and halt biodiversity loss
16	Promote peaceful and inclusive societies for sustainable development, provide access to justice for all and build effective, accountable and inclusive institutions at all levels
17	Strengthen the means of implementation and revitalise the Global Partnership for Sustainable Development

Political Forum on Sustainable Development (HLPF), a central platform for overseeing the follow-up and review of SDGs on a global level under the auspices of the UNGA and the Economic and Social Council (ECOSOC).[5] According to the latest HLPF review, many countries "nationalized ... targets for the 2030 Agenda in their national strategies and plans, including financing strategies and institutional mechanisms".[8] National legislation, policy frameworks and targets are critical for understanding and realising the SDGs in each country's own perspective for their people, enablers and actors.

1.3.2 The Circular Economy

The circular economy and the so-called 3Rs (Reduce, Reuse, Recycle) are very important concepts that have gained increasing prominence as a tool that presents solutions to some of the world's most pressing cross-cutting sustainable development challenges. The circular economy addresses the root causes, in which waste and pollution do not exist by design, products and materials are kept in use, and natural systems are regenerated, providing much promise to accelerate implementation of the 2030 Agenda. Given that massive changes to the way our societies and businesses are organised are essential to transition to a sustainable future, the circular economy holds particular promise for achieving the SDGs, including SDGs 6 on energy, 8 on economic growth, 11 on sustainable cities, 12 on sustainable consumption and production, 13 on climate change, 14 on oceans and 15 on life on land. A study reported that a circular economy development path could halve carbon dioxide emissions by 2030, relative to today's levels, across mobility, food systems and the built environment.[9,10]

The circular economy has deep-rooted origins, conceptually characterised as cradle-to-cradle.[11] The design philosophy behind the concept is to consider all materials involved in industrial and commercial processes to be nutrients, in which two main categories are the actors: (1) technical and (2) biological. Its practical applications have gained momentum from the late 1970s and early 1980s to modern economic systems and industrial processes.

Ghosh has defined the circular economy as:[12]

> ... *a systems-level approach to economic development and a paradigm shift from the traditional concept of linear economy model of extract-produce-consume-dispose-deplete (EPCD2) to an elevated echelon of achieving zero waste by resource conservation through changed concept of design of production processes and materials selection for higher life cycle, conservation of all kinds of resources, material and/or energy recovery all through the processes, and at the end of the life cycle for a specific use of the product will be still fit to be utilised as the input materials to a new production process in the value chain with a close loop materials cycles that improves resource efficiency, resource productivity, benefit businesses and the society, creates employment opportunities and provides environmental sustainability.*

The concept of EPCD2 is characterised as the traditional linear economy based on a model of mass production, mass consumption and mass disposal. The transformation from a linear to a circular economy has gone hand in hand with the acceptance for the need to ensure sustainable economic growth. For example, since the early 2000s, Japan has been advancing the 3Rs ahead of the rest of the world and has been making progress in a steady manner, reducing the amount of final disposal and improving recycling rates.

The correlation between SDGs and the circular economy is discussed in Chapter 2. Natural resource management is directly tied to at least 12 of the 17 SDGs, according to the International Resource Panel.[4] Concentration on natural resources to meet the Paris Climate Agreement commitments is so important that in its absence compliance to those commitments will be impossible. More efficient practices could cut greenhouse gas emissions by 60% by 2050.[13] Effective management of natural resources is also a key component of poverty eradication and resilient economic growth.

The circular economy model offers a new opportunity for innovation and integration among natural ecosystems, reengineering businesses, our daily lives and society as well as resource and waste management. For maximum traction, the circular model of resource management should be defined in a holistic manner that is internationally accepted.

1.3.3   The Role of Policy Support, Legislation and Standardisation

Currently available technologies can help to better manage energy demand, improve efficiency and support both low-carbon economic growth and job creation. But progress in the efficient deployment of these technologies faces a number of challenges, including non-technical barriers such as a lack of policy support, perceived and real project risk, lack of financial support for LMICs, lack of access to low-cost capital, long permitting timelines, community pushback, rebound effects and supply chain challenges.

While effective waste management may lead to the production of recycled products, there is generally a low awareness about the quality and characteristic of these products. The quality of recycled products depends on the competence and confidence levels of the supply chain of the waste starting from the source through to the use, disposal and resource recirculation stages. There should be standardisation of the recycling processes and recycled products. The effectiveness of the circular economy and the SDGs will depend to a great extent on whether national or international standards for a specific operation and products are available that can build customer and stakeholder confidence. Here the role of national standard bodies and the ISO is critical. It is noteworthy that the ISO technical committee 323 (ISO/TC 323) (Circular Economy) was formed in 2019 to develop standards relating to the circular economy (see Section 1.6).

1.4 Policy and Legislative Support to Achieve SDGs

1.4.1 Policy Tools

The Global Circularity Gap Report in 2019 reported that only 9% of the world economy was circular.[14] Due to the complexity of the sustainable development vision, most often its implementation needs to be supported by innovation designers, intermediaries and policymakers. They provide services and designs ushering appropriate radical changes in practices, policies and decision-making tools.[15,16] Although formulated for the agricultural sector, the UN Environment Programme's (UNEP's) compendium of policy tools to support implementation of the SDGs can be generalised as follows:[17]

- financing and technical assistance policies
- land access policies
- recognition of sustainability policies in master plans, urban zoning and instruments for territorial planning and land-use regulation
- policies for more sustainable water use and access
- policies and urban planning for local food production
- policies to strengthen markets for local producers
- policies that support research and data collection
- policy interventions for a circular economy – regulatory frameworks, fiscal frameworks, education, information and awareness creation, public procurement policies and innovation support schemes.

Policymakers should delineate up front the purpose/goal of their policies, to help optimise outcomes and to make explicit any trade-offs.

Because of the global nature of pollution, transnational agreements are often critical to achieve particular SDGs. Such is the case with marine pollution caused by discarded plastics, where studies have shown that plastic pollution is largely a regional issue with global implications.[18] The Osaka Blue Ocean Vision (Figure 1.1), a voluntary commitment of the G20 countries, aims to "reduce the additional pollution by marine plastic litter to zero by 2050 through a comprehensive life-cycle approach", with endorsements from 86 countries and regions (as of January 2021). The Vision addresses SDG 14 (Conserve and sustainably use the oceans, seas and marine resources) and in particular Target 14.1 (By 2025, prevent and significantly reduce marine pollution of all kind, in particular from land-based activities, including marine debris and nutrient pollution).

A compilation of policies relating to plastic pollution indicated that policies at the regional level were largely a European phenomenon (62% of regional policies in the inventory). At a national level, the upward trend in policy responses over the last decade largely reflected new policies introduced solely to address pollution from plastic bags, the instrument used most being a regulatory ban on plastic at some stage in the life cycle. National governments used regulatory instruments more frequently than economic

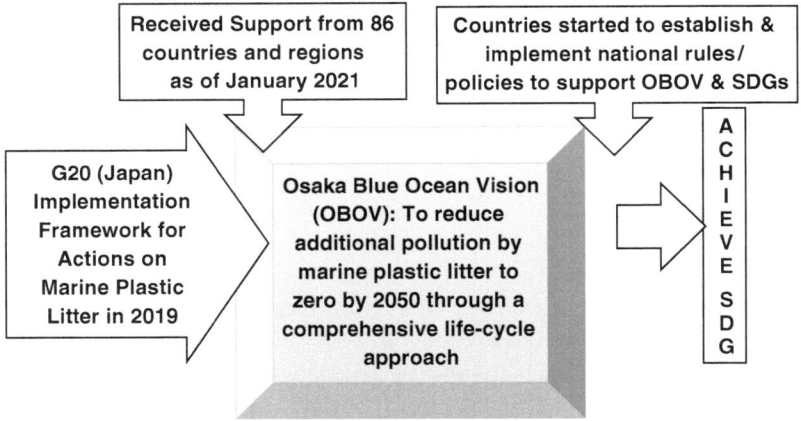

Figure 1.1   Osaka Blue Ocean Vision.

instruments and information and/or behaviour change instruments in the sample analysed.[19]

The adoption of SDG initiatives is illustrated in the following sections in the case of a regional entity (the European Union) and a national entity (India).

1.4.2   Achieving SDGs in the EU

As an active member of the UN, the Brundtland Commission's seminal report on sustainable development (see Chapter 2) was the foundation stone for policy initiatives of the European Communities (EC) prior to 1993, and of the European Union (EU) thereafter.[20] Article B of the Treaty on European Union (1992), otherwise known as the Maastricht Treaty, introduced an objective of the Union to "promote economic and social progress which is balanced and sustainable", with Article 2 undertaking:[21]

> *… to promote throughout the Community a harmonious and balanced development of economic activities, sustainable and non-inflationary growth respecting the environment a high degree of convergence of economic performance, a high level of employment and of social protection, the raising of the standard of living and quality of life, and economic and social cohesion and solidarity among Member States.*

The consolidated version of the Treaty published in 2012 stated at Article 3(3):[22]

> *[The Union] shall work for the sustainable development of Europe based on balanced economic growth and price stability, a highly competitive social market economy, aiming at full employment and social progress, and a high level of protection and improvement of the quality of the environment.*

In 2001 the EU published its first Sustainable Development Strategy (SDS) "to achieve a continuous long-term improvement of quality of life through the creation of sustainable communities able to manage and use resources efficiently, able to tap the ecological and social innovation potential of the economy and in the end able to ensure prosperity, environmental protection and social cohesion".[23] Renewed alongside a relaunch of the Lisbon Strategy in 2005, the SDS was updated in 2006 with the following priority areas for action to be achieved by 2010:

- climate change and clean energy
- sustainable transport
- sustainable consumption and sustainable production
- conservation and management of natural resources
- public health
- social inclusion, demography and migration
- global poverty and sustainable development challenges.

The 2001 SDS introduced an important objective to improve policy coherence, stating that "sustainable development should become the central objective of all sectors and policies", and that "All policies must have sustainable development as their core concern" (original emphasis).

Following a declaration in 2009 by the European Council that "Sustainable development remains a fundamental objective of the European Union" and that the SDS "will continue to provide a long-term vision and constitute the overarching policy framework for all Union policies and strategies", sustainable development as a policy imperative was incorporated in a new strategy termed Europe 2020, intended as a remedial to the 2008 economic crisis.[24] Particular emphasis was placed on combatting climate change and delivering clean and efficient energy. As the title indicates, the strategy extended the EU's policy and delivery horizon for sustainable development from 2010 to 2020.

The UN's adoption of its 2030 Agenda necessitated further reflection and review of the EU's programme for sustainable development, not least because of a new set of SDGs and a new delivery horizon of 2030 (see Chapter 2 for details). Responding to a reflection paper published by the Commission, the incoming presidency launched a revamped growth strategy, the European Green Deal (EGD) in December 2019 while continuing to mainstream the SDGs in EU policies. The EDG focused EU policy on the following themes:

1. increasing the EU's climate ambition for 2030 and 2050
2. supplying clean, affordable and secure energy
3. mobilising industry for a clean and circular economy
4. building and renovating in an energy- and resource-efficient way
5. accelerating the shift to sustainable and smart mobility
6. from farm to fork: designing a fair, healthy and environmentally friendly food system

7. preserving and restoring ecosystems and biodiversity
8. a zero-pollution ambition for a toxic-free environment.

Explicitly described as "the EU's strategy for reaching the 2050 goal [of achieving climate neutrality – net zero by 2050]", the EGD can be viewed as having a predominantly economic and environmental focus, despite its commitment to a "just transition" and for "sustainable competitiveness and social fairness and resilience". To address more fully the broader aspects of the SDGs, supporting policies were proposed, for example on human rights, gender equality, taxation, minimum wage, education and skills. Nevertheless, the general consensus among commentators is that the EU continues to take a piecemeal approach to actioning the SDGs, and that there is a lack of cohesion in its policymaking processes, not least because implementation is left to the discretion of each Member State. The open question is whether it is sufficient for the EU to embrace *sustainability* and *sustainable development* in a general sense in EU policymaking with touchpoints to some of the UN SDGs and their targets, which it undoubtedly does, as evidenced in the Commission's policy initiatives driving the eight themes of the EGD, or whether the EU should create more formal procedures to explicitly incorporate the UN SDGs and their indicators as a package into its policies, albeit tailored to suit regional circumstances (for instance, the EU tracks obesity as opposed to undernourishment under SDG 1).

There have been persistent calls from the Council of the EU (the Council of Ministers), the European Council (setting the overall political direction and priorities of the EU) and the European Parliament for the Commission to develop a comprehensive implementation strategy for the 2030 Agenda and the SDGs. The latest resolution of the European Parliament (June 2022) emphasises the urgency of the task, and "Calls for the Commission to adopt a new high-level EU 2030 Agenda implementation strategy ... building on EU resolutions and policy directives aimed at achieving the SDGs, given that there are fewer than eight years left to achieve the 2030 Agenda ...".[25,26] Included in the strategy should be: (a) a new governance framework, led by a single high-level Commissioner; (b) a revised set of concrete measurable, EU-wide, time-bound targets; (c) an updated monitoring system and indicators bearing in mind the voluntary nature of regional and local SDG reviews; (d) a single financial plan to achieve the EU's SDG objectives, linked to the above targets; and (e) a plan for the EU's SDG diplomacy and international cooperation.

It remains to be seen which direction the Commission will take to address the concerns expressed by its governing bodies.

1.4.3   Achieving SDGs in India

India's national development goals and commitment to "Sabka Saath, Sabka Vikas" (Collective Efforts, Inclusive Growth) chimes with the aims of the SDGs. The National Institution for Transforming India (NITI Aayog) was set up by the Government of India in 2014 with the twin mandate to oversee the adoption and monitoring of the SDGs in the country and promote competitive and cooperative federalism among States and the Union Territories. As reported by NITI Ayog, continuous engagement with the 29 States and 7 Union Territories during the course of construction of the SDG India Index in partnership with United Nations India reinforced the belief that the SDGs can be achieved by India if they are achieved at the subnational levels with appropriate national legislation and policy instruments.[27] In fact, the SDGs, with their focus on equality, inclusion and the principle of "leave no one behind", make the nations responsible to implement the same with significant roles. However, the wide diversities among the Indian states in localising the SDG agenda in their respective development planning–implementation–advocacy–evaluation strategies have created the need to document the varied localisation processes in different States and the lessons learnt, which will help to accelerate the country's drive to achieve the UN 2030 Agenda.

India launched a set of legislations and various schemes following a holistic approach towards its 2030 SDGs. There have been a number of instances where, with the establishment of national rules and policy framework, the implementation of SDGs has proved to have been successful. The effective implementation of the rules and policy instruments are an important factor. India's SDG Index Score ranges between 42 and 69 for States and between 57 and 68 for Union Territories. Kerala and Himachal Pradesh are the front runners amongst the States, with a score of 69; Chandigarh and Puducherry are the front runners with a score of 68 and 65 respectively among the Union Territories surveyed.

A number of policies and legislative initiatives have been established to achieve SDGs, for example:

- Swachh Bharat Mission (Campaign Clean India)
- Beti Bacho Beti Padhao (Save the girl child, educate the girl child)
- Pradhan Mantri Awas Yojana (Communal welfare scheme providing housing in urban areas)
- Smart Cities
- Pradhan Mantri Jan Dhan Yojana (Financial inclusion to ensure access to financial services)
- Deen Dayal Upadhyaya Gram Jyoti Yojana (Providing uninterrupted power to rural India)
- Pradhan Mantri Ujjwala Yojana (Providing LPG connection to women below the poverty line).

The Namami Gange Mission was launched in 2014 as a priority programme with the objectives of effective abatement of pollution, conservation and rejuvenation of the River Ganges – a key policy towards achieving SDG 6. With a budget outlay of INR 2000 million for 2015–2020, major components included sewerage project management, urban and rural sanitation, tackling industrial pollution, water use efficiency and quality improvement, ecosystem conservation and the Clean Ganga Fund, among others.

   The Nationally Determined Contribution (NDC) under the Paris Agreement
in 2015 was taken up by the Government of India. India has set a set of policy
instruments with ambitious targets to be achieved by 2030 to enable further
reduction in greenhouse gas emissions. In 2021 India continued exercising
significant climate leadership at the international stage under the International Solar Alliance (ISA), Coalition for Disaster Resilient Infrastructure
(CDRI) and the Leadership Group for Industry Transition (Lead IT Group). Initiative for sustainable finance have been adopted by various ministries, the
Reserve Bank of India (RBI) and the Securities and Exchange Board of India
(SEBI).

1.5   Policy and Legislative Support to Achieve a Circular Economy

1.5.1   International Perspective

Many countries have adopted the principles of the circular economy as part of
their future strategies. For example, the circular economy is promoted as a
top-down national political objective in India, the Republic of Korea and
China, while, in other areas and countries such as the EU, Japan and the USA,
it is a tool to design bottom-up environmental and waste management policies. The circular economy is a system that requires not only buy-in from governments to provide support with legislation and policies, but also the
involvement of other stakeholders such as industries, academia, plant operators, municipal administration, non-governmental organisations (NGOs) and
government departments.[28] Table 1.2 demonstrates the perspective of circular economy implementation globally.

   Rising interest has been observed in urban agriculture as a means to rebal-
ance urban–rural linkages, bring nature back into the city, address nutritional
insecurity of the urban and peri-urban poor and offer environmental benefits
of lower food miles travelled. These bring opportunities for resource circularity as well as for a circular economy, such as recycling nutrients from food
waste, utilisation of wastes generated from industrial, agricultural and municipal sectors, including wastewater treatment to farms. The signatory cities to
the Milan Urban Food Pact have been developing food action plans, such as
urban agriculture, emerging as a natural focal point with potential to advance
a number of SDGs including sustainable cities and communities (SDG 11),
good health and well-being (SDG 3), reducing inequalities (SDG 10) and zero
hunger (SDG 3), in addition to environmental dimensions related to land,
water and climate. Thus, the local and the national policy frameworks help in
achieving the aims of the SDGs and of the circular economy.

   A discussion is available on legislation pertaining to the implementation of
the circular economy in several countries: Afghanistan, Australia, Bhutan,
Canada, China, EU, Germany, India, Israel, Kenya, Laos, Mauritius, Malaysia,
Nigeria, Norway, Serbia, South Korea, Thailand, UK, USA and Vietnam.[12]

Table 1.2   A global perspective on circular economy implementation.

Regions/cities	EU	Circular Economy Action Plan
	Copenhagen, Paris, London, Amsterdam	Various initiatives
	Scotland	Circular Economy Investment Fund
	Wales	Circular Economy Fund
	Brussels	be.brussels
	Vancouver	2020 Goal
Countries	Cuba	Circular economy activities
	Columbia	EU trade mission
	Uruguay	Foro de Economia Circular
	Morocco	White paper on circular economy
	South Africa	EU dialogue on circular economy
	Rwanda	Working with the World Economic Forum on a circular economy model
	Turkey	European Bank for Reconstruction and Development (EBRD) initiative
	India	Strategy paper on resource efficiency
	China	Circular economy promotional law and 5-year plans UK–China circular economy taskforce
	Japan	Home appliance recycling law
	South Korea	Act on the promotion of saving and recycling of resources
	Vietnam	3R policies
	Laos	Government exploring a circular economy policy
Business coalitions	Europe	Ellen MacArthur Foundation Project Mainstream
	USA	US Business Council Sustainable Development
	UK	Circular Economy Taskforce

Legislation enforcing circular economy implementation in most of the countries in the EU is relatively robust (see following section) and has been in place for an extended period, whereas several countries in Asia and Africa have yet to initiate the same.

1.5.2   Promoting Circular Economy Principles in the EU

As noted in Section 1.4.2, "sustainable consumption and sustainable and production" and "conservation and management of natural resources", both desired outcomes of a circular economy, were recognised as priority areas for action in the EU's 2001 Sustainable Development Strategy, updated in 2006. Indeed, in its Report *On the implementation of the Circular Economy Action Plan*, the EU stated that "The circular economy is now an irreversible, global mega trend".[29] The Circular Economy Action Plan (CEAP), adopted in 2015, aimed to transition the EU's economy from a linear to a circular model through 54 actions and four legislative proposals on waste. With these actions largely completed by and in 2019, the Commission adopted a new CEAP in 2020, with the following objectives:[30]

- make sustainable products the norm in the EU
- focus on sectors that use most resources with high potential for circularity, such as: electronics and ICT, batteries and vehicles, packaging, plastics, textiles, construction and buildings, food, water and nutrients
- ensure less waste: reduce total waste generation and halve the amount of residual (non-recycled) municipal waste by 2030
- empower consumers and public buyers
- make circularity work for people, regions and cities
- lead global efforts on circular economy.

A resolution by the European Parliament in February 2021 required additional measures to achieve a carbon-neutral, environmentally sustainable, toxic-free and fully circular economy by 2050, including tighter recycling rules and binding targets for materials' use and consumption by 2030. The Commission responded with a first package of measures in March 2022 to speed up the transition towards a circular economy by boosting sustainable products, empowering consumers for the green transition (for example by introducing a right to repair), reviewing construction product regulation and creating a strategy on sustainable textiles. New rules on reducing packaging waste, improving packaging design and moving to bio-based plastics were proposed in November 2022.

The 2019 European Green Deal (EGD) referred to in Section 1.4.2 explicitly lists "Mobilising industry for a clean and circular economy" as a priority action. The EU Industrial Strategy of 2020 also emphasises the benefits of a circular economy for improved resource efficiency, reducing dependencies and strengthening the EU's resilience.

In 2021 the Commission prepared an EU Action Plan, *Towards Zero Pollution for Air, Water and Soil* (the Zero Pollution Action Plan (ZPAP)).[31] Described as "a cross-cutting objective contributing to the UN 2030 Agenda for Sustainable Development and complementing the 2050 climate-neutrality goal in synergy with the clean and circular economy and restored biodiversity goals", the aim of ZPAP is that by 2050, "air, water and soil pollution is reduced to levels no longer considered harmful to health and natural ecosystems and that respect the boundaries our planet can cope with, thus creating a toxic-free environment".

Cumulatively, the EU has enacted a range of legislative instruments actioning the circular economy – 10 Communications, 7 Directives and 8 Regulations by mid-2020.[32] However, each Member State transposes EU legislation into its own national legislative systems, and implementation is also a matter for individual Member States. In this regard, researchers have identified a so-called "two-speed Europe" in terms of their transition towards a circular economy. Countries most advanced in this transition include Germany, Belgium, Spain, France, Italy, the Netherlands and the United Kingdom. At a slower pace are countries of Central, Eastern and Southern Europe.[33]

1.5.3 Promoting Circular Economy Principles in the Asia Pacific Region

The popularity and awareness of the circular economy, supported by the implementation strategies of the SDGs, has been growing rapidly in several countries in the Asia Pacific region. Many of the countries in the region have adopted legislative instruments based on circular economy concepts and have been realising the benefits of implementation of those policies and targets. Countries such as Japan, India, China, the Republic of Korea and Australia have established specific legislation and have involved the cities, industries, the government and the academic fraternity in implementation. For example, the Japanese government is committed to transitioning to a circular economy and to becoming carbon neutral by 2050.

1.5.3.1 Promoting Circular Economy Principles in Sri Lanka

In recent years, the Government of Sri Lanka has recognised the importance of transitioning to a circular economy to address developmental challenges and promote sustainable development. This has led to the implementation of legislation, regulations and policies aimed at promoting circular economy practices.[34] Extended Producer Responsibility (EPR) regulations were introduced in 2018.[35] The EPR policy has been effective in managing plastic waste in Sri Lanka. Sri Lanka has also implemented waste management regulations to reduce waste generation and promote resource efficiency – the National Waste Management Policy, established in 2017. Sri Lanka has been able to achieve a recycling rate of around 20%, with plans to increase this to 50% by 2030.[36]

The Government of Sri Lanka has established green product design guidelines to encourage manufacturers to design products that are environmentally friendly, resource-efficient and easily recyclable. These guidelines were introduced in 2019 and apply to a range of products, including packaging materials, electronic goods and textiles.[37]

1.5.3.2 Promoting Circular Economy Principles in Australia

Australia's National Waste Policy Action Plan (NWPAP), agreed in late 2019, setting out 7 targets and 80 action items to implement its 2018 National Waste Policy. The plan was designed around five circular economy principles: avoid waste; improve resource recovery; increase use of recycled material and build demand and markets for recycled products; better management of material flows to benefit human health, the environment and the economy; and improve information to support innovation, guide investment and enable informed consumer decisions. The ACT Waste Management Strategy: Towards a Sustainable Canberra 2011–2025 was established with targets of decoupling waste generation from population growth, and expanding reuse of goods.[38] Table 1.3 presents the targets of the 2019 NWPAP and the actions realised by 2021.[39]

Table 1.3   The seven targets and implementation status of the Australian National Waste Policy Action Plan (NWPAP), up to 2021.

Targets	Progress summary report in 2021
Target 1	Ban on the export of waste plastic, paper, glass and tyres, commencing in the second half of 2020 The national Recycling Modernisation Fund (RMF) supported 78 new and upgraded infrastructure projects around the country. Glass, plastic, paper and tyres that were previously sent overseas will now be recycled here in Australia
Target 2	Reduce total waste generated by 10% *per capita* by 2030. In 2016–17, Australia *per capita* generated 2.7 tonnes of waste. A 10% reduction equates to *ca.* 300 kg *per capita* each year, reducing disposal through recycling, energy recovery National roadmap for reducing Australia's food waste by half by 2030. Stop Food Waste Australia was established. Planet Ark established the Australian Circular Economy Hub – a one-stop-shop website for circular economy inspiration, education and implementation in Australia. The Productivity Commission completed an inquiry into the 'Right to Repair'. APCO delivered the industry Action Plan for Problematic and Unnecessary Single-Use Plastic Packaging, presenting a pathway for industry to shift from single-use disposable plastic packaging to more durable, reusable and recyclable packaging that will support delivery of the 2025 National Packaging Targets
Target 3	80% average resource recovery rate from all waste streams following the waste hierarchy by 2030 23 grants were awarded under the National Product Stewardship Investment Fund (NPSIF) to enable new industry-led product stewardship schemes and the expansion of existing schemes. The 23 projects received AUD 17.5 million and are forecast to divert 1.5 million tonnes of waste from landfill and create approximately 560 jobs. Battery Stewardship Scheme (B-Cycle) is established. Australasian Recycling Label (ARL) was added to more products on supermarket shelves, where 600 brand owners are signed up to the ARL Program and the number of supermarket products carrying the ARL will continue to increase. The Australian Government ran a 'ReMade in Australia' campaign to encourage Australians to recycle by showing how recycled materials can be recirculated in the economy
Target 4	Significantly increase the use of recycled content by governments and industry The Commonwealth Sustainable Procurement Advocacy and Resource Centre has released a toolkit to assist Australian Government staff to include recycled content and environmental sustainability in the procurement process. As part of its Waste Optimisation Program, Defence reviewed the waste collection systems across its estates and found it needed around 32 466 new waste sorting bins. Market research revealed there were no existing multisort bins made in Australia from 100% Australian recyclate. Sustainable procurement by governments, businesses and individuals is helping create demand for recycled materials – supporting the long-term sustainability of our recycling sector

Targets	Progress summary report in 2021
Target 5	Phase out problematic and unnecessary plastics by 2025 In April 2021 Australia's environment ministers agreed to phase out 8 problematic and unnecessary single-use plastic product types nationally by 2025 (or sooner in some cases) under the National Waste Policy Action Plan. The 8 items agreed for phase-out are lightweight shopping bags, 'degradable' plastics (fragmentable/oxo-degradable), plastic straws, Plastic utensils and stirrers, plastic bowls and plates, Expanded Polystyrene (EPS) consumer food containers (*e.g.*, cups and clamshells), EPS consumer goods packaging (loose fill and moulded) and microbeads in personal health-care products
Target 6	Halve the amount of organic waste sent to landfill for disposal by 2030 Governments expanded Garden Organic (GO) and Food Organics and Garden Organics (FOGO) kerbside collection services to households and businesses in Australia. The SA Organics Sector Analysis summary, released in May 2021, found the sector highly circular with 83% of the 1.35 million tonnes managed annually turned into valuable products, contributing AUD 189 million to Gross State Product and providing over 1200 jobs
Target 7	Make comprehensive, economy-wide and timely data publicly available to support better consumer, investment and policy decisions Australia has improved the availability of high-quality waste and recycling information. This supports a clearer understanding of how waste moves around Australia, how kerbside recycling is processed and reused and how high-value recycled commodities are traded – all key measures of success. Victoria's Circular Economy Business Innovation Centre (CEBIC) provided AUD 6.3 million to 23 projects, which will design out over 41000 tonnes of waste each year, launch innovation to recycle over 27000 tonnes of waste each year, reduce yearly greenhouse gas emissions by over 41000 tonnes, create 72 new jobs and leverage AUD 10 million of private investment into Victoria's circular economy

1.5.3.3   *Promoting Circular Economy Principles in India*

In 2000 India established the Municipal Waste Management Rules. However, the effectiveness of the implementation was not encouraging, despite the scope for value creation out of waste in India with the generation of 62 million tonnes of municipal solid waste within a total waste generation of 1236 million tonnes. In 2014 the Government of India launched the Swachh Bharat Mission (SBM; see Section 1.4.3) and in 2016 established six sets of waste management rules based on the circular economy and 5Rs (Reduce, Reuse, Recycle, Recover and Remanufacturing) principles and subsequently, Extended Producer Responsibility (EPR) guidelines and battery waste management rules, which are as follows:

- Guidelines on Extended Producer Responsibility for Plastic Packaging 2022
- Plastic Waste Management Rules, 2016, as amended, 2022

- Construction & Demolition Waste Management Rules, 2016
- Biomedical Waste Management Rules, 2016
- Solid Waste Management Rules, 2016
- Battery Waste Management Rules, 2022
- E-waste Management Rules 2016, as amended in 2022
- Hazardous and Other Wastes (Management and Transboundary Movement) Rules 2016, Second Amendment Rules, 2021
- National Policy on Resource Efficiency 2019 (Draft), which is yet to be finalised as of June 2023.

The SBM includes waste management as one of six components. Swachh Survekshan (a survey to assess rural and urban areas for their levels of cleanliness) was introduced by in 2016 as a competitive framework to encourage cities to improve the status of urban sanitation while encouraging large-scale citizen participation. Swachh Survekshan assesses the status of implementation of waste management rules and sanitation in different municipalities. The assessment criteria also include the resource efficiency of waste utilisation. In 2016 the survey was attended by 75 cities, while in 2022, 4320 cities participated and were assessed for their efficiency of waste segregation at source, waste treatment producing recycled products and energy recovery and other aspects (see Figure 1.2). The assessment criteria become more stringent from year to year, leading to institutionalising good practice. The eighth edition of Swachh Survekshan has been launched in 2023 – SS 2023 under SBM Urban 2.0. The evaluation parameters of SS 2023 will be out of 9500 marks including 48% for service-level progress and 26% each for the certification and citizens' voice. The mission is being implemented under the overarching principles of "waste to wealth" and "circular economy".[40]

For industries a number of rules including EPR, the Zero Defect, Zero Impact (ZED) Scheme 2015, the Metals Recycling Policy and revised categories of industries were enforced and practised. A huge momentum has been created, with circular economy concepts adopted in many industries, municipalities and other sectors in India. In March 2021, 11 committees were formed, led by the concerned line ministries and comprising officials from the Ministry of Environment, Forest and Climate Change (MoEFCC) and NITI

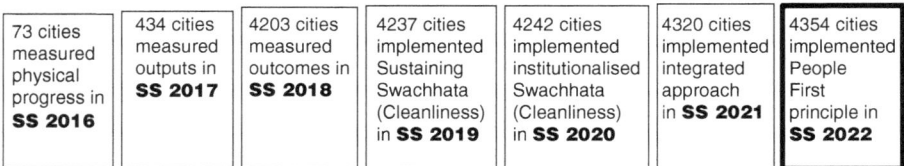

73 cities measured physical progress in **SS 2016**	434 cities measured outputs in **SS 2017**	4203 cities measured outcomes in **SS 2018**	4237 cities implemented Sustaining Swachhata (Cleanliness) in **SS 2019**	4242 cities implemented institutionalised Swachhata (Cleanliness) in **SS 2020**	4320 cities implemented integrated approach in **SS 2021**	4354 cities implemented People First principle in **SS 2022**

Figure 1.2   Progress of people participation in Swachh Survekshan (SS) in Indian cities for resource recovery and sanitation. Sustenance of ranking system from physical verification in 2016 to "People First" principle in 2022.

Aayog, domain experts, academics and industry representatives. The 11 focus areas identified by the Government of India are:[41]

- municipal solid waste and liquid waste
- scrap metal (ferrous and non-ferrous)
- electronic waste
- lithium-ion (Li-ion) batteries
- solar panels
- gypsum
- toxic and hazardous industrial waste
- used oil
- agricultural waste
- tyres
- rubber recycling and End-of-life Vehicles (ELVs).

The committees will prepare comprehensive action plans for transitioning from a linear to a circular economy in their respective focus areas. They will also carry out the necessary modalities to ensure the effective implementation of their findings and recommendations. The focus includes end-of-life products/recyclable materials/wastes that either continue to pose considerable challenges or are emerging as new challenge areas that must be addressed in a holistic manner.

1.5.3.4   *Promoting Circular Economy Principles in Indonesia*

The Government of Indonesia is in the process of strengthening its commitments and efforts in overcoming economic, social and environmental problems through low-carbon economy development and the implementation of circular economy concepts based on the SDGs and the target of reducing greenhouse gas emissions to satisfy the Paris Agreement, by 2030. Due to inefficient disposal practices in industrial sectors, namely, food and beverages, construction, electronics, textiles and plastics, the projected increases in waste percentage by 2030 are 54%, 82%, 39%, 70% and 40%, respectively.[42] Vision Indonesia 2045 and development policies and strategies in Indonesia incorporate circular economy concepts.

As an initial step, a study of the environmental, economic and social potential for the implementation of a circular economy in Indonesia was conducted in five industrial sectors, namely, food and beverages, construction, electronics, textiles and plastics. This will lead to the development of the National Action Plan and including the circular economy in the next National Medium Term Development Plan (RPJMN) 2025–2029. These five sectors represent 30% of Indonesia's Gross Domestic Product (GDP) and employed more than 43 million people in 2019.[43] The implementation of circular economy concepts in Indonesia contributes to an opportunity of economic growth, job creation and environmental sustainability. A study estimated that the circular economy could create 1.2 million new jobs and generate USD 10.2 billion in annual economic benefits for Indonesia by 2030.[42]

1.5.3.5   *Promoting Circular Economy Principles in Thailand*

Host to the Asia-Pacific Economic Cooperation (APEC) forum for 2022, Thailand overcame geopolitical, economic and pandemic-related head-winds to successfully guide the group to a consensus on a Joint Leaders Statement and the landmark Bangkok Goals on the Bio-Circular-Green (BCG) Economy. In January 2021 the Thai government announced a policy of positioning the BCG economy as a component of its "national agenda (key themes)". Thailand is still at an early stage of circular economy development while the country continues to make steady progress on the Global Green Economy Index, rising from 45th position in 2014 to 38th and 27th in 2016 and 2018, respectively.

The goal of a BCG economy is "to build a production and consumption system with less environmental impact through the promotion of agriculture and biotechnology-related industries, the encouragement of recycling, and the introduction of renewable energy".[44] The agriculture–food, bioenergy–biomaterial–biochemical, medical and wellness, and tourism and creative economies are the four focus areas of the BCG economy, which are closely related circular economy and green economy areas. Since 2018 the Thai government has focused on the production and trade of food trays, plastic bags, PET bottles[45] and automobile/power generation fuel made from bioethanol derived from agricultural waste under a resource circulation programme. The four strategies under the BCG economy are:[46]

- promote sustainability of biological resources by balancing conservation and utilisation
- strengthen communities and grassroots economy by employing resource capital, identity, creativity and advanced technology
- enhance sustainable competitiveness of Thai BCG industries
- build resilience to global changes.

The bioeconomy, circular economy and green economy share the common goal of environmental conservation with many overlapping aspects.

1.5.4   Promoting Circular Economy Principles in Canada

There are a number of policies and targets in Canada that encourage the implementation of a circular economy. Canada, specifically Ontario, became the first jurisdiction in the Americas to enact a comprehensive circular economy law, the Resource Recovery and Circular Economy Act 2016 (RRCEA). The Waste Diversion Act (WDA) was a government-managed scheme to oversee the diversion of target waste streams away from landfills in Ontario. RRCEA helps to implement circular economy principles that constitute resource recovery activities. The act encourages the extraction of useful materials or other resources from materials that might otherwise be waste, including through reuse, recycling, reintegration, regeneration or other activities.[47] Six sets of resource recovery Individual Producer Responsibility (IPR) obligations were imposed upon producers:

- designate materials
- define responsible parties
- set up a collection and management system for the end-of-life products and related packaging
- provide promotion and education
- register, report, auditing and recordkeeping
- reduce waste.

Canada has implemented IPR and the circular economy in plastics.[48]

1.6 ISO Standards on the Circular Economy

1.6.1 The ISO Process

A new International Organization for Standardization (ISO) technical committee was formed in 2019 to connect the dots in a circular economy and to introduce standardisation in the process, products and procedure for implementation of the concepts of circular economy. The ISO Technical Committee 323 (ISO/TC 323, Circular economy) is made up of experts from over 65 different countries and growing. The impetus behind the committee began with a seminar held by the Association Française de Normalisation (AFNOR), the ISO member for France, where business leaders from many sectors expressed the need to move from a linear to a circular economy model. What followed was a French standard, XP X30-901, *Circular economy – Circular economy project management system—Requirements and guidelines* published in 2018. The positive response spawned ISO/TC 323.

The ISO plans to develop a number of new standards based on circular economy concepts; many are in the developmental stage. Table 1.4 gives a glimpse of the standards under development through Technical Committee 323.[49-55]

In the ISO, the development of a standard follows a set path. A proposal for a new work item within the scope of an existing committee has to be submitted to the secretariat of that committee as a New Work Item Proposal (NWIP). Once the NWIP is accepted, responsibility of developing the standard is assigned to a working group. The working group develops the standard through various draft stages, to the final standard. In each of the stages voting of the member countries is conducted; the entire process can take a significant amount of time. As the circular economy is an important subject for all ISO TC 323 members, the development of the final standards should be expedited.

Table 1.4 List of the standards under development through ISO/TC 323 and the responsible working group.

ISO standard	Title	Scope
ISO 59004 – WD3 (3/2/2022): 2023 (Working Group 1)	Circular Economy – Terminology, Principles and Guidance for Implementation	Defines key terminology, establishes circular economy principles, and provides a framework for implementation and guidance regarding areas of action and helps to understand and contribute to a circular economy while aiming for sustainable development
ISO CD 59010 (Working Group 2)	Circular Economy – Guidance on the Transition of Business Models and Value Networks	Provides guidance for an organisation seeking to transition its business models and value networks from linear to circular
ISO CD 59020 (Working Group 3)	Circular Economy – Measuring and Assessing Circularity	Specifies a framework that provides guidance on how the circularity performance of an economic system can be objectively, comprehensively and reliably measured and assessed using circularity indicators and complementary methods enabling the organisations to contribute to sustainable development
ISO WD2 59040 (Working Group 5)	Circular Economy – Product Circularity Data Sheet	Provides a general methodology for improving the accuracy and completeness of circular economy-related information based on the usage of a Product Circularity Data Sheet when acquiring or supplying products. It also provides guidance for the definition and sharing of a Product Circularity Data Sheet, considering the type, content and format of information to be provided
ISO 59014 (ISO TC 207/SC 5/ISO TC 323 Joint Working Group 14)	Circular Economy – Secondary Materials – Principles, Sustainability and Traceability Requirements	Provides a framework and guidance for implementing the sustainability of activities and processes in providing/ obtaining/generating/managing/ capturing the secondary materials applicable to the environmental, social and economic aspects of such activities and processes. It also provides measures for enabling the traceability of secondary materials, hazardous wastes and resources not recovered as materials in the value chain

ISO standard	Title	Scope
Technical Report – ISO /TR-59031: 2021(E) (Working Group 4)	Circular Economy Performance-based Approach	Provides an analysis of cases for the implementation of specific aspects of a circular economy in organisations. This technical report focuses on performance-based approaches such as functional economy, service economy, product–service systems (PSS), Product as a Service (PaaS) and other performance-based approaches
Technical Report – ISO TR 59032 (Working Group 4)	Circular Economy – Review of Business Model Implementation	Provides a review of existing business model implementation relevant to the area of the circular economy. It presents the results of a survey to aid in the direction of the development of ISO 59010 for understanding the state of circular economy approaches

1.6.2 ISO/CD 59014:2022 – A Standard on Secondary Materials' Recovery Under Development

As noted previously, global waste generation is projected to increase to 2.59 billion tonnes by 2030 and 3.4 billion tonnes by 2050, a 70% increase from 2016 (2.01 billion tonnes); this continues to be an unexploited resource.

Secondary materials are embodied in trade activities. Next to the challenges connected with estimating the domestic extraction needed to enable the consumption of primary materials, there are also practical limitations to the accounting of secondary materials embodied in trade. Physical transaction data are scarce or unreliable, or outdated. Due to a lack of specific information, the estimate or forecasting of the available secondary materials in different sectors for exports and domestic use are not reliable, hindering effective planning for the use of these materials.

A standard entitled, *Environmental Management and Circular Economy – Sustainability and Traceability of Secondary Materials Recovery: Principles and Requirements* is under preparation by the members of the ISO TC 207/SC 5/ISO TC 323 Joint Working Group 14. The document is at present in the committee draft (CD) stage, as of June 2023, which will be elevated to the draft interim standard (DIS) and subsequently the final standard will be released after the voting process. The proposed ISO/CD 59014:2022 standard provides principles and requirements for enabling the sustainability and traceability of activities and processes in the recovery of secondary materials and addresses secondary materials, resources not recovered as materials and non-recovered wastes in the value chain of an organisation.

National bodies may also develop their own standards. The Bureau of Indian Standards has been carrying out a programme for developing a guideline standard for resource efficiency and secondary raw materials, which is being led by the author of this chapter. The standard, currently in the committee draft stage, focuses on the role of subsistence activities to ensure safe and healthy working conditions and continual improvement of well-being, livelihoods and professional practices while performing waste collection and treatment activities. The standard defines subsistence activity as activities conducted by an organisation involving an individual(s) or groups of individuals/families while earning below the living wage.

1.7   Conclusions

We need to halt and reverse nature loss by conserving more of the natural environment that we have left, restoring what is possible and sustainably managing the rest. Transitioning to a carbon-neutral and nature-positive society and economy is the only path to a safer and more equitable future for humanity. Material wastages need to be reduced and consumption patterns have to be redesigned, enhancing their life cycle. The combined strengths and capabilities of formal and informal sectors are required to improve the management of electronic, plastic, construction, demolition and biomedical waste. While hazardous wastes and nuclear wastes have attracted the attention of policymakers because of their destructive potential on human health and the environment, municipal wastes are the most neglected types of resources and have received relatively poor attention in many countries.

The circular economy and associated business models have emerged as a promising strategy, gaining increased attention and support in society and by different national governments around the world. Circular business models provide many opportunities that are more sustainable and restorative than existing linear models. The linear models based on extract–produce–consume–dispose–deplete (EPCD2) causes more extraction of natural resources, which is unsustainable. The circular economy offers opportunities worth trillions of dollars, employment generation, an enhanced economy, resource efficiency and, ultimately, environmental sustainability.

Implementing innovative technologies based on the 3Rs and circular economy principles are potentially more sustainable in the long term. Holistic zero-waste management strategies based on integrated tools, systems and technologies are required for the transition phase of a society. Selection or application of waste treatment technologies for zero-waste cities should consider holistic intergeneration resource recovery and product stewardship. In zero-waste city planning and design, the material flow of the city should be designed and managed in a balanced way, while taking into account differences in lifestyles, values and personal behaviours. More generally, the waste disposal behaviour and awareness of consumers needs to be addressed in the context of a country's complex socio-cultural and economic conditions, while committing all resource recovery sectors collectively to develop sustainable waste management systems.

The presence of the national legislation is the first step to provide a foundation for a country's activities related to the circular economy and sustainable development goals. International standards, working in tandem with national legislation and standards, will accelerate the transition to a more sustainable future.

Acknowledgements

The author acknowledges the support of the Going Global Partnerships – Industry Academia Collaborative Grant, funded by the British Council and Aston University, UK, to the International Society of Waste Management, Air and Water, India.

References

1. OECD, *Global Material Resources Outlook to 2060 – Economic Drivers and Environmental Consequences*, OECD, Geneva, 2018.
2. UNFPA, *Seeing the Unseen: State of World Population 2022*, New York, 2022.
3. World Bank, *What a Waste 2.0 A Global Snapshot of Solid Waste Management to 2050*, New York, 2020.
4. International Resource Panel, *Global Material Flows and Resource Productivity*, UNEP, Kenya, 2016.
5. United Nations, *Transforming Our World: The 2030 Agenda for Sustainable Development*, New York, 2015.
6. UN Conference on Sustainable Development, Rio+20, Rio de Janeiro, Brazil, 20–22 June 2012.
7. T. Janowski, *Government Information Quarterly*, 2016, **33**, 603. DOI: https://doi.org/10.1016/j.giq.2016.12.001.
8. ECOSOC, *Report of the High-level Political Forum on Sustainable Development Convened under the Auspices of the Economic and Social Council at its 2016 Session*, United Nations, New York, 2016.
9. Ellen MacArthur Foundation, The circular economy in detail: deep dive, https://ellenmacarthurfoundation.org/the-circular-economy-in-detail-deep-dive (accessed May 2023).
10. Ellen MacArthur Foundation, *Completing the Picture: How the Circular Economy Tackles Climate Change*, Cowes, 2019.
11. Ellen MacArthur Foundation, The circular economy model – brief history and schools of thought, https://ellenmacarthurfoundation.org/topics/circular-economy-introduction/overview (accessed June 2023).
12. *Circular Economy: Global Perspective*, ed. S. K. Ghosh, Springer Nature, Singapore, 2020.
13. P. Ekins and N. Hughes, *Resource Efficiency: Potential and Economic Implications*, UNEP, Kenya, 2017.
14. Circle Economy, *The Circularity Gap Report 2019: Closing the Circularity Gap in a 9% World*, https://docs.wixstatic.com/ugd/ad6e59_ba1e4d16c64f44fa94fbd8708eae8e34.pdf/.

15. P. Golinska, M. Kosacka, R. Mierzwiak, K. Werner-Lewandowska, *J. Clean. Prod.*, 2015, **105**, 28.

16. E. Küçüksayraç, D. Keskin and H. Brezet, *J. Clean. Prod.*, 2015, **101**, 38.

17. International Resource Panel, *Urban Agriculture's Potential to Advance Multiple Sustainability Goals*, UNEP, Kenya, 2022.

18. E. Napper and R. C. Thompson, *Global Challenges*, 2020, **4**, 1900081.

19. R. Karasik, T. Vegh, Z. Diana, J. Bering, J. Caldas, A. Pickle, D. Rittschof and J. Virdin, *20 Years of Government Responses to the Global Plastic Pollution Problem: The Plastics Policy Inventory,* Nicholas Institute for Environmental Policy Solutions, Duke University 2020.

20. S. Sabato and M. Mandelli, in *Social Policy in the European Union: State of Play 2020*, ed. B. Vanhercke, S. Spasova and B. Fronteddu, European Trade Union Institute (ETUI) and European Social Observatory (OSE), Brussels, 2021, pp. 113–132.

21. European Commission, *Treaty on European Union*, (92/C 191/01), OJ No C 191/1, 29.7.92, Brussels, 1992.

22. European Commission, Consolidated Version of the Treaty on European Union, OJ C 326/13, 26.10.2012, Brussels, 2012.

23. European Commission, *Sustainable Development in the European Union: A Statistical Glance from the Viewpoint of the UN Sustainable Development Goals*, Luxembourg: Publications Office of the European Union, 2016.

24. European Commission, *Europe 2020*, COM(2010) 2020, Brussels, 2010.

25. A. Widuto, *Sustainable Development Goals (SDGs) in EU Regions,* European Parliamentary Research Service, PE 659.415, Brussels, June 2022.

26. European Parliament, Resolution of 23 June 2022 on the implementation and delivery of the Sustainable Development Goals (SDGs), 2022/2002(INI); 2023/C 32/05, OJ 27.1.2023.

27. NITI Aayog, *Localising SDGs: Early Lessons from India*, New Delhi, 2019, https://sdghelpdesk.unescap.org/sites/default/files/2019-12/LSDGs_July_8_Web.pdf (accessed May 2023).

28. L. Wellesley, F. Preston and J. Lehne, *An Inclusive Circular Economy: Priorities for Developing Countries*, Chatham House, London 2019.

29. European Commission, Report from the Commission to the European Parliament, the Council, the European Economic and Social Committee and the Committee of the Regions on the implementation of the Circular Economy Action Plan, COM(2019) 190 final, Brussels, 2019.

30. European Commission, *A New Circular Economy Action Plan: For a Cleaner and More Competitive Europe*, COM/2020/98 final, Brussels, 2020.

31. European Commission, *Pathway to a Healthy Planet for All EU Action Plan: Towards Zero Pollution for Air, Water and Soil*, COM(2021) 400 final, Brussels, 2021.

32. M. C. Friant, W. J. V. Vermeulena and R. Salomone, *Sustainable Production and Consumption*, 2021, **27**, 337.

33. E. Mazur-Wierzbicka, *Environ. Sci. Eur.*, 2021, **33**, 111.

34. Ministry of Environment, *Sri Lanka Updated Nationally Determined Contributions*, Colombo, 2021, https://unfccc.int/sites/default/files/NDC/2022-06/Amendmend%20to%20the%20Updated%20Nationally%20Determined%20Contributions%20of%20Sri%20Lanka.pdf (accessed May 2023).

35. USAID, *Extended Producer Responsibility: Lessons learned from Sri Lanka*, Washington, DC, 2021.

36. N. Cooray, S. Peiris, K. Rasaputra, R. K. Singh, G. J. Premakumara and K. Onagawa, *National Action plan on Plastic Waste Management 2021–2030*, Ministry of Environment Sri Lanka, Colombo, 2021, https://ccet.jp/sites/default/files/2021-08/srilanka_report_web_fin_pw.pdf (accessed May 2023).

37. Green Building Council of Sri Lanka, *Green Labelling System For Sustainable Building Materials and Products Version 2.0*, Colombo, 2022, https://saicmknowledge.org/sites/default/files/publications/GREEN%20Labelling%20System%20(V%202.0).pdf (accessed May 2023).

38. P. S. M. Vaughan Levitzke, in *Circular Economy: Global Perspective*, ed. S. K. Ghosh, Springer Nature, Singapore, 2019, pp. 25–42.

39. Australian Government, *National Waste Policy Progress Summary Report 2021*, Department of Climate Change, Energy, Environment and Water, Canberra, 2021, https://www.dcceew.gov.au/environment/protection/waste/publications/national-waste-policy-progress-summary-report (accessed May 2023).

40. Ministry of Housing and Urban Development, *Swachh Survekshan Ranking Report 2022*, Government of India, New Delhi, 2022.

41. Press Information Bureau, Govt driving transition from linear to circular economy, Government of India, New Delhi, https://pib.gov.in/PressReleasePage.aspx?PRID=1705772 (accessed May 2023).

42. Kementarian/PPN, *The Economic, Social and Environmental Benefits of a Circular Economy in Indonesia*, UNDP, 2021.

43. The Indonesia Circular Economy Forum, *The Economic, Social and Environmental Benefits of a Circular Economy in Indonesia*, Embassy of Denmark, Jakarta, UNDP and the Indonesia Circular Economy Forum, January 2021.

44. S. Kumagai, *RIM Pacific Business and Industries*, 2022, **22**(84), 2.

45. N. Apinunwattanakul, *Challenges and Opportunities for Businesses under the Measure to Reduce Plastic Use*, SBC Economic Intelligence Center, Bangkok, 18 April 2019.

46. BCG, BCG in action: vision, https://www.bcg.in.th/eng/vision/ (accessed May 2023).

47. J. Cocker and K. Graham, in *Circular Economy: Global Perspective*, ed. S. K. Ghosh, Springer Nature, Singapore, 2020, pp. 87–122.

48. Smart Prosperity Institute, *A for a Circular Economy for Plastics in Canada*, Ottawa, 2019.

49. ISO, ISO/DIS 59004 – WD3 (3/2/2022): 2023 Circular economy – terminology, principles and guidance for implementation, https://www.iso.org/standard/80648.html (accessed May 2023).

50. ISO, ISO/DIS 59010 Circular economy – guidance on the transition of business models and value networks, https://www.iso.org/standard/80649.html (accessed May 2023).

51. ISO, ISO/DIS 59020 Circular economy – measuring and assessing circularity, https://www.iso.org/standard/80650.html (accessed May 2023).

52. ISO, ISO/CD 59040 Circular economy – product circularity data sheet, https://www.iso.org/standard/82339.html (accessed May 2023).

53. ISO, ISO/CD 59014 Secondary materials – principles, sustainability and traceability requirements, https://www.iso.org/standard/80694.html (accessed May 2023).

54. ISO, ISO/CD TR 59031 Technical report – circular economy – performance-based approach – analysis of cases studies, https://www.iso.org/standard/81183.html (accessed May 2023).

55. ISO, ISO/CD TR 59032.2 Technical report – circular economy – review of business model implementation, https://www.iso.org/standard/83044.html (accessed May 2023).

CHAPTER 2

Sustainable Development and the Circular Economy: Concepts, Progress and Prospects

GEV EDULJEE*

Resource Futures, Create Centre, Smeaton Road, Bristol BS1 6XN, UK
*E-mail: geduljee@gmail.com

2.1 Introduction

Historically, sustainability and the circular economy evolved from different foundational principles and have generally been applied as mutually independent paradigms – the former integrating economic, environmental and social considerations within the rubric of sustainable development, the latter more technical and process-oriented considerations typically confined to the economic sphere. However, with the recognition that the circular economy can play an important role in delivering the United Nations' Sustainable Development Goals (SDGs), the focus in recent years has been to study more rigorously the similarities and differences between the two paradigms and how they can be made to work in tandem to fulfil the SDGs.

Chapter 2 introduces the concepts behind the two paradigms, their progress to date, the challenges they face and the prospects for further implementation and for meeting SDG target dates. The chapter is intended as an

Issues in Environmental Science and Technology No. 51
The Circular Economy: Meeting Sustainable Development Goals
Edited by Sadhan Kumar Ghosh and Gev Eduljee
© The Royal Society of Chemistry 2024
Published by the Royal Society of Chemistry, www.rsc.org

overview – subsequent contributions to this book explore particular sustainable development and circular economy themes in more detail.

2.2 Sustainable Development

2.2.1 Sustainability

Concerns relating to the impact of humans on the quality of our environment and on its conservation are apparent as far back as the ancient civilisations of India, China, Mesopotamia and Egypt.[1,2] Its emergence in modern times has been traced from the late-romantic idealisation of and nostalgia for a passing age of "ecological innocence" as intensive industrialisation and urbanisation took hold in the 17th and 18th centuries, through to the "scientific conservation" of the 19th century with a more explicit articulation of the consequences of the actions of humans and the setting up of formal institutions to manage the environment, and finally to the "ecology of affluence" from the mid-20th century onwards, when environmental management became a mainstream policy imperative, backed by the maturing environmental sciences.[2] Much of the early concerns focused on preserving wilderness and forest resources, wood being a key domestic and industrial resource. Whereas prehistoric and early-modern concerns tended to identify and ameliorate localised impacts, this last phase of ecological awareness is additionally characterised by an increasing appreciation of impacts at a global, planetary scale, climate change being a prime example, and by the birth of citizen-based environmental movements.

The terms "sustainability" and "sustainable" have Latin, French and German roots dating from the 14th and 17th centuries, but their modern idiom and usage is relatively recent, introduced in a global sense in the report *The Limits to Growth* in 1972, with its exhortation to create "a world system … that is sustainable".[3,4,5] Also in 1972, the United Nations Conference on the Human Environment culminated in the Stockholm Declaration, enunciating a set of "common principles to inspire and guide the peoples of the world in the preservation and enhancement of the human environment". The declaration introduced the concept of environmentally sound development, later characterised as eco-development, which argues for "a different, environmentally prudent, sustainable and socially responsible growth".[5,6] The declaration set an important milestone in the context of future interpretations of sustainability, by explicitly linking social improvement with economic development:

In the developing countries most of the environmental problems are caused by under-development. Millions continue to live far below the minimum levels required for a decent human existence, deprived of adequate food and clothing, shelter and education, health and sanitation. Therefore, the developing countries must direct their efforts to development, bearing in mind their priorities and the need to safeguard and improve the environment.

In the chapter titled "Towards Sustainable Development", the strategy states that:

Development and conservation operate in the same global context, and the underlying problems that must be overcome if either is to be successful are identical.

One particular "difference" in the declaration was the recognition that humans' use of the environment should be mindful of the needs of future generations. This was most clearly articulated in the International Union for Conservation of Nature (IUCN) World Conservation Strategy of 1980:[7] "… we have not inherited the earth from our parents, we have borrowed it from our children".

These two principles, namely that social improvement goes hand in hand with economic development and that concurrently the needs of future generations must be respected, are embodied in the most prevalent and widely quoted enunciation of sustainability, that of the Brundtland Commission.[8] Under the aegis of the United Nations (UN), the so-called Brundtland Report emphasised the need for economic growth, especially in the developing world, reaffirming the importance of sustainable development, namely "[economic] growth that is forceful and at the same time socially and environmentally sustainable". Sustainable development was defined as:

Development that meets the needs of the present without compromising the ability of future generations to meet their own needs.

The Brundtland Report galvanised global interest and especially action on sustainability and sustainable development:

- In 1992 the Earth Summit in Rio de Janeiro gave rise to the Rio Declaration to guide sustainable development, and Agenda 21, "a comprehensive plan of action to be taken globally, nationally and locally … in every area in which human impacts on the environment". Subsequent progress summits were held in 1997, 2002 and 2012.
- In 1993 the Commission for Sustainable Development was set up to monitor and promote implementation of Agenda 21. In 1996 the Commission published 130 indicators to measure progress. A revised set was published in 2001.
- In 2000 at the Millennium Summit in New York, the UN agreed its Millennium Declaration and Goals (MDGs), a set of eight goals ranging from poverty alleviation to providing universal primary education, all by 2015. Goal 7 specified "environmental sustainability".
- In 2002 the UN World Summit on Sustainable Development, held in Johannesburg, adopted a political declaration and implementation plan to achieve development respectful of the environment in areas such as water, energy, health, agriculture and biodiversity.

- In 2012 at a conference on sustainable development in Rio de Janeiro, the UN adopted *The Future We Want,* a set of broad sustainability objectives to be negotiated and adopted as internationally agreed Sustainable Development Goals (SDGs) by the end of 2014. The Commission for Sustainable Development was replaced by the High-level Political Forum on Sustainable Development to monitor progress on the SDGs and on Agenda 21.
- In 2015 at a summit in New York, the UN adopted *Transforming Our World: The 2030 Agenda for Sustainable Development*; this contained 17 SDGs that subsumed the MDGs adopted in 2000 and moved the target date from 2015 to 2030.

Representation of the business community at the 1992 Earth Summit was organised through a new group, the Business Council for Sustainable Development, renamed the World Business Council for Sustainable Development (WBCSD). The WBCSD has developed sector roadmaps to guide businesses in achieving the SDGs (also see Section 2.2.3).

2.2.2   The Pillars of Sustainable Development

A particular construct of sustainable development that recurs in UN declarations and related documents and has received common acceptance, especially with policymakers, is that based on the "three pillars": economic, environmental and social, though it has been pointed out that neither their origin in terms of a defining document, nor their theoretical underpinning, are clear-cut.[5,9,10] For example, while Venn diagram representations of the three pillars imply that they are of equal importance with a balanced three-way sweet spot at the centre, it is often the case that the economic pillar dominates policymaking, presumably reflecting the notion that economic development, and in particular economic growth, is at the heart of poverty alleviation (see Section 2.2.1) and that we can afford environmental protection measures only if the economy is thriving. This thinking is implied, for example, in the definition of the World Bank:[11]

> *Sustainable development means basing developmental and environmental policies on a comparison of costs and benefits and on careful economic analysis that will strengthen environmental protection and lead to rising and sustainable levels of welfare.*

Whatever the merits of these arguments, the fact remains that a rigorous interrogation and analysis of the terms sustainability and sustainable development is lacking, together with a common consensus on precisely what development means, or who development is for.[5] While the three pillars and the Brundtland Report definition remain the mainstream interpretations, this ambiguity has engendered other paradigms and approaches. Thus, the European Union's (EU's) proposal for a directive on corporate sustainability has an implied, more wide-ranging definition of sustainability as it relates to business:[12]

... to foster ... the respect of the human rights and environment in their own operations and through their value chains, by identifying, preventing, mitigating and accounting for their adverse human rights, and environmental impacts, and having adequate governance, management systems and measures in place to this end.

The IUCN's *World Conservation Strategy* defined sustainable development in ecological terms:

Sustainable development is about maintenance of essential ecological processes and life support systems, the preservation of genetic diversity, and the sustainable utilization of species and ecosystems.

A follow-up meeting to monitor the IUCN's *World Conservation Strategy* offered a definition that arguably emphasises social elements over definitions that focus on economic aspects:

Sustainable development ... seeks ... to respond to five broad requirements: integration of conservation and development; satisfaction of basic human needs; achievement of equality and social justice; provision for social self-determination and cultural diversity; and maintenance of ecological integrity.

Caring for the Earth, the update on the *World Conservation Strategy*, based development around nine "interrelated and mutually supporting [principles of a] sustainable society".[13] Other researchers have suggested additional pillars such as institutional, governance, time, cultural and technical.[5,14,15] The UN's 2030 Agenda states that the SDGs are based on five pillars (the 5Ps): people, planet, prosperity, peace and partnership.

A number of alternative definitions have been proposed, as collated by Pezzy until the early 1990s, and considerably more have appeared since.[16,17] The rationale behind these diverse and divergent viewpoints has been explored.[18] Notwithstanding these ambiguities, the conventional three pillars have solidified in policymaking, even though, as in the case of the EU's proposed directive and the IUCN (see above), definitions of sustainability are sometimes qualified by including other attributes judged to be important for implementation.

2.2.3 The Triple Bottom Line

The buy-in and participation of the corporate sector was considered to be crucial in implementing sustainable development (see Section 2.2.1 relating to the Earth Summit). Supplementing the traditional profit-related focus of a business, the three pillars were re-imagined into the concept of the Triple

Bottom Line (TBL), coined and developed in 1994. Termed "people, planet, profit", the concept "focuses corporations not just on the economic value that they add, but also on the environmental and social value that they add – or destroy".[19] Described as a "sustainable capitalism transition", a concept open to challenge,[20] TBL aims to integrate seven sustainability drivers into corporate culture and practices: markets, values, transparency, life-cycle technology, partnerships, time and corporate governance.

Of the three pillars, that of the environment was considered to be most familiar to corporations, not least as a result of long-standing mandatory environmental legislation relating to emission and discharge controls, pollution abatement, clean-up, *etc.*, dating from the early 20th century and particularly from the mid-1950s. Companies in effect grafted the remaining two pillars of sustainability onto the concept of Corporate Social Responsibility (CSR) which, from the 1960s, aimed to incorporate societal goals into the behaviours and actions of corporations. The TBL approach formalised these extra-financial aspects by instituting more rigorous and standardised tools for analysis and reporting, especially requiring openness and public disclosure of a corporation's activities and progress towards its TBL goals and targets.

Whether the TBL approach has been effective in directing corporate behaviour more towards societal and environmental wellbeing is a matter of debate. Amid accusations of greenwashing, studies have sometimes characterised TBL as empty rhetoric.[5,21] Yet despite its perceived shortcomings, *not* creating formalised structures to operate alongside a focus solely on profit merely serves to preserve the *status quo*. The efforts of policymakers to create regulatory instruments to further embed environmental and societal considerations into corporate behaviours should be seen as a welcome development. Government organisations such as the Competition and Markets Authority in the UK and the Authority for Consumers and Markets in the Netherlands have introduced codes that inform businesses of their obligations when making environmental and sustainability claims.[22,23]

2.2.4   Sustainable Development Goals

The SDGs introduced in Section 2.2.1, together with their associated targets and indicators, form the basis of the present book. The system comprises 17 goals, 169 targets and 231 discrete indicators (with 13 indicators repeated under two or three different targets).[24,25] Adopted by 193 UN member states, the SDGs came into force on 1 January 2016, with a target implementation date of 2030. The UN Division for Sustainable Development Goals provides support, with annual SDG progress reports presented by the Secretary General. The UN-sponsored Sustainable Development Solutions Network published a detailed discussion on the SDG system and its indicators.[26]

A detailed exposition on each SDG, target and indicator is beyond the scope of this book – subsequent chapters will pick up on those relevant to their theme. In this introductory chapter some general remarks will be presented.

The general architecture of the SDG system is summarised in Table 2.1.

Table 2.1 SDGs, targets and indicators. Data from ref. 24 and 25.

No	SDGs (with accepted abbreviations)	No	SDG
1	End poverty in all its forms everywhere (No poverty). 7 targets, 12 indicators	10	Reduce inequality within and among countries (Reduce inequalities). 10 targets, 11 indicators
2	End hunger, achieve food security and improved nutrition and promote sustainable agriculture (Zero hunger). 8 targets, 14 indicators	11	Make cities and human settlements inclusive, safe, resilient and sustainable (Sustainable cities and communities). 10 targets, 15 indicators
3	Ensure healthy lives and promote wellbeing for all at all ages (Good health and wellbeing). 13 targets, 26 indicators	12	Ensure sustainable consumption and production patterns (Responsible production and consumption). 11 targets, 13 indicators
4	Ensure inclusive and equitable quality education and promote lifelong learning opportunities for all (Quality education). 10 targets, 11 indicators	13	Take urgent action to combat climate change and its impacts (Climate action). 5 targets, 7 indicators
5	Achieve gender equality and empower all women and girls (Gender equality). 9 targets, 14 indicators	14	Conserve and sustainably use the oceans, seas and marine resources for sustainable development (Life below water). 10 targets, 10 indicators
6	Ensure availability and sustainable management of water and sanitation for all (Clean water and sanitation). 8 targets, 11 indicators	15	Protect, restore and promote sustainable use of terrestrial ecosystems, sustainably manage forests, combat desertification, and halt and reverse land degradation and halt biodiversity loss (Life on land). 12 targets, 14 indicators
7	Ensure access to affordable, reliable, sustainable and modern energy for all (Affordable and clean energy). 5 targets, 6 indicators	16	Promote peaceful and inclusive societies for sustainable development, provide access to justice for all and build effective, accountable and inclusive institutions at all levels (Peace, justice and strong institutions). 12 targets, 23 indicators
8	Promote sustained, inclusive and sustainable economic growth, full and productive employment and decent work for all (Decent work and economic growth). 12 targets, 17 indicators	17	Strengthen the means of implementation and revitalize the Global Partnership for Sustainable Development (Partnership for the goals). 19 targets, 25 indicators
9	Build resilient infrastructure, promote inclusive and sustainable industrialisation and foster innovation (Industry, innovation and infrastructure). 8 targets, 12 indicators		

The 2030 Agenda emphasises the importance of treating the SDGs as a single, coherent sustainability framework: "all goals and targets are equally important". Given the complexity of the system, the issue arises as to how individual goals and targets interact with each other in both a positive and a negative sense – achieving a particular set of goals/targets may enhance or, conversely, adversely impact progress towards others. These so-called synergies and trade-offs between SDGs and their targets have been extensively studied.[27,28,29,30] A number of approaches have been suggested to examine these linkages and their interactions. One such study for example, found "positive developments with notable synergies" for SDGs 1, 3, 7, 8, and 9, but "notable trade-offs" between SDGs 11, 13, 14, 16, and 17 "as well as non-associations with the other goals".[31] Another study, employing principal component analysis, concluded that the SDGs were "largely compatible in the sense that for each country the level of attainment in one SDG tends to be correlated with the level of attainment in the SDG agenda as a whole", but that SDG 10 was negatively correlated with most of the other principal components.[29] A further study, researching the question "can the 2030 Agenda objectives be considered homogeneous, and can they adequately measure the concept of sustainability?" answered both in the negative.[32] In these and other studies, the key conclusion was that while the 2030 Agenda should be implemented as an indivisible package, countries should be mindful of potential synergies and trade-offs between various goals and targets – some mutually reinforcing, others potentially in conflict.

Another issue arising from the configuration of the SDG system is the cross-cutting nature of some general themes – a feature which has influenced the structure of this book. For instance, protection of public health has been correlated with SDG 1/Target 1.4; SDG 3/Targets 3.2, 3.3, 3.9; SDG 11/Targets 11.1, 11.6; and SDG 12/Target 12.4.[33] Other examples are readily identified.[34] It has been suggested that to focus implementation, a virtual SDG might be created for such cross-cutting themes, corralling relevant targets from disparate goals into a new consolidated SDG.[33]

A further overarching theme spanning biodiversity and cultural diversity was discussed at the UN Conference of the Parties (COP) 15 at Montreal in December 2022. Arising out of the conference was a draft agreement as a contribution to the UN 2030 Agenda, placing "biodiversity, its conservation, the sustainable use of its components and the fair and equitable sharing of the benefits" central to sustainable development.[35] The longer-term vision is that "by 2050, biodiversity is valued, conserved, restored and wisely used, maintaining ecosystem services, sustaining a healthy planet and delivering benefits essential for all people". The framework contains 23 targets to be achieved by 2030.

2.2.5   Financing the SDGs

Aside from politically related factors, adequate financing (as implied previously) is seen as a major obstacle to achieving SDGs expeditiously. Following the formal adoption of the Paris Agreement and the SDGs, the so-called

Addis Ababa Action Agenda (AAAA) set out a seven-point plan for financing the SDGs that relied on "both domestic actions and a commitment to create an enabling international environment that supports national efforts". Especially targeting low- and middle-income developing countries, actions range from mobilising public and private sector resources to stimulating official development assistance (ODA).[36] However, the Organisation for Economic Co-operation and Development (OECD) noted that even pre-COVID-19 SDG financing was deficient, to the extent of an annual gap of USD 2.5 trillion. This gap was expected to increase by a further USD 1.7 trillion (including spending on COVID-19) in 2020, together with a fall in external financing to developing countries.[37] The UN estimated that an annual global spend of between USD 3–5 trillion was required to meet the SDGs by 2030, with the COVID-19 pandemic increasing this estimate by an additional USD 2 trillion annually.[38]

While national and international public development funding is vital for financing SDG commitments, it is recognised that reliance on this funding stream alone will not be sufficient. It is reported that total ODA fell by 4.3% in 2018; although ODA increased in 2020, cumulatively, ODA investment totalled USD 16.1 billion over 2012–17, a small proportion of the total funding gap that needs to be bridged.[39] Private sector finance is acknowledged to play a key role in delivering finance. It is estimated that the current investment gap of USD 2.5 trillion per year in developing countries can be achieved by mobilising 7.76% of global assets under management each year, which amounts to less than the capital exchanged on a single day in the world's financial markets.[40] Put another way, the funding gap represents over 25 times the total of ODA funds, but only 1.1% of global financial assets as valued in 2020 at more than USD 378.9 trillion.[41]

Comparing the Environmental, Social and Governance (ESG) investing performance of 8550 companies in the Morgan Stanley Capital International (MSCI) All Country World Index across a four-point scale, the World Economic Forum estimated that some 38% of companies were aligned to the SDGs whereas almost 55% were misaligned or neutral. The ESG investment performance of only 0.2% of companies was "strongly aligned" to the UN SDGs.[42] These figures suggest considerable scope for gearing up private sector ESG investment in support of actioning the SDGs, even among companies and investors that have publicly declared their intent.

Given the importance of private sector involvement in financing the SDGs, considerable effort has been put into creating platforms, mechanisms and financial instruments to foster public–private partnerships and to encourage direct ESG financing. For example, the Sustainable Development Goals Fund (SDG Fund) was created by the UN in 2014 to bring together UN agencies, national governments, academia, civil society and business to achieve the SDGs. The fund was preceded by the Millennium Development Goals Fund (2007) and superseded by the Joint SDG Fund in 2017.

The World Bank issued its first ever set of equity-index linked green bonds in 2017, directly linking financial returns to companies meeting the standards and aims of the SDGs. The bonds created a USD 175 million funding pot sourced by institutional investors.[43] In 2019 the World Bank launched a EUR 1.5 billion 10-year Global Sustainable Development Bond listed on the Dublin and Luxemburg Stock Exchanges, focused on mobilising action towards achieving the SDGs.[44]

In 2020 the SDG500 platform was launched, a coalition including UN entities, other private and public sector organisations, non-governmental organisations (NGOs) and private equity firms. The investment platform targets businesses across the agriculture, finance, energy, education and healthcare sectors, with investors committing USD 500 million to help developing nations achieve SDGs.[45] Also in 2020, a Chief Financial Officers (CFO) Leadership Group was convened by the UN Global Compact (UNGC), publishing a set of principles guiding companies seeking to mobilise corporate finance and investments to achieving the SDGs.[46] In March 2022 the UNGC launched the CFO Coalition for the SDGs, in order "to develop frameworks for ensuring that corporate finance is aligned with the SDG framework". The coalition aims to create a USD 10 trillion market for corporate SDG-directed finance by 2030, up from USD 500 billion in 2020.[47]

UN-backed initiatives targeting specific issues such as climate change can contribute to the achievement of the SDGs. For example, the UN-convened business grouping The Net Zero Asset Owner Alliance reported in September 2022 that its members collectively managed USD 10.6 trillion of assets, USD 3.3 trillion of which would be directed towards investments in climate action/low-carbon solutions (SDG 13).[48] Other UN-backed finance initiatives include the Net Zero Banking Alliance and the Net Zero Asset Managers Initiative. The latter, launched in 2020, is reported to have USD 61.3 trillion of assets under management, USD 16 trillion of which are targeting low-carbon solutions.[49]

2.2.6   Measuring Sustainability

Owing to the multifaceted nature of the concepts, there is no settled methodology to measure sustainability or sustainable development. Sustainable development has been described as a "contested concept" given the multiplicity of definitions (see Section 2.2.2) and it has even been suggested that stakeholders could be "engaging in a futile exercise of measuring the immeasurable".[18,50] However, a more pragmatic view might be that sustainability and sustainable development should be regarded as intrinsically malleable concepts. Even a relatively straightforward definition of sustainable development such as that of the Brundtland Report raises legitimate issues of interpretation, for example, precisely what the policy objectives are, what needs should be addressed and whether these needs are universal or nationally/culturally influenced, and over what timescales should the needs of future generations be respected, to which there are no single right answers. Moreover, the sheer range and diversity of problems and scenarios that sustainability and sustainable development are required to address suggests that there is no one-size-fits-all definition; correspondingly, approaches to measurement need to be tailored to the problem under consideration. In other words, a case can be made for a more nuanced approach – some elements would be more important in determining the sustainability of a particular TBL-based corporate business model than, say, of food and farming or of public health practices at a national or an international scale. This is not to advocate for a methodological free-for-all. Without some standardisation and consensus on which indicators/indices are important under which scenarios, how they should be measured and how individual measurements should be aggregated to provide an overall estimate of sustainability/sustainable development, estimates of progress, comparisons and meaningful interventions – technical, political, financial, funding – cannot be made.

A meta-analysis of the published literature on sustainability measurements identified eight main areas of inquiry: sustainability disclosure and performance, determinants of sustainability disclosure, critical environmental accounting, sustainability metrics, sustainable operations and supply chain management, carbon accounting, diffusion of sustainability standards, and assurance of sustainability reporting. Issues such as corporate sustainability and disclosure received greater attention than, for example, carbon accounting, while topics such as ecosystem valuation and biodiversity accounting were emerging areas of inquiry. According to the study, the field has developed in an *ad hoc* manner, akin to "the moral of the blind men and the elephant", with divergent conceptualisations and research outcomes.[51] Given our previous remarks, the finding of a lack of high-level coherence across the broad spectrum of sustainability and sustainable development is not unexpected.

Building on a framework intended for environmental sustainability,[52] a taxonomy of indicators and indices is helpful in structuring the hundreds of metrics introduced in Section 2.2.4 and Table 2.1, as well as providing a framework for creating new indicators pertinent to the issue under consideration:

- **Descriptive indicators:** those ones that describe the real situation pertaining to a particular issue, expressed in quantified units (*e.g.*, Gross Domestic Product (GDP), water usage, atmospheric emissions, crime rate, conflicts).
- **Performance or effectiveness indicators:** the result achieved relative to its pre-established objective, that is, distance to target (*e.g.*, percentage of waste treated, water quality, diversity and inclusion, crime, employment, gender inequality, levels of education).
- **Efficiency indicators:** the ratio between the result achieved and the economic/other resources used (*e.g.*, cost–benefit analysis, economic and social return on investment, manufacturing emissions per unit of product).
- **Subjective indicators:** mainly wellness indicators that measure general wellbeing, for example *via* population surveys (*e.g.*, standard of living, health achievements, feelings of security and safety, social connectedness).

While the dividing lines between the categories are blurred – it is possible, for instance, for wellness to be expressed as performance indicators – a matrix for each SDG target allows appropriate indicators to be allocated and interrogated, amended or augmented relative to their contribution to each of the three pillars of sustainable development.

Among issues open to debate (such as generational time frames), two in particular affect which indicators to apply, and how they should be interpreted: scale and integration. The Sustainable Development Solutions Network identified four levels across which indicators can be monitored and measured: national, global, regional and thematic.[26] Indicators for assessments at national level will reflect particular political proclivities and policies, and therefore will likely differ from nation to nation, though there will generally be some coherence such that inter-nation comparisons and aggregations can be effected. This also applies to regional sustainability assessments, such as within the EU, BRICS (Brazil, Russia, India, China and South Africa) or Association of Southeast Asian Nations (ASEAN) blocs. Indicators at global level would include, for example, sea-level change, atmospheric greenhouse gas concentrations, fish stocks, biodiversity and aggregated national-level metrics such as poverty and health statistics. Thematic assessments refer to specific sectors, industries or operations, for example the corporate, health, agricultural and education sectors. Indicator selection is likely to be specific to each theme, as discussed at the start of Section 2.2.6.

How the suite of indicators should be interpreted is an issue of ongoing debate. Indicators can either be left disaggregated, perhaps with a selection displayed as a dashboard, or combined into individual indexes. Indexes can be further combined and expressed as a single aggregated number that can subsequently be used in ranking systems. Two examples where a broadly similar set of indicators is expressed in these two ways are provided by the Wellbeing model and the Doughnut model. In the former, wellbeing and sustainability is expressed as a combination of human wellbeing and ecosystem wellbeing. The various indicators and sub-indicators representing each of these factors are quantified, expressed as ratios reflecting distance to target, and then aggregated into a national Human Wellbeing Index and an Ecosystem Wellbeing Index, which are plotted as axes on a graph colour-coded in five bands from bad to good.[53] The Doughnut model argues for a political economy operating within a "safe and just" (doughnut-shaped) space between an inner social and an outer planetary boundary. Indicators parallel to those used in the Wellbeing model are quantified and then individually graphed onto the doughnut, leaving them disaggregated and without gradations.[54]

Given the large number of indicators across diverse SDG targets, and the interconnectedness of the three pillars ("in general economic sustainability has an ecological cost and sustainability has an economic cost",[55] some form of integration is inevitable if only to better focus policy and other types of interventions. This might take the form of developing indexes representing particular themes, or of combining weighted indicators into a single overall score to facilitate ranking.[55,56] Examples of the latter are provided in Section 2.2.7. From the array of indexes proposed, a selection is presented in Table 2.2.[57,58]

Table 2.2 A selection of sustainability indexes.

Theme	Selected indexes
Innovation, knowledge and technology	Innovation Index
	Technology Achievement Index
Development	Corruption Perception Index
	Democracy Index
	Education Index
	Human Development Index
	Life Expectancy Index
	Index of Sustainable and Economic Welfare
	Sustainable Society Index
	Water Poverty Index
Market and economy	Energy Sustainability Index
	Gross National Product
	Internal Market Index
	Business Climate Indicator
Ecosystem	Air Quality Index
	Biodiversity Index
	Ecological Footprint
	Environmental Protection Index
	Living Planet Index
	Water Quality Index
Products and services	Eco-innovation Index
	Life Cycle Index
	Product Sustainability Index
Cities	City Development Index
	Compass Index of Sustainability
	Green Score Cities Index
	The Sustainability Cities Index
	Urban Sustainability Index
Industry	COMPLIMENT – Environment Performance Index for Industries
	Eco-points
	Eco-compass
	Eco-indicator 99
	Sustainable Circular Index
	Sustainable Process Index
Social	Gender Parity Index
	Gender Empowerment Measure
	Gini Coefficient/Index
	Gross National Happiness
	Happy Planet Index
	Health System Performance Attainment
	Index for Sustainable Society
	National Health Care Systems Performance
	Physical Quality of Life Index
	Social Progress Index
	Wellbeing Index

In summary, the measurement of sustainability is an evolving, vigorous discipline and a subject of considerable academic and institutional interest, which accounts for the volume, range and diversity of approaches proposed. A number of fundamental methodological issues remain to be resolved, such as a standardised approach when applying multicriteria analysis to aggregate indicators for ranking purposes.[55,56] As noted previously, consistency of approach is vital for robust trend analysis and inter-country comparisons.

To this end efforts to introduce consistency are evident, in general coalescing around various thematic strands. For example, the Global Reporting Initiative (GRI) has developed standardised bespoke sustainability measurement and reporting frameworks that have been widely adopted by major corporations and by industries such as agriculture, aquaculture and fisheries.[59] As discussed in Chapters 1 and 12, national and international standards bodies are developing formal standards for measurement and reporting. Coupled with diffusion of best practice by regulatory agencies, trade bodies and institutions, there are encouraging signs that standardisation is beginning to emerge.

2.2.7   Progress Towards SDGs

The UN publishes an annual update on progress towards meeting the 2030 Agenda. In addition to these updates, signatories to the 2030 Agenda are required to produce at least one Voluntary National Review, assessing progress towards achieving the SDGs. The UK, for example, presented its first review to the UN High-level Political Forum in 2019, while Norway reports annually. Other studies have assessed regional and global compliance. Progress towards SDG goals was found to be patchy across the EU, while a meta-analysis of over 3000 scientific studies published over 5 years found that, globally, although the goals "have had some political impact ... this impact has been largely discursive", with "more profound ... impact ... remaining rare".[32,60] Another study covering 156 countries for which adequate information was available supported the conclusion that compliance was uneven.[34]

Methodologies have been developed to rank countries in relation to their progress towards attaining SDGs (see Section 2.2.6). For example, financial market data provider Refinitiv mapped 242 detailed indicators feeding into each of the 17 SDGs against 8.5 million active economic indicators over the past 70 years, available through its macroeconomic time series database Datastream. Scored out of ten, the top ten and bottom ten countries in terms of progress since 2015 are listed in Table 2.3.[61] Also listed in the table are the SDG indexes for top and bottom ten countries (scored out of 100) according to a comprehensive analysis published in 2022.[62]

While relative positions vary according to the different methodologies and year of assessment, the pattern is clear: challenged states and economies, conflict zones and other restraining factors inhibit actioning SDGs. The COVID-19 pandemic has exacerbated the implementation challenge, as the bleak picture painted in the UN's latest assessment for 2021 reports: owing to the pandemic, "2021 will be decisive as to whether or not the world can make the transformations needed to deliver on the promise to achieve the SDGs by 2030 ... [y]ears, or even decades, of progress have been halted or reversed".[63]

Table 2.3 Progress towards SDGs: Refinitiv country scores and SDG indexes. Data from ref. 61 and 62.

Top 10 countries		Bottom 10 countries	
Refinitiv (2020)	SDG Index (2021)	Refinitiv (2020)	SDG Index (2021)
Norway	Finland	Somalia	South Sudan
UK	Denmark	Central African Republic	Central African Republic
Switzerland	Sweden	Eritrea	Chad
Sweden	Norway	Yemen	Somalia
Netherlands	Austria	Libya	Sudan
Iceland	Germany	Haiti	Liberia
Germany	France	Papua New Guinea	Democratic Republic of the Congo
Denmark	Switzerland	Chad	Madagascar
Finland	Ireland	Guinea Bissau	Djibouti
France	Estonia	Liberia	Angola

In other words, the target date is in danger of slippage. This trend has been noted by others, for example in 2022: "For the second year in a row, the world is no longer making progress on the SDGs ... partly due to slow or non-existent recovery in poor and vulnerable countries ... Before the pandemic ... the world was progressing on the SDGs at a rate [too slow] to reach the 2030 deadline ... restoring and accelerating SDG progress in all countries ... should be a major priority of recovery plans and reforms to the international development finance system".[62]

A useful compendium of progress to date towards achieving individual SDGs has been published.[64] Highlights from the report are provided in Table 2.4.

In summary, the challenge is that "sustainability [cannot be achieved] at any scale without fundamental and radical transformations in human activities and supporting financial, legal, political, and governance systems". Five principles of transformation have been proposed: rethinking growth; rethinking efficiency, rethinking the state, rethinking the commons and rethinking justice.[65]

2.3 The Circular Economy

2.3.1 Concepts and Definitions

As with sustainability and sustainable development, the origin of the term "circular economy" is unclear – there is no seminal text to which the term can unequivocally be attributed. Some commentators claim Boulding was the first to introduce the concept, though he did not use the term itself.[66] Others point to Pearce and Turner, while they in turn refer back to the academic literature of the 1970s and 1980s, in which the term was first used to describe a

Table 2.4 SDGs – progress to date and prospects. Data from ref. 64.

SDG	Progress to date	Prospects
1	The number of people living in extreme poverty has fallen by 36% between 1990 and 2015. The COVID-19 pandemic has reversed these gains – extreme poverty levels rose in 2020	SDG 1 was not on track prior to the pandemic. According to the UN, "without immediate and significant action, it will remain beyond reach"
2	The number of people falling into hunger has increased since 2015 to ≈690 million, equivalent to ≈9% of the global population	The UN warns that this number could surpass 840 million by 2030, and that the underlying factors are predominantly human-made (e.g., poorly performing economies)
3	The UN claims "significant strides" have been made, especially in terms of increasing life expectancy and responding to disease outbreaks, but that the pandemic has reversed recent gains	Positive signals are evident, for example an increase in the number of health professionals, a decline in the mortality rate for those aged under 5 years, etc. A post-pandemic action plan needs to keep the gains on track
4	SDG 4 has progressed positively. Before COVID-19 the percentage of children who were out of primary and secondary school education had declined from 26% in 2000 to 17% in 2018	More children have access to primary and secondary education; more teachers are being trained. Expenditure in areas such as education has increased between 2019 and 2021
5	Strong progress has been made in education and employment. The "average gender gap" is currently ≈31%, the aim being to reduce this to zero	Steady progress is being made to reduce gender inequality, though still too slow to meet the 2030 target date
6	Globally, access to drinking water sources has increased from 76% to 90% since 1990. However, other indicators are concerning, for example more than 80% of wastewater is discharged without pollution control	More than 2.2 billion people are without access to safely managed drinking water, plus a further two billion without access to any sanitation. Water scarcity is one of the most severe threats facing the global population
7	Strong progress has been made, with access to electricity for an additional 1.7 billion people between 1990 and 2010	The price of renewables continues to fall, providing more of the global population with affordable, cleaner energy
8	Progress has slowed, with global annual GDP growing by ≈2% between 2010 and 2018, slowing to 1.5% between 2018 and 2019. Unemployment rates are stagnating	The UN has urged nations to deliver a financial boost to markets through policies, and invest in community-led initiatives, to reverse the COVID-19 economic slowdown
9	Research and development investment increased from 1.5% in 2000 to 1.7% in 2015, flatlining thereafter. Manufacturing Gross Value Added (GVA) increased from 15% in 2005 to over 16% in 2017. Online connectivity and 3G internet access has increased dramatically	The UN urges the least developed countries to increase their manufacturing capacity and capabilities, and to scale up investment in innovation and research

10	Economic inequality is increasing for over 70% of the global population. In 2015, 40% of global income went to the richest 10%, while the poorest 10% received less than 7%. Health outcomes were worse in the poorest 20% of the population	While acknowledging the efforts made to reduce economic inequalities, the UN urges policies aimed at an inclusive post-pandemic recovery, including access to better healthcare, welfare support and economic opportunities for persons with disabilities
11	The global building stock growth is failing to keep pace with the growing urban population, resulting in urban sprawl, slums, air pollution and poor basic services	SDG 11 requires a complete rethink on how cities and living spaces are planned and expanded. Signs are positive – more countries with urban strategies and better city services
12	According to the UN, current production and consumption is outpacing the earth's natural resources and ability to replenish. This applies to food sources as well as to drinking water	A global reappreciation for consumption is required, with more focus on decoupling of economic growth from resource use, that is, moving with greater speed towards circular closed-loop economies
13	The Intergovernmental Panel on Climate Change (IPCC) warns of slow progress on adaptation with global warming trending towards exceeding 1.5 °C on pre-industrial temperatures, potentially leading to serious consequences to economies and to public wellbeing	The COP26 Glasgow Climate Pact provides a platform and vehicle for action, urging developed countries "to at least double their collective provision" of climate finance for adaptation to developing nations from 2019 levels by 2025
14	The UN warns that one-third of the world's fisheries are overfished. Global warming and pollution from human activities on land is leading to marine health and productivity being jeopardised	Significant economic benefits can accrue if pollution into waterways is reduced. The One Ocean Summit (2022) pledged EUR 4 billion of ocean finance by 2025 towards marine conservation and restoration
15	Deforestation and desertification are increasing. The UN states that land degradation has reduced productivity by 23%, and up to USD 577 billion in annual crop output is currently at risk	Despite more countries committing to protecting forests, other natural ecosystems and biodiversity and setting appropriate targets, action has been slow. More than half of global GDP is at risk from nature loss
16	According to the UN, 100 civilians are killed in armed conflict daily. The number of people fleeing conflict and persecution in 2019 reached a new high. Climate change and unsustainable resource management will exacerbate conflict and mass migration	While access to information and legislative approaches to justice are improving, overall significant challenges remain to agree and implement a global compact on administering justice and the rule of law, and reducing marginalisation
17	The UN states that the COVID-19 pandemic has reduced both national and international partnering and aid-giving. ODA continues to fall and donor countries are failing to deliver on pledges to increase and support global finance developments	The number of cross-sector partnerships, especially with the corporate sector, has increased dramatically, potentially committing significant finance to SDGs to offset the funding shortfall. The challenge is to convert these commitments into concerted action

closed-loop economy.[67] It has been noted that the "linear economy" was coined and popularised at that time in opposition to the "circular economy", contrasting two different world views: the former a "cowboy economy" (Boulding's phrase) leading to pollution and environmental degradation, the latter mimicking the earth's natural biological and geochemical cycles whereby feedback loops eliminate waste and maintain a sustainable equilibrium across all environmental systems.[66,68,69] This dichotomy has been used to characterise the circular economy as standing in opposition to a linear economy: the latter assumes "there is an unlimited supply of natural resources and that the environment has an unlimited capacity to absorb waste and pollution".[70]

Typically, the circular economy is seen as a portmanteau term:[71]

The circular economy model synthesises several major schools of thought. They include the functional service economy (performance economy) of Walter Stahel; the Cradle to Cradle design philosophy of William McDonough and Michael Braungart; biomimicry as articulated by Janine Benyus; the industrial ecology of Reid Lifset and Thomas Graedel; natural capitalism by Amory and Hunter Lovins and Paul Hawken; and the blue economy systems approach described by Gunter Pauli.

Given the slipperiness of the concept, reviews of the circular economy literature have highlighted a number of ways in which the circular economy is defined, identifying between eight and 114 different concepts and definitions.[72,73]

The United Nations Environment Programme regards a circular economy as one which:

… balances economic development with environmental and resource protection, with the aim of designing out waste, circulating technical nutrients to delay or eliminate their entry into the environment, returning biological nutrients safely, and using renewable energy to power the economy.

The restorative and regenerative potential of a circular economy is its main attribute.[74]

Attempting to formalise the concept of restoration and regeneration, another definition states that the circular economy is.[75]

… a regenerative system in which resource input and waste, emission and energy leakage are minimised by slowing, closing, and narrowing material and energy loops. This can be achieved through long-lasting design, maintenance, repair, reuse, remanufacturing, refurbishing, and recycling.

More elaborate definitions attempting to capture a broader range of aspects have been suggested – see Section 2.3.2.[76,77]

Undoubtedly the most common representation of the circular economy has been in the context of the so-called waste hierarchy (prioritising various forms of value retention and creation over disposal) together with the 3Rs – Reduce, Reuse, Recycle – which has been expanded to 5Rs, 6Rs (Rethink, Refuse, Reduce, Reuse, Recycle, Replace), 9Rs, 10Rs and even 12Rs.[78,79] Particularly in the early stages of its development, this has had the effect of focusing its application on the waste management/recycling/repair and associated production/manufacturing sectors. It has, however, been pointed out that this view emasculates the true potential of the concept – rather than regarding the circular economy as a *sector* it should be treated as a *system*, with far more wide-ranging applicability, as the ensuing chapters of this book demonstrate.[80]

Beyond the depiction of circularity as the integration of two halves of the economy – the production and consumption cycle with the waste cycle into a seamless whole – with no formal, settled definition, the notion has become a catch-all for an array of policy objectives. Is it about resource conservation? Resource efficiency? Resource substitution? Landfill diversion? Waste prevention? Ecodesign? Remanufacture? Recycling? Resource security? Jobs and sustainable growth? Combating climate change? Changing consumption patterns?

The general response has been "all of the above", and certainly they are all inter-related. But many of these policy strands are open to different interpretations. For example, it is possible to improve resource efficiency at a process level (more output for less input) while the business as a whole remains within the linear economy (products are discarded after use, rather than entering a regenerative system). The majority of project commitments within the European Commission's Raw Materials Initiative (a feeder into its Circular Economy Package) fall into this category.[81]

2.3.2 The Circular Economy and Sustainability

As the name suggests, the circular economy has most often been thought of in terms of economic development – specifically, given the language of recycling, repair and reuse, of the (industrial) production economy, with the corporate sector as its principal actors. However, as other chapters in this book demonstrate, the concept has wide-ranging applicability, including to natural resources (biotic and abiotic materials), water and land.[82]

Regarding the impact of the circular economy on the social and environmental pillars of sustainability (see Section 2.2.2), studies have generally given qualified acknowledgement of a positive contribution. For example, while job creation is undeniably a feature of the circular economy (for example in refurbishment and repair), whether this inevitably enhances social equity is by no means obvious.[10,68,75] Similarly, while the take–make–dispose philosophy of the linear economy is intuitively unsustainable in the long run, poorly designed circular economy solutions, while conferring individual environmental benefits such as resource conservation and reduced carbon emissions, may lead to sub-optimal environmental consequences when assessed in net terms across all impact categories.[10,32,83] In the words of one study, the circular economy concept should be "reworked to acknowledge the reality that closed-loop recycling is not always more beneficial than open-loop recycling and that remanufacturing is not always preferable to recycling from an environmental perspective".[77] We prefer the phrase that as a matter of course an environmental analysis of circular economy solutions should be undertaken against alternatives, to identify and quantify net impacts, ultimately to identify environmentally optimal solutions.

Considered in the round, the circular economy is most often viewed as a vital operational/technical tool to drive sustainable development, which needs to incorporate systems such as life-cycle analysis and corporate social responsibility into a more rounded and integrated sustainability paradigm. With this *proviso*, alternative definitions to those presented in Section 2.3.1 have been suggested to capture more fully the link between the circular economy and the three pillars of sustainability. One such definition is as follows:[73]

> *[The circular economy] is an economic system that is based on business models which replace the end-of-life concept with reducing, alternatively reusing, recycling and recovering materials in production/distribution and consumption processes, with the aim to accomplish sustainable development, which implies creating environmental quality, economic prosperity and social equity, to the benefit of current and future generations.*

Another definition that highlights sustainability is as follows.[68]

> *[The circular economy] is an economic model wherein planning, resourcing, procurement, production and reprocessing are designed and managed, as both process and output, to maximise ecosystem functioning and human well-being.*

A more extreme definition that places the circular economy firmly within the ambit of sustainability is expressed as "seven pillars":[84]

- materials are cycled at continuous high value
- all energy is based on renewable sources
- biodiversity is supported and enhanced through human activity
- human society and culture are preserved
- the health and wellbeing of humans and other species are structurally supported
- human activities maximise generation of societal value
- water resources are extracted and cycled sustainably.

The social and environmental sustainability benefits of a circular economy approach have received more recent attention, primarily through publication of case studies.[85,86,87] The extent of convergence of the concepts underpinning the circular economy and sustainability was examined in a bibliographical study identifying similarities and differences.[75] Similarities included:

- emphasis on intra- and intergenerational commitments
- a global perspective
- multi- or interdisciplinary approaches
- integrating non-economic aspects into development
- cooperation between stakeholders as imperative
- regulation and incentives as core implementation tools
- the central role of the corporate sector, due to resources and capabilities.

Differences included the emphasis placed on each of the three pillars – economic, social and environmental – owing to the different origins and development paths of the concepts; the time frames to achieve particular goals; the perception as to who the primary beneficiaries are intended to be; and the agencies through which change might be effected, for example regulation and incentives *versus* education and engagement.

A recent study analysed three strategies – circular resource use, operational eco-efficiency (more return per unit of resource used) and sufficiency (less resource used by limiting the consumption of products and services) – within a single model to examine the overall effect of their interactions in reducing resource consumption to a sustainable level when they are applied in tandem.[88] Using example scenarios, the study illustrated that applied singly, each strategy could have either a positive or a negative effect on the others, indicating that policy implementation should develop an optimised collective approach rather than applying any one individual strategy to the exclusion of the others.

2.3.3 The Circular Economy and SDGs

Alongside the release of the 2030 Agenda in 2015 came the recognition that the circular economy could play an important part in achieving particular SDGs: it has been "commonly accepted … that the linear economy cannot lead to sustainable development".[10] For example, the EU stated in a 2015 action plan that "the circular economy [will be] essential … for the implementation of … the UN 2030 Agenda for Sustainable Development … and … will be instrumental in reaching the SDGs by 2030, in particular Goal 12 of ensuring sustainable production and consumption patterns".[89] Responsible consumption is discussed in Chapter 6.

Searching the literature for keywords, one study mapped the circular economy (CE) onto the SDG targets using a common framework:[83]

- direct, strong contribution of CE practices to achieving the goal
- indirect contribution (*via* another SDG being targeted)
- progress on target supports uptake of CE practices
- cooperation opportunity to promote CE practices
- weak or no link.

The study concluded that circular economy practices can contribute directly to achieving 21 of the 169 targets, can contribute indirectly to 28 out of 169, support 52 out of 169 and promote 33 out of 169; finally, the study identified 35 targets with weak or no links to the circular economy. A more detailed target-wise representation broken down by sustainability pillar has been provided, as summarised in Table 2.5.[77,83]

Overall, the strongest (direct and indirect) linkages were between the circular economy and targets in SDGs 6, 7, 8, 12 and 15. Targets in SDGs 1, 2 and 14 were indirectly affected by circular economy practices. SDGs 4, 9, 10, 13, 16 and 17 "show a relationship insofar as progress on the targets would positively contribute to uptake of circular economy practices globally".[83] Targets in SDGs 3, 5, 10, 11 and 16 showed no or only weak links. The strong links identified in the study can be compared with the recommendations of the UN as to where "the circular economy holds particular promise for achieving multiple SDGs", namely SDGs 6, 8, 11, 12, 13, 14 and 15.[90]

These positive and negative relationships chime with the predominantly economic/production focus of the circular economy, with a relatively weaker link to the environmental pillar and, weakest of all, to the social pillar of sustainability. To that extent, the findings are perhaps self-fulfilling, but

Table 2.5   The impact of the circular economy on SDG targets. Data from ref. 77 and 83.

Impact type	Economic pillar	Social pillar	Environmental pillar
Direct (21)	7.1, 7.2, 7.3, 8.2, 8.4, 9.2, 9.4	2.4, 3.9, 6.1, 6.2, 6.4, 11.6, 12.2, 12.4, 12.5	6.3, 14.1, 15.1, 15.2, 15.3
Indirect (28)	1.1, 7b, 8.1, 8.5, 8.6, 10.1, 12.6	1,2, 1.8, 1b, 2.1, 2.2, 2.3, 11.1, 11.2, 11.4, 12.1, 12.3, 12.8, 12b, 16.1	6.6, 13.1, 14.2, 14.3, 15.4, 15.5, 15.7
Contribution/ support (52)	1.4, 5.5, 5a, 8.3, 8.8, 8.9, 8.10, 9.1, 9.3, 9.5, 9a, 9b, 9c, 10.4, 10.6, 10a, 12.7, 12c, 17.4	2.5, 2b, 2c, 4.3, 4.4, 4.5, 4.6, 4.7, 10.2, 11.3, 11a, 11b, 16.5, 16.7, 16.8, 16.10, 17.10, 17.11, 17.12, 17.13, 17.15, 17.18	13.2, 13.3, 13a, 14.5, 14.6, 14b, 16.6, 16b, 17.14, 17.19
Cooperation/ promotion (33)	7a, 8a, 8b, 10b, 17.1, 17.2, 17.3, 17.5	1a, 2a, 3d, 4b, 4c, 5b, 6.5, 6a, 6b, 11c, 12a, 17.6, 17.8, 17.9, 17.17	13b, 14.7, 14a, 14c, 15.6, 15a, 15b, 15c, 17.7, 17.16
No/weak link (35)	5.1, 5.2, 5.3, 5.4, 8.7, 10.5, 10c	1.3, 3.1, 3.2, 3.3, 3.4, 3.5, 3.6, 3.7, 3a, 3b, 3c, 4.1, 4.2, 4a, 5.6, 5c, 10.2, 10.3, 10.7, 11.5, 11.7, 16.2, 16.4, 16.9, 16a	14b, 15.8, 16.3

also emphasise the point that the circular economy's impact should best be measured against *targets and indicators* as opposed to the overall *goals*. Municipal waste management provides an example – despite the identification of a weak link or no link between the circular economy and SDG 11 (see above), the circular economy is in fact particularly strongly linked to Targets 11.6 and 12.5 and their associated indicators, as would be expected given the discussion in Section 2.3.1 on the dominance of the waste hierarchy and the 3Rs in conceptual thinking.[33,91]

It has been noted that the circular economy does not figure in the 2030 Agenda, nor in the annual SDG progress reports of the UN Secretary General.[77] Furthermore, the absence of settled theoretical underpinnings for either the circular economy or for sustainability, and the fact that the two concepts have developed from different starting assumptions, suggests that the early literature is likely to converge on goals and targets relating to the economic pillar and will tend to diverge on goals and targets in the two remaining pillars. As this lacuna is increasingly recognised and the circular economy concept in the context of sustainable development evolves as a consequence (see for example the expanded definition presented in Section 2.3.2 explicitly incorporating environmental and social aspects) it would be interesting to update the analysis with post-2017/18 literature in which SDGs regarded as having no or weak links with the circular economy receive more attention.[92] Indeed, more recent thinking has emphasised the wider benefits the circular economy can bring in implementing the SDGs, benefits that are not limited to the technical resource/material cycle.[93]

2.3.4 Financing the Circular Economy

Expanding on the new business models alluded to in Section 2.3.2,[73] five such models along circular economy lines have been proposed:[94]

1. circular supplies
2. resource recovery (waste as a resource)
3. product life extension
4. sharing
5. product as a service.

While the circular economy was seen as a growing market estimated to generate 1% to 4% economic growth over 10 years, financing of these business models came with some challenges:

- the changing nature of the cash flow of the firm (for example, from outright sale of a product to leasing or servitisation)
- increased up-front capital needs (for example, replacing production equipment still within its operating life)
- legal issues surrounding collateral and its value (for example, accepting contractual cover instead of the right of legal ownership over assets).

Fundamental barriers to investing in circular economy solutions exist, characterised as market imperfections, that are more difficult to overcome because they rely on measures such as regulation, fiscal adjustments and industry standards to level the playing field. These include unpriced positive and negative externalities not acknowledged when costing a linear business model, but which would put at a disadvantage a circular investment that takes into consideration life-cycle social and environmental costs; short-termism in decisions involving allocation of capital; subsidised sectors offering protection to linear models; and regulatory uncertainty, for example concerning the status of a product made from waste.[95] These barriers and proposed de-risking options are discussed in more detail in Section 2.6. To support the transition to circularity, financial institutions will need to develop valuation and risk models that suit the characteristics of circular business models, for example by capturing the value of upcycling/recovering a discarded product in business cases, rather than writing down its value to zero.[94]

Since the late 2010s the number of financial instruments targeting circular economy outcomes has grown dramatically. Managed assets in public equity funds with a circular economy focus are reported to have grown 26-fold in less than two years from USD 0.3 billion to USD 8.0 billion in 2021; corporate and sovereign bonds have grown around five times from USD 4.5 billion to USD 21 billion in 2021.[95] Different numbers but comparable growth rates have been reported by others.[92,96] Specialised circular equity funds are being designed, for example, the ECPI Group's Circular Economy Equity Leader Index comprising categories such as circular supplies, resource recovery, product life extension, sharing platform, and product as a service.[95]

It is, however, salutary to note that the USD 800 billion of annual corporate spend with a circular economy focus accounts for a little over 2% of the USD 35 trillion that the private sector spends on the take–make–dispose linear economy. The global annual government closed-loop spend of USD 510 billion amounts to 4% of the total global public sector spend of USD 13 trillion.[92]

2.3.5   Measuring the Circular Economy

As with sustainability and sustainable development, the measurement of circularity depends on its definition and the specific context to which it applies. Relatively circumscribed definitions based on closed-loop economic models (see Section 2.3.1) tend to focus on measurements relating to resource flows and the resource cycle (recovery, recycling, renewable energy, *etc.*) whereas more expansive definitions such as that of the "seven pillars" in Section 2.3.2 will include, along with resource considerations, measurements of environmental and social impact.[97] Along this continuum lies a range of possible measurement frameworks and indicators depending on the context.

The European Environmental Agency and the Italian Institute for Environmental Protection proposed a set of principles called the Bellagio Declaration for developing circular monitoring systems:[98]

1. monitor the circular economy transition
2. define indicator groups
3. follow indicator selection criteria
4. exploit a wide range of data and information sources
5. ensure multilevel monitoring
6. allow for measuring progress towards targets
7. ensure visibility and clarity.

A classification of circular economy indicators has been proposed based on the extent to which life-cycle considerations are included in the analysis, together with a system reflecting the scale of application:[97,99]

- **Scope 0:** indicators restricted to measuring physical properties from the technological cycle, with no life-cycle considerations.
- **Scope 1:** indicators measuring physical properties from the technological cycle with full or partial life-cycle considerations.
- **Scope 2:** indicators measuring environmental, economic and/or social burdens/benefits from the technological cycle.
- **Micro level:** circular economy indicators applied to a specific service, operation, product or substance at a business or a local level.
- **Meso level:** circular economy indicators applied at a larger scale (industrial parks, cities, regions) involving agglomerations of diverse processes, operations, activities with a variety of consumption behaviours.
- **Macro level:** circular economy indicators applied at country level with emphasis on exchanges of materials between economy and environment, international trade and material accumulation in national economies.

Examples of indicator frameworks applied in these taxonomies are provided in Table 2.6.[97,100,101]

It is apparent that the majority of indicators fall within Scope 0 regardless of scale of application and, although expressed in a variety of ways, they are broadly compatible at different scales, implying that resource flow indicators applied at micro level (such as company recycling rates) can be aggregated up to meso (regional) or macro (country) level, taking care to avoid double-counting of resource flows. Indeed, China's circular economy indicator system is applied at all scales. This further implies that aggregation of judiciously selected Scope 0 resource flow indicators up to a dashboard and further to a single numerical ranking or expression of circularity (see Section 2.3.6) can have some methodological validity and is arguably less contentious than aggregation of SDG indicators across diverse impact categories, as discussed in Sections 2.2.6 and 2.2.7. Some Scope 1 and

Table 2.6 Examples of circular economy indicator frameworks. Data from ref. 97, 100 and 101.

Scope	Micro scale	Meso scale	Macro scale
	WBSCD circular transition indicators	EU circular economy monitoring framework	China's circular economy indicator system
0	1: Percentage material circularity 2: Percentage water circularity 3: Percentage renewable energy 4: Percentage critical materials 5: Onsite water circulation	1: Self-sufficiency for raw materials 2: Green public procurement 3a–c: Waste generation 4: Food waste 5a–b: Overall recycling rates 6a–e: Recycling rates (various packaging materials, biowaste, recovery of construction and demolition waste)	1: Resource productivity 2: Recycling rates pf principal waste streams (agricultural, industrial, renewables, urban food waste, *etc.*) 3–7: Comprehensive utilisation (recycling and reuse) rates of straw stalk, industrial solid waste, building material waste, kitchen waste, *etc.* 8: Reuse rate of industrial water 9: Recycling rate of renewable resources 10–13: Discharge of waste (solid waste, industrial sewage, landfill, discharge of main pollutants) 14: Energy productivity 15: Water productivity 16: Construction land productivity 17: Gross value output of recycling industries
1	6: Percentage recovery type (recycling, repair, remanufacturing, *etc.*) over lifetime 7: Circular material productivity	6f: Recycling rate of e-waste 7a–b: Contribution of recycled materials to raw materials demand	
2	8: Circular transition revenue 9: Greenhouse gas impact (future elements will cover other impacts)	8: Trade in recyclable raw materials 9a–c: Private investments, jobs, GVA 10: Number of patents	

Scope 2 indicators in Table 2.6 would also appear to be amenable to such treatment if metrics and time frames are coordinated (for example, circular economy jobs, revenue from circular activities and greenhouse gas emission reductions).

2.3.6   Progress Towards Circularity

For the circular economy to replace the linear economy as a policy objective and, ultimately, as an operational reality, the relevant implementing measures, together with the tools to measure and track this transition, need to be in place. Unsurprisingly for such a wide-ranging concept and the entrenched nature of the linear economy, this has proved a challenge. Indeed if the business case for the circular economy was as strong and self-evident as is often claimed, then economic actors would spontaneously adopt this as their preferred business model. Save for small pockets of excellence and individual examples of good practice, the global economy is still firmly wedded to the linear economy. An estimate of circularity taking into account resource use across seven sectors (housing, services, nutrition, mobility, healthcare, consumables and communication) found that in 2018 the global economy was only 9.1% circular, falling to 8.6% in 2020.[102] Raworth claims that a "truly circular economy belongs to the fantasy of perpetual motion machines" owing to inevitable leakage from an industrial loop: in the case of Japan, 98% domestic recycling of metal *versus* a 2% loss from the system.[54] Certainly, no industrial system is perfect – an element of redundancy is inherent in every activity – but the present 91–92% leakage from a global circular loop indicates that, even if theoretically pure circularity is unattainable, there is considerable room for improvement in reducing waste and in putting unavoidable discards and secondary resources to better use before a residual 2% leakage gives cause for concern.

A UNGC survey of 1000 leading Chief Executive Officers (CEOs) found that their greatest problem was the lack of a link between sustainability and business value – "in many cases, business leaders feel that given the structures, incentives and demands of the market, they have taken their companies as far as they can".[103] Eighty-four per cent of CEOs interviewed called for active intervention by governments and policymakers to align public policy with sustainability, preferring hard interventions such as regulations, standards and tax measures, realigning the market and price signals through appropriate policy interventions.

Obstacles to achieving a circular economy have been identified, ranging from fiscal and financial to institutional to societal. A summary of the findings of one study is provided in Table 2.7,[104] which is broadly in line with other research.[69] The daunting list of obstacles is no doubt influenced by the global reach of markets – raw materials sourced from one part of the world, products made in another and sold in a third.

A range of measures have been suggested to aid the transition towards a circular economy, as shown in Table 2.8.[92,104]

Table 2.7   A summary of obstacles to achieving a circular economy. Data from ref. 104.

Theme	Obstacles
Financial	Major up-front investment costs
	Environmental costs (externalities are not taken into account)
	Shareholders with short-term agenda dominate corporate governance
	Recycled materials are often more expensive than virgin ones
	Higher costs for management and planning
Institutional and government policies	Unlevel playing field created by current institutions
	Financial governmental incentives support the linear economy
	Circularity is not effectively integrated in innovation policies
	Competition legislation inhibits collaboration between companies
	Recycling policies are ineffective to obtain high-quality recycling
	Governance issues concerning responsibilities, liabilities and ownership
Infrastructure within the value chain	Limited application of new business models
	Lack of an information exchange system
	Confidentiality and trust issues hamper exchange of information
	Exchange of materials is limited by capacity of reverse logistics
Societal and value-related benefits of recycling material	Lack of awareness and sense of urgency, also in businesses
	GDP does not show the real progress or decline of our society
	Resistance from powerful stakeholders with large interests in *status quo*
Technology, knowledge and data	Limited attention for end-of-life phase in current product designs
	Limited availability and quality of recycling material
	New challenges to separate the bio- from the techno-cycle
	Linear technologies are deeply rooted

Some headway has been made in terms of fiscal and regulatory policy development, metrics and indicators, circular business models and measurement and tracking tools, as the examples below illustrate:

- In 2020 Sweden introduced tax breaks on repairs for consumer goods, cutting the Value Added Tax (VAT) rate on repairs from 25% to 12%. Residents can claim back from income tax half the labour cost of appliance repair. Changes to the terms of warrantees have been enacted in countries such as France, Austria and Norway. Producer obligations *vis-à-vis* product take-back have been strengthened.

Table 2.8 Measures to enable a circular economy. Data from ref. 92 and 104.

Theme	Measures
Supporting measures	Set up a simple index for circular performance
	Encourage experimentation, innovation and redesign
	Gather and spread successful business examples
	Integrate the circular economy principles in education and training programmes
	Initiate and stimulate stakeholder forums about the circular economy
Corporate strategies and markets	Develop a long-term company vision identifying linear risks and circular economy opportunities
	Search for material pooling opportunities
	Promote circular products using modern marketing techniques and social media
	Prepare roadmaps for established economic sectors
Transforming the economy	Replace traditional financial reporting by mandatory and accountable integrated reporting
	Develop the concept of True Value
	Create a tax shift from labour towards resources
	Implement a new economic indicator beyond GDP that steers towards circularity
	Establish international independent systems to organise material flows
	Adjust national and international government policies

- Fiscal and regulatory measures such as landfill tax and bans, aggregates tax, peat tax, plastic packaging tax, plastic bag levy, bans on certain single-use plastic products, carbon tax, *etc.*, are moving tax reform away from labour and towards externalities and beneficial environmental outcomes of materials, products and emissions, a key policy enabler to further circularity and sustainability.[105,106]
- Organisations such as the Ellen MacArthur Foundation, C40 Cities and the WBCSD have taken great strides in developing, promoting and supporting circular sectoral strategies, providing communities of stakeholders with platforms for sharing success stories.
- Tools have been developed for corporations to measure and transition to circularity – such as Circulytics (Ellen MacArthur Foundation), the EU Product Environmental Footprint (PEF) and Organisation Environmental Footprint (OEF) Guides, the Circular Transition Indicators Tool (WBCSD) and the Circular Economy Playbook (Sitra/Accenture).

Nevertheless, the findings on the circular economy noted previously, of overall poor global performance, emphasise the need for a paradigm shift from the linear economy, that in turn suggests a more radical realignment of resources, markets and consumption.[65,107]

While the role of corporations and of the public sector in driving the circular economy has been acknowledged and analysed in some detail, the role of the third sector and of NGOs "as key agents of change in the shift towards the concepts of resource efficiency and circular economy, at the community level" has received less recognition or attention.[108,109] The third sector offers a unique entry point into the circular economy, reaching and engaging with communities that are often bypassed by other circular economy actors, closing a resource loop that would otherwise remain linear (for example, supporting repair and repurposing activities on discarded products that would otherwise be consigned to landfill) while offering opportunities for environmental improvement, economic regeneration, job creation and poverty alleviation (see Chapter 3).

2.4   Conclusions

Considerable strides have been made, with generally positive trends, in progressing the majority of the SDGs, especially since the Paris Agreement of 2015, until the start of the COVID-19 pandemic in 2019. However, much remains to be done if the 17 SDGs are to be achieved by the target date of 2030. The same applies to the goal of increasing circularity. Financing of and investment in SDGs is well below the level required to sustain the required trajectory, while the contributions of both the public and private sectors are still overwhelmingly directed towards the linear economy. Progress to date is starkly divided, with developed economies and stable polities scoring well overall relative to challenged economies and unstable regions, exacerbated by chronic underinvestment in low- and middle-income countries, and more recently by the negative legacy of the COVID-19 pandemic. Furthermore, the private sector in particular has called for appropriate regulatory and fiscal realignment to level the playing field between the linear and circular economies, without which neither the circular economy nor, by implication, many of the SDGs will be fully realised.

Sustainability and circularity have complementary goals; as concepts they have evolved from different starting principles and in different directions – the former with a greater emphasis on environmental protection and social equity and the latter more towards a closed-loop economy – however, in developmental terms, both emphasise the importance of managed economic growth sensitive to the environment. The circular economy has direct relevance to a number of SDGs, especially those with an economic and resource efficiency focus, as would be expected given the former's conceptual provenance. It may however be stated that the circular economy as currently defined and applied is a necessary but not sufficient condition to achieve sustainable development as articulated by the SDGs. It is one tool in a toolbox that must contain supporting tools to action and optimise the social and environmental agendas. The challenge as to how to embed all three sustainability pillars into a formal operational circular economy framework that is consistently applied has yet to be fully resolved.

Nevertheless, it is recognised that applying circular economy principles and solutions will be essential for the implementation of the SDGs. A key factor has been an appreciation that the circular economy is not confined to a stand-alone sector centred around discarded products and the waste management system, but it has much wider applicability as a conceptual model linking elimination of waste with retention of value. As such, its principles are applicable to almost any area of human endeavour. The economic benefits and jobs that circular economy practices can create feed directly into the aspirations underlying the SDGs.

There are encouraging signs of a convergence, at least in conceptual thinking, between the aspirations of the circular economy and those of sustainable development, with phrases such as "the circular economy: a new sustainability paradigm?", "circular economy model – the sustainable approach" and "circular economy – a way forward to sustainable development" being increasingly seen in the academic and grey literature and in the language of policymakers, though a robust theoretical assimilation and underpinning of the two concepts has yet to be attempted. Whether the circular economy is merely a "tool" or a fundamental conceptual framework integrated into that for sustainable development will determine how and with what speed the former will be systematically applied to achieving the SDGs.

References

1. J. A. Du Pisani, *Environmental Sciences*, 2006, **3**, 83.
2. R. Guha, *Environmentalism*, Penguin Books India, Gurugram, 2014.
3. Club of Rome, *The Limits to Growth*, Universe Books, New York, 1972.
4. J. L. Caradonna, *Sustainability: A History*, Oxford University Press, Oxford, 2014.
5. B. Purvis, Y. Mao and D. Robinson, *Sustainability Science*, 2019, **14**, 681.
6. L. B. Sohn, *Harvard International Law Journal*, 2019, **14**, 423.
7. International Union for Conservation of Nature, *World Conservation Strategy*, Gland, 1980.
8. World Commission on Environment and Development, *Our Common Future*, Oxford University Press, Oxford, 1987.
9. V. Spaiser, S. Ranganathan, R. B. Swain and D. J. Sumpter, *Int. J. Sustain. Dev. World Ecol.*, 2017, **24**, 457.
10. N. Millar, E. McLaughlin and T. Borger, *Ecological Economics*, 2019, **158**, 11.
11. World Bank, *Findings, Africa Region*, World Bank, Washington, DC, 1995.
12. European Commission, Proposal for a Directive on Corporate Sustainability Due Diligence and amending Directive (EU) 2019/1937, COM(2022) 71 final, Brussels, 2022.
13. IUCN, *Caring for the Earth: A Strategy for Sustainable Living*, Gland, 1991.
14. E. Huttmanova, R. Novotny and T. Valentiny, *Eur. J. Sustain. Dev.*, 2019, **8**, 409.
15. V. Prieto-Sandoval, C. Jaca and M. Ormazabal, *Memoria Investigaciones En Ingenieria*, 2017, **15**, 85.

16. J. Pezzy, *Sustainable Development Concepts: An Economic Analysis*, World Bank Environment Paper No. 2, World Bank, Washington, DC, 1992.
17. C. A. Ruggerio, *Sci. Total Environ.*, 2021, **786**, 147.
18. S. Bell and S. Morse, *Sustainability Indicators: Measuring the Immeasurable?*, Earthscan, New York, 2nd edn, 2008.
19. J. Elkington, *Cannibals with Forks: The Triple Bottom Line of 21st Century Business*, Capstone, Oxford, 1997.
20. G. Feola, *Environmental Innovation and Societal Transitions*, 2020, **35**, 241.
21. D. Sarokin, *Corporate Sustainability: Does it Make a Difference?* Primedia eLaunch LLC, 2022.
22. Competition and Markets Authority (UK), *CMA Guidance on Environmental Claims on Goods and Services*, London, 2021.
23. Netherlands Authority for Consumers and Markets, *Guidelines: Sustainability Claims*, The Hague, 2021.
24. United Nations, *Transforming Our World: The 2030 Agenda for Sustainable Development*, New York, 2015, https://sdgs.un.org/2030agenda (accessed June 2023).
25. United Nations, *Report of the Inter-Agency and Expert Group on Sustainable Development Goal Indicators* (E/CN.3/2016/2/Rev.1), New York, 2016.
26. Sustainable Development Solutions Network, *Indicators and a Monitoring Framework for the Sustainable Development Goals*, UN, New York, 2015.
27. P. Pradhan, L. Costa, D. Rybski and J. P. Kropp, *Earth's Future*, 2017, **5**, 1169.
28. N. Weitz, H. Carlsen and C. Trimmer, *SDG Synergies – an Approach for Coherent 2030 Agenda Implementation*, Stockholm Environment Institute, Stockholm, 2019.
29. H. Hegre, K. Petrova and N. von Uexkull, *Sustainability*, 2020, **12**, 8729.
30. F. Renaud, X. Zhou, L. Bosher, B. Barrett and S. Huang, *Sustainability Science*, 2022, DOI: https://doi.org/10.1007/s11625-022-01209-9.
31. C. Kroll, A. Warchold and P. Pradhan, *Palgrave Communications*, 2019, **5**(140).
32. J. M. Rodriguez-Anton, L. Rubio-Andrada, M. S. Celemin-Pedroche and S. M. Ruiz-Penalver, *Int. Environ. Agreem*, 2022, **22**, 67.
33. D. C. Wilson, in *The Routledge Handbook of Waste, Resources and the Circular Economy*, ed T. Tudor and C. J. C. Dutra, Routledge, London, 2021, pp. 54–68.
34. J. M. Rodrigues-Anton, L. Rubio-Andrada, M. S. Celemin-Pedroche and M. D. M. Alonso-Almeida, *Int. J. Sustain. Dev. World Ecol.*, 2019, **26**, 708.
35. UNEP, Kunming-Montreal Global Biodiversity Framework, Non-Paper on Item 9A, UNEP, Nairobi, 2022, https://www.cbd.int/doc/c/7a5e/1d9a/f8718d1a5dd9828dba764053/cop-15-item9a-nonpaper-president-en.pdf (accessed May 2023).
36. Inter-agency Task Force on Financing for Development, Financing for development: progress and prospects, UN, New York, 2017, https://www.un.org/esa/ffd/wp-content/uploads/2017/03/2017-IATF-Report_Key-messages-and-recommendations.pdf (accessed May 2023).

37. OECD, Launch of the 2021 global outlook on financing for sustainable development, https://www.oecd.org/about/secretary-general/global-outlook-on-financing-for-sustainable-development.htm (accessed May 2023).

38. UN Global Compact, Sustainable finance, https://www.unglobalcompact.org/sdgs/sustainablefinance (accessed May 2023).

39. P. Schröder and J. Raes, *Financing an Inclusive Circular Economy*, Chatham House, London, 2021.

40. International Institute for Sustainable Development, The evolution of private sector action in sustainable development, 2022, https://www.iisd.org/articles/evolution-private-sector-action-sustainable-development (accessed May 2023).

41. OECD, *Framework for SDG Aligned Finance*, OECD, Paris, 2020.

42. World Economic Forum, UN Sustainable Development Goals: how companies stack up, 2021, https://www.weforum.org/agenda/2021/03/how-aligned-are-un-companies-with-their-sustainable-development-goals/ (accessed May 2023).

43. apolitical, World Bank creates first bond for Sustainable Development Goals, 2017, https://apolitical.co/solution-articles/en/world-bank-creates-first-bond-sustainable-development-goals (accessed may 2023).

44. World Bank, World Bank announces Euro 1.5 billion 10-year Sustainable Development Bond in Ireland, 2019, https://www.worldbank.org/en/news/press-release/2019/05/16/world-bank-announces-euro-15-billion-10-year-sustainable-development-bond-in-ireland (accessed May 2023).

45. UN Capital Development Fund, The SDG500 platform investment opportunity, https://www.uncdf.org/article/5311/the-sdg500-platform-investment-opportunity (accessed May 2023).

46. UN Global Compact, *CFO Principles on Integrated SDG Investments and Finance*, New York, 2020.

47. UN Global Compact, UN Global Compact launches CFO coalition for the SDGs to drive more private sector investment towards sustainable development, https://www.globenewswire.com/en/news-release/2022/03/29/2411460/0/en/UN-Global-Compact-launches-CFO-Coalition-for-the-SDGs-to-drive-more-private-sector-investment-towards-sustainable-development.html (accessed May 2023).

48. UNEP, *Advancing Delivery on Decarbonisation Targets*, Nairobi, 2022.

49. Net Zero Asset Managers Initiative, https://www.netzeroassetmanagers.org/ (accessed May 2023).

50. M. Jacobs, Sustainable Development as a Contested Concept, in *Fairness and Futurity: Essays on Environmental Sustainability and Social Justice*, ed. A. P. Dobson, Oxford University Press, Oxford, 1999, ch. 1, p. 21.

51. M. Mura, M. Longo, P. Micheli and D. Bolzani, *Int. J. Manag. Rev.*, 2018, **20**, 661.

52. Bearing-news.com. Environmental sustainability and measurement's tools, 2022, https://www.bearing-news.com/environmental-sustainability-and-measurements-tools/ (accessed May 2023).

53. R. Prescott-Allen, *The Wellbeing of Nations*, Island Press, Washington, DC, 2001.
54. K. Raworth, *Doughnut Economics*, Random House Business, London, 2017.
55. G. Munda, *Environment Development and Sustainability*, 2005, **7**, 117.
56. P. Allin and D. Hand, *The Wellbeing of Nations*, John Wiley, Chichester, 2014.
57. R. K. Singh, H. R. Murty, S. K. Gupta and A. K. Dikshit, 2009, *Ecological Indicators*, **9**, 189.
58. P. Ghadimi, N. M. Yusof and M. Z. M. Saman, *Pertanika J. Sci. & Technol.*, 2013, **21**, 303.
59. Global Reporting Initiative, https://globalreporting.org/ (accessed May 2023).
60. F. Biermann, T. Hickmann, C. A. Senit, M. Beisheim, S. Bernstein, P. Chasek, L. Grob, R. E. Kim, L. J. Kotzé, M. Nilsson, A. O. Llanos, C. Okereke, P. Pradhan, R. Raven, Y. Sun, M. J. Vijge, D. van Vuuren and B. Wicke, *Nature Sustainability*, 2022, DOI: https://doi.org/10.1038/s41893-022-00909-5 with a correction at https://doi.org/10.1038/s41893-022-00938-0.
61. Refinitiv, Refinitiv debuts country sustainable development scores to measure how extensively a country meets UN SDGs, 2020, https://www.refinitiv.com/en/media-center/press-releases/2020/october/refinitiv-debuts-country-sustainable-development-scores-to-measure-how-extensively-a-country-meets-un-sdgs (accessed May 2023).
62. J. Sachs, G. Lafortune, C. Kroll, G. Fuller and F. Woelm, *Sustainable Development Report 2022*, Cambridge University Press, Cambridge, 2022.
63. UN Statistics Division, *The Sustainable Development Goals Report 2021*, New York, 2022.
64. Edie, *Achieving the SDGs: A Blueprint for Business Leadership*, Edie Insight, 2022, https://www.edie.net/achieving-the-sdgs-a-blueprint-for-business-leadership/ (accessed May 2023).
65. T. McPhearson, C. Raymond, N. Gulsrud, C. Albert, N. Coles, N. Fagerholm, M. Nagatsu, A. Olafsson, N. Soininen and K. Vierikko, *Urban Sustainability*, 2021, **1**(5).
66. B. K. E. Boulding, in *Environmental Quality in a Growing Economy*, ed H. Jarrett, Johns Hopkins University Press, Baltimore, 1966, pp. 3–14.
67. D. Pearce and R. K. Turner, *Economics of Natural Resources and the Environment*, Harvester Wheatsheaf, London, 1990.
68. A. Murray, K. Skene and K. Haynes, *J. Bus. Ethics*, 2017, **140**, 369.
69. P. Ekins, T. Domenech, P. Drummond, R. Bleischwitz, N. Hughes and L. Lotti, *The Circular Economy – What, Why, How and Where*. OECD, Geneva, 2019.
70. T. Cooper, *J Sustain. Prod. Des.*, 1999, **8**, 7.
71. Ellen MacArthur Foundation, What is the circular economy?, https://guides.co/g/mv5ue63s0a/165170 (accessed May 2023).
72. D. Masi, S. Day and J. Godsell, *Sustainability*, 2017, **9**, 1602.

73. J. Kirchherr, D. Reike and M. Hekkert, *Resource Conservation Recycling*, 2017, **127**, 221.
74. UNEP, *Circular Economy: An Alternative for Economic Development*, UNEP, Paris, 2006.
75. M. Geissdoerfer, P. Savaget, N. Bocken and E. J. Hultink, *J. Clean. Prod.*, 2017, **143**, 757.
76. B. Saurez-Eiroa, E. Fernandez and G. Mendez-Martinez, *J. Clean. Prod.*, 2019, **214**, 952.
77. J-M. Valverde and C. Aviles-Palacios, *Sustainability*, 2021, **13**, 12652.
78. J. Potting, M. Hekkert, E. Worrell and A. Hanemaaijer, *Circular Economy: Measuring Innovation in the Product Chain*, Netherlands Environmental Assessment Agency, The Hague, 2017.
79. Amsterdam Economic Board, Levels of circularity, https://amsterdam economicboard.com/en/10-rs-2/ (accessed May 2023).
80. ReLondon, *The Circular Economy at Work: Jobs and Skills for London's Low Carbon Future*, ReLondon, London, 2022.
81. European Commission, *Communication from the Commission to the European Parliament and the Council – The raw materials initiative: meeting our critical needs for growth and jobs in Europe* (SEC(2008) 2741), Brussels, 2008.
82. European Environment Agency, EEA Report No. 2/2016, Copenhagen, 2016.
83. P. Schroder, K. Anggraeni and U. Weber, *J. Ind. Ecol.*, 2018, **23**, 77.
84. E. Gladek, The seven pillars of the circular economy, Metabolic, 2019, https://www.metabolic.nl/news/the-seven-pillars-of-the-circular-economy/ (accessed May 2023).
85. Netherlands Enterprise Agency and Holland Circular Hotspot, *Circular Economy and SDGs*, The Hague, 2020.
86. T. N. da Silva and E. A. Pedrozo, in *The Routledge Handbook of Waste, Resources and the Circular Economy*, ed T. Tudor and C. J. C. Dutra, Routledge, London, 2021, pp. 243–251.
87. A. B. Sutherland and I. Kouloumpi, *More than Just SDG 12: How Circular Economy can Bring Holistic Wellbeing*, International Institute for Sustainable Development, Winnipeg, 2022.
88. F. Figge and A. S. Thorpe, *Ecological Economics,* 2023, **204**, 107692.
89. European Commission, *Closing the Loop: An EU Action Plan for The Circular Economy*, COM(2015)614 final, Brussels, 2015.
90. UN, *Circular Economy for the SDGs: From Concept to Practice*, General Assembly and ECOSOC Joint Meeting, 10 October 2018.
91. A. Whiteman, M. Webster and D. C. Wilson, *Waste Management & Research*, 2021, **39**, DOI: 10.1177/0734242X211035926.
92. P. Schroder and J. Raes, *Financing an Inclusive Circular Economy*, Chatham House, London, 2021.
93. A. B. Sutherland and I. Kouloumpi, How the circular economy can help us achieve the Sustainable Development Goals, Circle Economy, 2022, https://www.circle-economy.com/blogs/how-the-circular-economy-can-help-us-reach-the-sustainable-development-goals (accessed May 2023).

94. ING Bank, *Rethinking Finance in a Circular Economy*, 2015, https://www.ing.com/MediaEditPage/Financing-the-Circular-Economy.htm (accessed May 2023).

95. GIZ, *Financing Circular Economy: Insights for Practitioners*, Eschborn, 2022.

96. Ellen MacArthur Foundation, *Financing the Circular Economy*, Cowes, 2020.

97. G. Moraga, S. Huysveld, F. Mathieux, G. A. Blengini, L. Alaerts, K. Van Acker, S. de Meester and J. Dewulf, *Resources, Conservation & Recycling*, 2019, **146**, 452.

98. UN Economic Commission for Europe, *Measuring and Monitoring the Circular Economy and use of Data for Policy-making*, Geneva, 2021, https://unece.org/sites/default/files/2021-12/BACKGR~1.PDF (accessed May 2023).

99. M. I. Khan and N. Akhtar, Indicators for Circular Economy, in *Asian Circular Economy for Tertiary Education, UN Environment*, Bangkok, 2020, p. 111, https://www.researchgate.net/publication/344121488_Indicators_for_Circular_Economy (accessed May 2023).

100. WBCSD, *Circular Transition Indicators V3.0, WBCSD*, New York, 2022, https://www.wbcsd.org/Programs/Circular-Economy/Metrics-Measurement/Resources/Circular-Transition-Indicators-v3.0-Metrics-for-business-by-business (accessed May 2023).

101. B. Zhu, Bing Zhu discusses the development of China's Circular Economy Indicator System, 2018, https://ec.europa.eu/newsroom/env/items/618580/en (accessed May 2023).

102. Circle Economy, *The Circularity Gap Report 2022*, https://www.circularity-gap.world/global (accessed May 2023).

103. UN Global Compact, *The UN Global Compact-Accenture CEO Study on Sustainability 2013*, New York, 2013.

104. L. Kok, G. Wurpel and A. Ten Wolde, *Unleashing the Power of the Circular Economy*, Report by IMSA Amsterdam for Circle Economy, 2013.

105. UCL Green Economy Policy Commission, *Greening the Recovery*, University College London, London, 2014.

106. The Ex'Tax Project, *The Taxshift: An EU Fiscal Strategy to Support the Inclusive Circular Economy*, 2022, https://ex-tax.com/taxshift/ (accessed May 2023).

107. G. Cole, *Radical Policies to (Finally) Enable the Circular Economy*, Circular, 2022.

108. I. D. Williams, T. Curran and F. Schneider, *Waste Management*, 2012, **32**, 1739.

109. J. Dururu, C. Anderson, M. Bates, W. Montasser and T. L. Tudor, *Waste Management & Research*, 2015, **33**, 284.

CHAPTER 3

The Circular Economy in Low- and Middle-income Countries – A Tool for Sustainable Development?

MIKE WEBSTER*

Systemiq, Gran Rubina Business Park 16th Floor, Jl. H. R. Rasuna Said, RT.2/RW.5, Karet Kuningan, Setiabudi, South Jakarta City, Jakarta 12940, Indonesia
*E-mail: Mike.Webster@systemiq.earth

3.1   Introduction

Consider the scene. The clang of hammer on metal as a mechanic fixes a tired-looking motorbike with nothing more than a few spanners and pot of used oil. Opposite is the clothes shop and dressmaker – they will take your trousers in for a few pennies but also have racks of imported second-hand T-shirts at low prices. There's the furniture maker giving chairs made from local teak a final sand, next door to the mobile phone repair shop where they will fix your cracked screen and promise to unlock any brand (but forget about your warranty!). Beyond that is one of the few remaining local paddy fields amongst the peri-urban sprawl, whose irrigation system is managed by an ancient form of local democracy, scrupulously fair and sharing out a precious and vital resource – water – ensuring both those at the top of the hill and

Issues in Environmental Science and Technology No. 51
The Circular Economy: Meeting Sustainable Development Goals
Edited by Sadhan Kumar Ghosh and Gev Eduljee
© The Royal Society of Chemistry 2024
Published by the Royal Society of Chemistry, www.rsc.org

those at the bottom get what they need for another successful rice harvest. Finally, an overladen scooter putters past, carrying carefully folded cardboard and bags of sorted metals, rigid plastics and glass bottles – an informal recycler with their pickings, off to the waste aggregator to negotiate for a good price.

But then we reach a patch of scrubland apparently belonging to no one. A pile of waste, some garden clippings but mainly noodle packets, plastic bags, sachets and other flexible plastics and sachets gently smouldering amidst the pervading scent of burning plastic. A cow grazes, a plastic bag drooping from its mouth. A flooded drain, blocked with waste, pours filthy water across the street. Above this stagnant pond of sewage hangs a cloud of mosquitoes, promising dengue or malaria for an unlucky passer-by.

This would describe a typical neighbourhood in a mid-sized city in Indonesia, representative of this part of the world and a familiar scene for those billions living in towns and cities across the Global South. In many ways it also encapsulates a few features and challenges of the circular economy – the repair economy of the mechanics and the design-for-repair of their vehicles; dressmakers fixing and adjusting, the international used-clothes trade; entrepreneurs refurbishing and repairing electronics thus preventing e-waste; local fabrication of durable furniture from local, sustainable timber; the age-old system sharing precious common irrigation resources. All features that circular economy enthusiasts from the Global North wax lyrical about and try to emulate with their various reports and pilot schemes.

The flip side of this apparently bucolic scene is a waste management system completely overwhelmed, and widespread open dumping and burning of waste, increasing every day as ever more unrecyclable, low-value plastics enter the system with few apparent answers as to how this increasing flow of unwanted materials can best be managed (see also Chapter 7).

This illustrates the challenges facing many Lower- and Middle-income Countries (LMICs) as their economies grow, as their populations urbanise and educate themselves and as they continue the journey from subsistence to consumer societies. The aim of Chapter 3 is to illuminate some of the issues and their solutions:

- What happens to the circularity of a society as it develops? What have we seen from those countries that have seen significant economic development in recent decades in terms of the waste and resources they produce and how the repair and reuse economies change?
- What are key linkages with LMICs between circularity, United Nations Sustainable Development Goals (SDGs) and the broader development agenda?
- What are the roles of different actors in achieving a circular economy? How can we influence the often seemingly relentless grind towards a system of take, make and dispose, and to what must we simply adapt?
- What are the challenges faced by policymakers in this field? How can we avoid a loss of circularity (and a move to linearity) as a country develops?

The learnings and success points from case studies that have attempted to increase circularity within LMICs will be discussed, including a case study in Indonesia creating a more formalised, comprehensive and circular solid waste management system.

Background reading for this chapter indicated how esoteric discussions in this field can be, and how far removed they can seem from the noise and fumes of a mechanics workshop – but it is in the latter that the much-lauded repair economy, with all its grittiness and real-world pay-offs, can be truly witnessed.

3.2 LMICs: Development, Waste, Resources and Circularity

The last half-century has seen billions lifted from poverty, as some of the world's largest countries by population have transformed their economies.[1] The unprecedented change has provided so much for so many in terms of opportunity and quality of life, but it has also brought challenges as consumption levels rocket.[2] It also underscores an important point – economies and countries are in constant flux, with, of course, a general effort towards, and desired outcome being, improvement in quality-of-life indicators and a push in the direction of economic and social development, even with bumps in the road and reversals along the way. To understand how the circularity of LMICs will change as they develop, and what interventions policymakers should make, the connections between economic development, waste and resource management, and circularity must be considered.

The circular economy predates recent interest by Global North-based thought leaders such as the Ellen McArthur Foundation. It fundamentally reflects the resource-conservation ethic of the urban poor.[3] Simply put, poor people do not consume much, they do not throw much away and they are careful to repair what they can. This raises the following questions:

- "What" new aspects does this modern concept of a circular economy bring to the development debate?
- "Who" should be the players in it, particularly in those places character-ised by weak governance (often a feature of economies with a large informal element)?
- "How" can interventions thus be made?

In the following discussion the idea of circularity is broken down into different elements, roughly correlated with the outer loops (where the product is disassembled or destroyed and the materials made into similar or different products, *e.g.*, recycling) and inner loops (where most value can be captured because they retain more of the embedded value of a product by keeping it whole, *e.g.*, sharing, maintaining and reusing) of the Ellen MacArthur Foundation circular economy "butterfly" diagram[4] (also see Chapter 8, Figure 8.2).

3.2.1   Waste, Resource Management and Development

As economies develop and populations grow, ever-increasing volumes of waste are produced at both total and per capita levels. Composition diversifies from one dominated by organic waste to include increased volumes of "dry" packaging – paper, metals, glass, e-waste and, notably, given current concerns, plastics. The amount of waste produced globally is projected to double by 2050 from 2017 levels.[5] The drivers are as follows:

- A shift from a subsistence to consumer lifestyle. The relationship between income levels and proportion of population engaged in subsistence agriculture is well established,[6] leading to generally low levels of consumption, low levels of waste arisings and a high proportion of organic waste. As cash economies emerge, foodstuffs and other consumption items are purchased rather than self-grown or made, and with an increase in discretionary purchases, packaging and other non-organic waste streams also increase, notably of plastics.[5]
- Urbanisation: By 2050 almost 7 in 10 people will live in towns and cities, many in megacities in the Global South.[7] The shift to a consumer lifestyle occurs earlier and more quickly in towns and cities. Hence urbanisation leads to increases in the volumes and complexity of the solid waste generated. This can be seen in the increased consumption of electronics; the growth of the internet as a cheap and effective method of communication, accessed *via* mobile devices, has been a particularly notable aspect of life in LMICs since the turn of the century and hence smartphone-type devices have become near ubiquitous.[8] Such devices are of complex, multi-material construction presenting challenges to repair, reuse and recycling. Furthermore, they contain potentially toxic substances and when burnt release hazardous substances.

As urban areas develop economically, the manner in which waste and recycling collections are organised also tends to change. Policymakers grapple with the problems caused by the generated waste. Regulation strengthens and the capacity of the government to deliver and manage services increases.[9] There tends to be process of formalisation of waste collection,[10] as public health concerns lead to the organisation of municipal collection and banning of open burning and dumping, followed by steps to improve waste disposal and ensuring containment in sanitary landfills. Once material is collected and contained in a satisfactory manner, and society has the resources to move beyond basic disposal techniques (dumpsites and landfill), resource efficiency and hence circularity again becomes a focus.[10] Resource efficiency and global environmental issues direct policy, leading to a focus on segregating waste, producer responsibility and novel approaches to incentivising positive behaviours, including financial and behaviour change strategies. For example, in the UK this developed in the 1990s and early 2000s with the introduction of statutory recycling targets.

Observing several stages in this process, countries were divided into nine Development Bands (DBs), with commonalities as to how they manage waste, with common challenges faced by each, as shown in Table 3.1.[11] Whilst the full table can be viewed in the publication, many (if not all) LMICs are situated within the first four DBs, hence the truncated version in Table 3.1.

A lack of comprehensive waste collection tends to be a feature of many lower-income countries, with around 3 billion noted as not enjoying the benefit of waste collection and controlled disposal[12] and hence within DBs 1–3. Reaching DB 4a and 4b, universal controlled waste collection and disposal can be seen as a key development milestone.

3.2.2   Challenges as Waste Systems Develop

Approximately 15 million people work in the informal waste sector, in some cases providing low-income cities in LMICs with higher recycling rates than those in the Global North.[13] Usually, the poorest and most marginalised informal recyclers offer collections, generally for free or even paying for materials and can often physically reach places that formal services cannot, in some cases working where there is no organised waste or recycling collection.[14] Within DBs 1–3 as outlined in Table 3.1, they are often the only form of resource recovery, providing important environmental services in the form of collection and recycling.

The challenges they face are multiple: there is a lack of economic mobility with little chance to scale up their activities or move out of poverty; the work is dirty and life expectancy is often relatively low; and they are often excluded as economically developing cities formalise their waste collection and face confrontation from registered waste managers. This formalisation process (often taking place on the approach to DB 4a and 4b) inevitably leads to conflicts with the informal sector, as the latter depends on a level of deregulation to access waste. Such conflict could be in the form of access to dumpsites or transfer sites where they can pick over collected waste for material; the ability to collect or purchase and then transport materials from householders or businesses without a licence, and fear of harassment as a result.

Development, as outlined previously, can lead to a period of reduction in recycling rates. However, given that the aim of the transition to a circular economy is resource efficiency, of which recycling is a large part, there are obvious crossovers with the aims of the informal recycler, namely to divert material for reprocessing from disposal, contingent on the challenges faced by the informal recycler. Firstly, the informal sector targets only those materials with the highest value within secondary markets. This is a potential source of income for formal waste and recycling collections and the exclusion of the informal recycling sector is sometimes a condition of contract. Secondly, they are, by their nature, undocumented, often not paying tax or licence fees, hence being seen as unfair competition by formal waste contractors. Getting this integration right is a key challenge for countries looking to provide more comprehensive waste collection coverage.

Table 3.1 As countries develop, they tend to go through different stages of waste management. Adapted from ref. 11 with permission from SAGE Publishing, Copyright 2021.

Development band	System characteristics	Common challenges	Pressure point	Example locations
DB 1: new beginnings	Most waste self-managed, uncontrolled dumping and open burning the norm Collection coverage: <30% Managed in a controlled facility: 0% Anything with value reused, repaired or recycled, at home or by informal sector	Introduce basic collection systems	Operator	Many towns and cities in the least developed countries; areas recently affected by conflict or natural disaster; refugee camps; peri-urban and slum areas in cities in many LMICs
DB 2: early movement	Collection coverage: 30–60% Some collected wastes disposed at designated sites Managed in a controlled facility: up to 20% Active informal recycling	Expand collection coverage Introduce basic operational management practices at disposal sites	Municipal capacity to assume responsibility for service provision (*i.e.,* client/employer function)	Many cities in LMICs that are growing rapidly due to influx from rural areas. Includes many secondary cities.
DB 3: service extension	Collection coverage: 60–80% Managed in a controlled facility: up to 50% Informal recycling often well established for a limited range of materials	Further expand collection coverage Introduce some engineered control and upgrade operational management practices at recovery and disposal sites	Planner	Many cities and megacities in LMICs

			Revenue collector (Environmental regulator)	Diverse situations across the world, in cities of all different sizes, and in most continents Includes many small islands Residual pockets may persist for some time after a country progresses to the higher DBs
DB 4: consolidating control	Collection coverage: 80–95%+ Managed in a controlled facility: moving towards 95%+	DB4a: Extend collection service coverage in cities to 95%+ DB4b: extend controlled disposal in cities to 95%+		
DB 4a: universal collection	As collection and disposal costs rise, diversion of waste from landfill by extending recycling moves up the municipality's agenda	Introduce gate fee or distinct budget line for disposal – but avoid illegal dumping Build on existing informal recycling sector to enhance recovery system performance (*e.g.,* by more separation at source)		
DB 4b: controlled disposal				

Two different, contrasting approaches from Cairo, Egypt,[15] and Pune, India,[16,17] are considered in the following sections, with very different outcomes.

3.2.2.1   Unsuccessful Integration: The Zabbaleen, Cairo

The Zabbaleen, which translated means garbage people, are the traditional waste collectors of Cairo. Originally migrants from upper Egypt but long settled in Cairo, they have one of the world's most efficient recycling systems. Waste collection started with organic waste to be fed to their pigs in return for a small monthly fee paid by residents. Over time this included dry recyclables as well and the system has mechanised, moving from donkey carts to trucks; their supporters claim they have greatly improved the capacity of Cairo to manage its waste at minimal cost or effort to the city administration.[15]

However, this system is under threat. Since 2003, the Cairo governorate privatised waste collection. While the Zabbaleen had previously recycled 80% of the waste they collected, the new contractors were required to recycle only 20%, the rest of which would go into landfill. The Zabbaleen were permitted to keep their jobs as wage workers with these companies, and they would also be responsible for street sweeping and placement of garbage bins. The Zabbaleen claim, however, that the salaries offered are less than what they used to make independently, and that they used to earn 90% of their income from recycling rather than from the collection fee. Citizens preferred the traditional door-to-door collection method of the Zabbaleen. Furthermore, the large trucks of the companies cannot go into the narrow streets of Cairo, requiring the placement of bins in central collection points, to the dismay of residents and leading to large amounts of open dumping.[15]

Recycling rates collapsed and the amount of waste sent to landfill increased. This was compounded by the ending of organic waste collections by the Zabbaleen following pig culls due to swine flu. With their main processor of organic waste gone, the Zabbaleen refused to collect organic waste from Cairo, leaving piles of garbage in the streets, replacing the threat of swine flu with the threat of typhus. Their situation was made worse following the official policy of moving their activities, including sorting, recovery, trading and recycling, outside the city under the guise of making their neighbourhoods cleaner and healthier, increasing the Zabbaleen's travel distance and, consequently, the cost of services delivered to residents.

The citizens prefer the Zabbaleen system with its cheaper fees. They reject the government's plan to pay extra fees to private companies. As such the Zabbaleen still collect municipal solid waste alongside companies and local municipalities.

3.2.2.2   Successful Integration of Informal Recyclers in Pune, India

In Pune, India, a cooperative of 3000 waste pickers joined forces with municipal authorities to collect waste door to door. In 1993 they organised themselves into a collective that eventually won union (the KKPKP) and official recognition as workers. In 2007 they set up a wholly-owned worker cooperative of waste pickers, itinerant waste buyers, waste collectors, named SWaCH, an acronym for Solid Waste Collection Handling, that also means "clean" in the local language. They provide front-end waste management services to Pune City, with support from the Pune Municipal Corporation. KKPKP argued that collection of recyclable materials constituted work and sought recognition for the workers from the corporation. A series of collective actions that protested abuse and discrimination of aggrieved waste pickers followed.[16,17]

3.2.3   Reuse, Repair and Refurbishment and Development

Maintenance and repair are important ways of keeping resources from being thrown away and for prolonging a product's lifespan. Repair slows down the flow of materials from production to recycling, which is important as, even when at the end of the product life cycle all the materials do get recycled, recycling itself requires additional resources, for example energy.[18]

In recent decades there has been a widely accepted shift away from a reuse and repair culture in the Global North. A 2016 European Parliament study noted that the number of specialised firms in electronics repair in the European Union reduced from 4500 to 2500 over a 10-year period, while in Poland it decreased by 16% between 2008 and 2010,[19] with a similar shift in the USA.[20] This has been matched by a measured increase in inbuilt obsolescence.[21] To address this, various pieces of legislation have been enacted, such as the Right to Repair across the EU.[22]

Is this decline repeated in LMICs? Whilst the literature is more limited, there is apparently a much stronger repair and reuse economy. Much of the evidence focuses on the reuse and repair of electrical and electronic items. The e-waste sector is estimated to have a global annual worth of USD 62.5 billion,[23] employing 18 million people worldwide in 2010.[24] Examples include the following:

- In the West African state of Ghana, the value of the industry is approximately USD 105–268 million annually, supporting up to 200 000 workers in the informal sector.[25,26]
- In Nigeria, it has been estimated that e-waste repair and refurbishing is responsible for 30 000 jobs in both Accra and Lagos,[27] with a total of 52 000 e-waste refurbishers across Nigeria, 80 000 e-waste collectors and "several thousand" recyclers.[24] In terms of contribution to the national economy, the electrical reuse and repair sector in Lagos generates USD 50.8 million per year, equivalent to 0.015% of Nigeria's gross domestic product.[24]
- In Brazil between 0.4 and 1.0 million people benefit from the e-waste sector as their source of income.[28,29]

- In India, the e-waste industry is estimated to be worth about USD 3 billion annually[30] involving nearly 25 000 informal workers in Delhi alone[31] and more than 30 000 in Seelampur,[32] resulting in 450 000 direct jobs and 180 000 indirect jobs nationwide.[33]
- In China the e-waste industry employs an estimated workforce of 700 000.[34]

In terms of employment, refurbishing and repair activities offered "significant employment opportunities in Lagos",[27] providing jobs that are mostly attractive for medium- and high-skilled workers. Collection and recycling provides employment opportunities to a lower-skilled workforce – an interesting shift away from a narrative that often focuses on low-income, low-quality jobs.

A corollary of this thriving second-hand market is the much greater availability of cheap laptops and similar devices due to the availability of second-hand components, leading to much greater access by students to laptops.[35] Outside the electronics sector, examples such as the vehicle repair and remanufacturing cluster in Kumasi in the Ashanti district of Ghana have provided new and different kinds of skills and jobs.[36] Over more than 30 years, the Suame/Kumasi cluster of micro and small to medium-sized enterprises has grown and now employs 200 000 workers – an increase from 40 000 in the early 1980s. The cluster consists of more than 12 000 businesses working on repair and remanufacture at the edge of the international automotive industry and is bigger than anything found in Europe.

It might seem paradoxical that reuse and repair thrive in LMICs whilst they are in need of policy intervention to prevent their decline in Europe. In lower-wage economies, the relative cost of repair is much lower than the cost of new goods, making it much cheaper to repair items across the value spectrum, from mobile phones and electronics to cars and clothing. This leads to a thriving repair sector visible across towns and cities in lower-income countries – mechanics, mobile phone and computer repair, tailors. Simply finding someone to fix something is apparently much easier and much cheaper than in most of Europe or the USA. This sector is entirely what might be described as "informal": the businesses are set up purely to respond to local demand for such services, and are often small and locally run independent businesses; convenient and affordable to the local population; and leading the way in reuse, redistribution and refurbishment.[37]

Whilst repair and reuse certainly predates the internet and recent advances in communications, businesses have adapted. In the poorest places they tend to be simple, focusing on mechanics, tailors and possibly carpentry. In wealthier and more urban areas there is a greater emphasis on electronics (notably the ubiquitous smartphone). As the cost of labour increases, repair and reuse tends to become marginalised, and even for those willing to pay a premium to have consumer goods repaired, simply finding a service becomes harder[20] as it often becomes cheaper and more convenient to replace items entirely.

3.2.4 Waste Colonialism

Is second-hand second best? The concept of 'waste colonialism' is one that has gained recent traction in some quarters, with the current global second-hand trade an example of how reuse and recycling efforts in the Global North can have negative impacts, undermining the quality of jobs and the (circular) economies of lower-income countries.[38,39] The non-governmental organisation (NGO) Tearfund has described the global second-hand trade as a "a double-edged sword for developing countries", noting that, on the one hand, access to imported second-hand electronics increases access to laptops, mobile phones and hence engagement with the internet whilst, on the other hand, the eventual fate of e-waste and associated recycling practices create major health and environmental problems.[40] Practices such as open burning of residual waste, burning off plastic insulation to obtain copper wires and acid bath recycling by the informal sector have serious impacts on local public health and the environment surrounding e-waste dumps, with a particular negative impact on child health.[41]

3.2.5 Weaker Product Standards

The informality, and the weaker governance that this implies, that stimulates a thriving recycling sector and a network of independent repair outlets in LMICs has its downside when it comes to products standards. They tend to be weaker, if not non-existent, with fake brands using sub-standard parts more prevalent than the *bona fide* product.[42] This directly affects the quality and hence longevity of manufacturers' items and parts, but can also be directly addressed by improved regulation. The literature notes the need for more policy support to develop inner-cycle activities (repair, reuse, refurbishment, remanufacturing) and to design more durable products.[43,44]

3.3 Contributions of a Circular Economy to the SDGs in LMICs

3.3.1 Global Waste Targets and the SDGs

Chapter 2 provides an overview of linkages between the circular economy and the SDGs; however, the priorities of LMICs and of the Global North are not necessarily aligned.[45] There are several emerging-economy issues that can be addressed by the circular economy:[46,47,48]

- Poverty reduction and job creation. In the LMICs this is often the greatest focus whilst in the Global North the circular economy is often focused primarily on environmental and economic impacts.
- Waste management, disposal of electronic waste, materials recycling, remanufacturing and repair, anaerobic digestion and biogas production.
- Energy needs, including the use of renewable energy.

Clear linkages are apparent with SDG 1 (No poverty), SDG 8 (Decent work and economic growth), in particular with SDG 1, Target 1.1 – to halve the number of those living in poverty by 2030.[49] Whilst it has been noted in Chapter 2 that the circular economy is not specifically mentioned in the SDGs, without a transition to the former, the SDGs will not be achieved. Furthermore, SDG 12 (Ensuring sustainable consumption and production patterns – a partial definition of the circular economy) does reflect the needs of a circular economy.[50]

Focusing on the outer loops of the circular economy (see Section 3.2 and Chapter 8, Figure 8.2), it is notable that the lack of improved solid waste management, including collection and controlled disposal targets, is not a primary goal. Does this mean that it is a low priority? An optimistic view is that improved Waste and Resource Management (WaRM) is recognised as a foundation stone of a circular economy and is considered as an entry point to address the SDGs.[50] However, in recognition of the lack of a specific SDG on waste, the seminal UN Environment Programme (UNEP) Global Waste Management Outlook set out five Global Waste (GW) targets that effectively indicate how the provision of improved WaRM would impact on the SDGs, as shown in Figure 3.1.[51]

The conclusions drawn by Figure 3.1 align with the conclusions of others who indicate strong relationships between circular economy practices and the targets of SDG 6 (Clean water and sanitation), SDG 7 (Affordable and clean energy), SDG 8 (Decent work and economic growth), SDG 12 (Responsible consumption and production) and SDG 15 (Life on land).[52]

Figure 3.1 and the collation of SDGs into the Global Waste Targets points to areas with relevance for LMICs, as discussed in Sections 3.3.2–3.3.4.

3.3.2   Improving Public Health

GW 1 (ensuring access for all to waste collection) has public health as its initial driver. Historically, this has been the case for the institutionalisation of solid waste management. For example, 19th century London witnessed a series of public health epidemics, directly linked with poor sanitation and a lack of waste management, with over 250 000 people dying from cholera between 1848 and 1854, leading directly to the provision of waste collection and disposal by municipal authorities.[53] Globally this has been the primary driver where comprehensive WaRM has been introduced. In addition, there is a clear linkage between improved WaRM and clean water and sanitation (SDG 6). Indeed, the division between sanitation (*i.e.,* the management of wet, human-sourced waste) and solid waste management (*i.e.,* the management of organic and inorganic, non-human-sourced waste) might be seen as artificial.

3.3.3   Jobs and Poverty Reduction

Jobs and poverty reduction relate to SDGs 1 and 8. The provision of jobs, and the equitable distribution of economic benefits, is a necessity when implementing a circular economy, if only because anything that causes large-scale

Sustainable Development Goals	GW1	GW2	GW3	GW4	GW5	GOV
1 NO POVERTY	1.4					
2 ZERO HUNGER						
3 GOOD HEALTH AND WELL-BEING						
4 QUALITY EDUCATION						
5 GENDER EQUALITY						
6 CLEAN WATER AND SANITATION		6.3				
7 AFFORDABLE AND CLEAN ENERGY						
8 DECENT WORK AND ECONOMIC GROWTH						
9 INDUSTRY, INNOVATION, AND INFRASTRUCTURE						
10 REDUCED INEQUALITIES						
11 SUSTAINABLE CITIES AND COMMUNITIES	11.1, 11.6	11.6	11.6			
12 RESPONSIBLE CONSUMPTION AND PRODUCTION			12.4	12.5	12.3	
13 CLIMATE ACTION						
14 LIFE BELOW WATER						
15 LIFE ON LAND						
16 PEACE, JUSTICE AND STRONG INSTITUTIONS						
17 PARTNERSHIPS FOR THE GOALS						

Key:
(1) Global Waste Target 1: Access for all to basic waste collection services.
(2) Global Waste Target 2: Stopping uncontrolled dumping and open burning.
(3) Global Waste Target 3: Managing all waste properly, particularly hazardous waste.
(4) Global Waste Target 4: Reducing waste and creating recycling jobs.
(5) Global Waste Target 5: Halving food waste from markets, shops, homes; reducing food losses in the supply chain.
(6) GOV: Governance factors that underpin sustainable waste management.
(7) Dark green = direct link.
(8) Light green = indirect link.
(9) Number = target that explicitly requires a basic level of waste management.

Figure 3.1   The contribution of waste management towards achieving the SDGs. Reproduced from ref. 51 with permission from by WasteAid UK, Copyright 2017.

unemployment and social disruption is unlikely to be politically and socially acceptable.

Views on the impact of circular economy-linked employment levels vary. Circular economy practices may displace some jobs following declines in new product sales and loss of manufacturing jobs resulting from increased sharing, self-service technologies and extended product life and use.[49] Other studies are more optimistic, noting many opportunities from the development of local circular business models including the creation of local reuse and repair markets contingent on access to repair, recycling and remanufacturing equipment. Access to circular services such as digital software and expertise in product–service systems' trade in second-hand goods can also stimulate demand for local repair, refurbish and remanufacturing jobs.[54] Trade in secondary raw materials (and waste intended for recovery) provides a low-cost feedstock for domestic industry, helping to boost competitiveness in global markets. It also improves resource security and protects economic growth from external resource price increases and volatility, whilst reducing the polluting impacts of growth.[47]

How a circular economy develops is also important. An inclusive circular economy that ensures the informal recycling and repair sector is not sidelined in favour of large, possibly multinational, recycling and waste collection firms is key when considering the role sustainable livelihoods play in poverty eradication programmes and policies, and the kind of decent circular job creation strategies assumed in SDG 8.[54] The informal recycling sector often provides either the majority or all of recycling and waste services in many lower-income communities. Ensuring the inclusive provision of jobs and that informal recyclers are not excluded from the source of their livelihoods is a matter of justice, of maintaining social order (as illustrated in the case study from Cairo in Section 3.2.2.1) and of ensuring that generations of experience are not lost. However, if not properly regulated, this could incentivise dangerous and low-paid work, such as in informal recycling and repair hubs, exposing workers to toxic substances and other hazards.[40,49,55] This could also lead to dependency on inefficient second-hand goods (such as old electronics or diesel cars), which may lock the importing country into a more inefficient and costly mode of consumption.

3.3.4 Environmental Protection

Environmental protection can easily be blurred with public health. For example, ending open burning and open dumping and preventing leakage into waterways not only have local public health benefits but also global environmental benefits.

Open burning of waste, which releases dioxins, furans, arsenic, mercury, polychlorinated biphenyls (PCBs), lead, carbon monoxide, nitrogen oxides, sulphur oxides and hydrochloric acid,[56,57,58] leads to higher levels of stunting, infertility, asthma and several hormonal and neurological conditions, and an estimated 270 000 deaths per year globally.[59]

Burning waste is also a leading cause of black carbon. With a global warming potential 2000 to 5000 times greater than carbon dioxide, black carbon contributes, according to some estimates, as much as 20% of the planet's warming, making it the second highest contributor to climate change after carbon dioxide.[60] Addressing this practice would have both health and environmental benefits. Furthermore, the benefits in terms of avoided methane emissions from landfill and marine leakage of plastics are well understood.

3.4 Case Studies

This section presents case studies from different parts of the world looking at various elements of the circular economy as they apply to LMICs.

3.4.1   Project STOP in Muncar, Indonesia

This case study examines the movement from no waste management to a circular waste system. Indonesia is, by population, the world's fourth largest nation. Identified as the world's largest source of marine litter,[61] only ≈40% of the population receive a formalised waste collection and around 139 million tonnes of waste leaks into the environment annually.[62] Of the waste that is collected, around 25% is recycled, 50% goes to managed disposal sites and 25% to dumpsites. Indonesia has 2.0–3.7 million pickers, some organised into cooperatives.[62]

Muncar is a sub-district of Banyuwangi regency in the extreme east of Java, the world's most populous island. The economy is based on fishing and agriculture. In 2018 Muncar had a population of 134 780 with 40 485 households.

Baseline studies included waste characterisation, local attitudes and behaviours towards waste, an assessment of the local governance and legislative environment and opportunities for recycling end-markets.[63] The following key characteristics were noted:

* Waste generation: 0.37 kg per capita per day, 47 tonnes per day, 75% organic, 13% plastic.
* Waste collection responsibility of the village government. Waste collection coverage is ≈20%, significantly lower than the 39% national average and 45% in similar cities.
* In Muncar, 80% of waste leaked into the environment: about 20% was buried or dumped on land, 20% was open burned and 40% was dumped in waterways.
* Pre-existing recycling. Recycling collection is entirely in the informal sector. Pickers collect ≈6 tonnes of recyclables per day, transported 300 km to Surabaya, East Java.
* Uncollected waste is managed mainly by women at household level.

Of the barriers to improved WaRM, funding is a major constraint. In principle, waste collection and disposal are funded largely from national, regency or village budgets, but it is not a high priority and on average only 0.7% of local government budgets is spent on waste management. Meanwhile waste budgets are limited and hard to spend with little capital funding and restrictions on operating spend. Local politicians are reluctant to set waste fees at a level that makes universal collection sustainable, to improve worker productivity or to enforce ethics and discipline.[64]

The first national waste policy in Indonesia was introduced in 2007. A new National Solid Waste Management Policy and Strategy was launched 2017 with responsibility spread across multiple ministries and local governments. Responsibility for waste collection is set at village level. Within the areas of the STOP Muncar project, there are ten villages with a population of roughly 10 000 per village, although this varies significantly. Solid waste management staff are political appointees, generally without technical expertise, and can be replaced suddenly. Inter-village resentment limits cooperation on larger disposal facilities.

The Indonesian government has prioritised solid waste management and in 2018 adopted ambitious plans to reduce the amount of plastics entering the oceans by 70% by 2025. Project STOP, founded by chemical company Borealis and system change company Systemiq, assists in meeting that target. Muncar was the first of three cities in East Java and Bali to pilot the project (April 2018–June 2023), ultimately offering waste services to a total population of 450 000 people once fully rolled out.

Targeted interventions focused on four key areas:

- **Establishing a waste collection function.** Various approaches were considered; ultimately, a village-based organisation, an autonomous non-profit waste collection body called a BUMDES, was formed. Capital finance, along with technical support, was made available for new services, as was support to facilitate behaviour change in local communities, both to present waste for collection and to separate at source. One Materials Recovery Facility (MRF) was selected acknowledging the existing informal recycling networks and the need to avoid disruption. A second MRF was developed as a cooperative village-level social enterprise.
- **Strengthening municipal capacity around waste and recycling** at the sub-regency and then regency level so that, once handed over the to the community, the local government was able to support, advise and advocate. Local legislation was formulated to forbid open dumping and burning and support payment of collection system fees. This was also supported by developing autonomous, financially secure and technically capable municipal solid waste management units within the sub-district government. The focus included the client function, managing the operators to ensure services delivered, stakeholder engagement at the village and household levels, and policy engagement at the regency and national levels.
- **System development.** Building two MRFs and implementing a motorised tricycle-based collection system along with support for a longer-term waste planning process.
- **Developing waste retribution systems** and ensuring long-term financial sustainability is a key focus. Revenue from the sub-district budget is important, but during the COVID-19 pandemic, this was diverted elsewhere. Collection fees at USD 0.70 per household per month are designed to be affordable to all and material sales also provide additional revenue, although these are typically low value as informal recyclers have generally removed higher-value materials. The three year programme also aims to engage international premium markets for plastics diverted from the oceans and financial support from packaging and product producers (extended producer responsibility).

STOP Muncar was handed over to the community in January 2022. The programme is now being rolled out across the regency as part of the Banyuwangi

Hijau programme, supported by efforts to improve the economics of waste collection by accessing premium international markets and support with carbon and plastic credits. At the time of writing, notable achievements across the three STOP cities include establishment of five separate waste collection systems in East Java and Bali; collection of almost 35 000 tonnes of waste, mainly in communities that were previously open dumping and burning; and the creation of 318 full-time equivalent jobs.

3.4.2 Automotive Clusters – West Africa

Many developing countries, including those in West Africa, have flourishing car repair and refurbishing industries because they cannot afford new cars. Instead, these countries import used vehicles from industrialised countries. For example in Nigeria, 95% of cars are second-hand.

To service and support this industry, large-scale automotive clusters have grown in Ghana and Nigeria – areas for vehicle repair and manufacturing. In Ghana, the largest cluster for vehicle repair and manufacturing in Africa is in a suburb of the second city Kumasi, at Suame Magazine. Suame employs an estimated 200 000 people who provide symbiotic skills in dismantling, refurbishing, repurposing and remanufacturing for the local automotive industry. It has evolved into a self-organised system capable of building almost anything using car parts, for example fences, swings, water pumps, carbide compressors, welding machines, but principally vehicles adapted to the African market. Everything is recycled. Even the unrepairable car parts have a destination: the blast furnace, which melts broken engine blocks into new iron products.[65]

In Nigeria, the Nnewi in the southeast of the country automotive cluster started as a local scheme over 40 years ago and now generates an estimated 80% of all locally fabricated automotive spare parts in Nigeria. Known locally as the 'Taiwan of Africa', the clusters provide jobs for over 30 000 people and handle over 560 000 tonnes of automotive materials annually.[66] The Anambra State Government has recently recognised this impact and has committed to providing support to the Nnewi Automotive Industrial Park with the provision of asphalted roads, electricity and water.

3.4.3 Eliminating Plastics in the Agricultural Supply Chain in Indonesia

Founded in 2015, Kecipir is a programme developed by Dutch company Enviu that connects horticultural farmers with urban consumers *via* a plastic-free, circular delivery system.[67]

Customers engage through an app to order and pay for their fresh food products. Farmers then harvest the produce, while Kecipir oversee the packaging, sorting and distribution. Farmers must fulfil certain duties, including implementing an organic farming system, and a commitment to applying the principles of zero-waste packaging. Fresh products travel no more than 60 km and delivery time is less than 24 hours, thus avoiding the need for refrigeration.

Kecipir is currently increasing its range of circular products and features. It recently launched a new refill category for staple products, such as rice and cooking oil, reducing single-use packaging. Customers can also exchange their used cooking oil for shopping vouchers, reducing waste from kitchens – currently by about 0.5 tonnes per year, serving 1200 orders per month. Following growth projections, the amount of plastic that can be avoided will rise to over 116 tonnes per year by 2024.

Benefits are not only environmental. Farmers can realise a 40% increase in profit due to a fair and direct market. Customers get better and fresher produce and pay less for organic produce than they would in the supermarket. Plastic waste is eliminated, and carbon emissions associated with food waste are avoided.

3.5 Transitioning to a Circular Economy: Learnings from the Case Studies

Drawing on the case studies presented in Sections 3.2 and 3.4, this section addresses the following questions in relation to making the transition to a circular economy in LMICs: Are there commonalities between the case studies? What learnings can we apply elsewhere? What pitfalls should be avoided?

3.5.1 Common Features and Challenges

As countries get wealthier and more urbanised, there are clear commonalities as societies shift from a subsistence to a consumer society. Waste arisings increase, and types of waste change, with more non-organic waste, particularly low-value film plastics. We see an increase in tonnages collected by the informal sector of those materials they can easily valorise, but we also see an increase in residual waste. As pressure grows to manage all waste, there is often conflict between informal and formal, as the informal collectors tend to collect the highest-value material, which negatively affects the economics of collection for those collecting all materials. There are a few key pressure points during this process:

- Widespread normalisation of open dumping and burning – often a habit normalised in earlier, less wasteful times – presents a major local pollution and public health risk. A notable recent impact has been the increased awareness around leakage of plastics into the ocean, mainly caused by a lack of waste management (see also Chapter 7). However, there is a growing body of evidence that open burning is a more prevalent activity than dumping in waterways (for example in Indonesia some estimates are that half the population burn their waste as their preferred disposal route).[68] This practice must now be addressed with similar vigour as the transition to a circular economy.

- Inadequate collection systems to collect and contain waste – hand in hand with traditional attitudes towards open dumping – causing problems as use of novel inorganic materials increase. This is exacerbated by a lack of local markets for novel materials. Flexible plastics are on ongoing challenge and ultimately those putting such difficult-to-recycle packaging on the market must be engaged to design out such materials.
- A shift away from a thriving reuse and repair sector as this becomes less economic and as there is growing consumer demand for new products.

In addressing these concerns, the issue is not only about what local and international policymakers *should* do to encourage a circular economy but also what they *can* do. In countries characterised by weak governance and often patchy policy implementation, this is moot. Key challenges can be summarised thus:

- How can we ensure waste systems can cope with the inevitable increase in waste arisings?
- How can cultures of repair and reuse be maintained? Is there a way the relatively low levels of waste arisings of low-income countries can be maintained whilst enjoying the large waste collection coverage of wealthier countries?
- How can existing recycling systems (which tend to be almost exclusively of the informal sector) be maintained as a country develops economically and socially? How can they integrate with systems that collect all materials – can they still have access to collection and disposal sites? Can their 'grey' status be maintained?
- How can we avoid the rapid shift to linearity we have seen in many middle-income countries, especially with the introduction of various difficult-to-recycle single-use items? How can multinational companies be engaged to design out, and design for recycling, such tricky packaging?

Some strategies to address the above challenges are discussed in the following section.

3.5.2   Policy and Fiscal Tools

A circular economy transition requires a fiscal and policy environment that means the shift away from linear models (or indeed maintaining and supporting pre-existing circular models) makes business and legal sense. The fiscal tools tend to focus on reducing taxes on income and replacing them with taxation of non-renewables and of the generation of waste and emissions, thus encouraging the recovery of materials over extraction of non-renewables.[69] Goods and Services Tax (GST) or Value-added Tax (VAT) is a widely used tax, simple to collect in societies with high levels of informal employment.

This can be set up to favour reuse and repair activities, the use of recycled content and secondary materials, and for the second-hand goods markets.[37]

Support for an inclusive approach to materials' recovery is key, not only for reasons of social equity but also from a practical standpoint. Excluding the informal sector not only pushes people who are often some of the most marginalised in society out of work but also excludes the most knowledgeable and effective operators in this space (see Section 3.2.2.1). Whilst some authors do not consider this feasible by the informal sector given their lack of adherence to social and environmental standards,[70] there needs to be a recognition that in many places they are the only option, and that there would be significant welfare, justice and livelihood impacts from their exclusion, as well as environmental impacts from reduced resource recovery.

3.5.3   Behaviour Change and Education

To ensure that shifts to a circular economy are politically and socially acceptable and sustained, there must be large scale buy-in from the population at large. There are several areas to focus on:

- Ensuring buy-in for controlled waste collection and disposal. Setting up new schemes for those without any current options requires new fiscal instruments and ultimately this means the householder pays – whether through central taxation, household fees or *via* other novel approaches such as inclusion on utility bills. This requires an understanding of the need for such systems and an acceptance to pay such fees.[62] The experience of Project STOP presented in Section 3.4.1 supports this view – new systems and new approaches often incur new costs or new ways of paying. There must be buy-in at household, local and national governmental levels for these to succeed.
- New skills to provide the workforce with the tools to design, repair and recycle safely and effectively.[69] Given the challenges, noted previously, associated with e-waste recycling, there is a clear need to develop safe, low-cost and realistic approaches that work for the informal recycling sector.

3.5.4   Promotion of Relevant Business Models

The spread of the internet and increasing accessibility in LMICs provides an opportunity for simple, convenient access to repair, reuse and take-back delivery services. There are opportunities to access take-back and repair schemes that are equal to single-use systems in terms of convenience. However, consumer buy-in is fundamental to ensuring that these move beyond pilot and niche schemes and ensuring that consumers understand why new business models (for example, take-back schemes) are preferable to existing linear models.[71]

Tearfund notes five interventions that would support increased circularity in electronics[40] – they demonstrate the multifaceted nature of interventions required to transition to a circular economy and indicate what might work elsewhere:

- design interventions to allow for easy disassembly and reduced toxicity
- appropriate environmental, health and safety regulation of e-waste recycling
- appropriate and well calibrated incentives for informal operators to participate in formalised recycling sectors
- capacity building with the informal repair and recycling sectors to reduce occupational health and safety issues
- engagement with industry, especially large multinational electronics manufacturers, on effective schemes for sustainable e-waste management, such as enhanced producer responsibility.

3.5.5   Mainstreaming the Circular Economy into Aid

To ensure that this fits into the aid sector, the circular economy should be mainstreamed into existing aid programmes and embedded in existing multilateral capacity-building programmes. Key areas for circular capacity building could include investing in infrastructure to enable domestic circular activities such as repairing, remanufacturing and recycling; trade infrastructure; customs systems and enforcement measures to counter illegal waste shipment; circular production skills and training; and policy development.[54]

3.6   Conclusions

When advocating for a circular economy, especially by thought leaders from the Global North, there needs to be clarity on the direct, immediate benefits for populations in the developing world. There should be recognition that many of these benefits may be overlooked in the Global North, or indeed taken for granted, because they have been implemented many generations ago. Basic public health, with its direct benefit to those engaged in the activity, should not be overlooked as a key driver for spreading improved solid waste management to the 3 billion people currently existing without it,[72] above other issues with less direct appeal to those affected communities, such as ocean plastics or even climate change. There clearly needs to be a focus on ensuring decent and inclusive employment – good jobs from the circular economy will not appear on their own and the risks of worker exploitation and environmental harm need to be worked against.

Multiple indicators of circularity are better in LMICs than the Global North – lower consumption, lower levels of waste arisings, the presence of a thriving reuse and repair sector. This raises the pertinent question of what, if any, capacity building can be done by the Global North for the Global South and whether the former has anything to tell the latter. The key areas where assistance could be provided are as follows:

- Assistance around the planning, financing and provision of skills to increase the roll-out of comprehensive waste collection and recovery and formalised disposal. This requires enabling legislation, the political will to support such initiatives and a financial framework to support capital expenditures (CAPEX) and operating expenses (OPEX).
- Support for the informal sector (recycling, repair, remanufacture) to ply their trades in a safer, more just and more effective way, for example, providing skills and techniques that allow safer recover of materials; access to personal protective equipment; credit lines to allow for economic and social mobility along value chains; research into simple techniques that allow recovery of valuable secondary materials more safely and more healthily.[70]
- Pressure on multinational corporations, often headquartered in the Global North, to take responsibility for the end-of-life of their packaging materials post consumer, particularly in those places without comprehensive waste collection.

Ultimately, the challenge for policymakers in the Global South is to understand what they can do to avoid a loss of circularity and a shift to linearity as incomes increase. Although at the lower end of the waste hierarchy and on the periphery of the circular economy, the case studies identify points within the system where governmental and international donors can make meaningful and thought-through interventions fundamental to enjoying the wider benefits a circular economy might bring.

References

1. E. Sucking, Z. Christensen and D. Walton, Economic poverty trends: global, regional and national, Factsheet 10, Development Initiatives, November 2021, https://devinit.org/resources/poverty-trends-global-regional-and-national/ (accessed May 2023).
2. T. Jackson, *Post Growth: Life after Capitalism*, Polity, London, 2021.
3. D. Sharma and J. Joshi, in *The Circular Economy and the Global South: Sustainable Lifestyles and Green Industrial Development*, ed. P. Schröder, M. Anantharaman, K. Anggraeni and T. J. Foxon, Routledge, London, 2019, part 4, ch. 10.
4. Ellen MacArthur Foundation, The butterfly diagram: visualising the circular economy, https://ellenmacarthurfoundation.org/circular-economy-diagram (accessed May 2023).
5. S. Kaza, L. Yao, P. Bhada-Tata and F. van Woerden, *What a Waste 2.0: A Global Snapshot of Solid Waste Management to 2050*, World Bank, Washington, DC, 2018, http://datatopics.worldbank.org/what-a-waste/ (accesssed May 2023).

6. L. J. Larsen, *Western Economic Journal*, 1968, **6**(4), 304.
7. United Nations Department of Economic and Social Affairs, *Revision of World Urbanization Prospects*, UN, New York, 2018, https://www.un.org/en/desa/2018-revision-world-urbanization-prospects (accessed June 2023).
8. Global System for Mobile Communications Association (GSMA), The Mobile Economy, Sub-Saharan Africa, 2022, https://www.gsma.com/mobileeconomy/sub-saharan-africa/ (accessed May 2023).
9. C. A. Velis, D. C. Wilson, Y. Gavish, S. Grimes and A. Whiteman, *Socio-Economic Development Drives Solid Waste Management Performance in Cities: A Global Analysis Using Machine Learning*, 2022, http://dx.doi.org/10.2139/ssrn.4254784 (accessed May 2023).
10. D. C. Wilson, *Waste Manag. Res.*, 2007, **25**(3), 198.
11. A. Whiteman, M. Webster and D. C. Wilson, *Waste Manag. Res.*, 2021, **39**(10), 1218.
12. D. C. Wilson, Global Waste Management – The Way Forward, in *Global Waste Management Outlook*, UNEP-ISWA, Paris, 2015, ch. 6, p. 268.
13. C. A. Velis, *Waste Manag. Res.*, 2015, **33**(5), 389.
14. M. Webster, in *The Routledge Handbook of Waste, Resources and the Circular Economy*, ed. T. Tudor and C. J. C. Dutra, Routledge, London, 2020, ch. 24, pp. 252–261.
15. Environmental Justice Atlas, *Multinational takeover threatens the livelihood of the Zabbaleen, Egypt*, EJAtlas, 2019, https://ejatlas.org/conflict/cairos-zabbaleen-continue-facing-hardships-after-the-multinational-waste-management-contracts-have-to-an-end-in-2017 (accessed June 2023).
16. Women in Informal Employment: Globalizing and Organizing (WIEGO), *Integrating Waste Pickers into Municipal Solid Waste Management in Pune, India*, Policy Brief (Urban Policies) No. 8, WIEGO, Manchester, 2012.
17. P. Chikarmane, in *Putting Public in Public Services: Research, Action and Equity in the Global South*, Cape Town, 2014, http://municipalservicesproject.org/userfiles/Chikarmane_Of_Users_Providers_and_the_State_Solid_Waste_Management_in_Pune_India.pdf (accessed May 2023).
18. European Commission, *Circular Economy Action Plan*, Brussels, 2020, https://ec.europa.eu/environment/circular-economy/pdf/new_circular_economy_action_plan.pdf (accessed May 2023).
19. European Parliament, Briefing Note – Consumers and repair of products, Brussels, 2019, https://www.europarl.europa.eu/RegData/etudes/BRIE/2019/640158/EPRS_BRI(2019)640158_EN.pdf (accessed May 2023).
20. J. McCollough, *Int. J. Consum. Stud.*, 2009, **33**, 619.
21. P. Siddharth, G. Dehoust, M. Gsell, T. Schleicher and R. Stamminger, *Influence of the Service Life of Products in Terms of their Environmental Impact: Establishing an information base and developing strategies against obsolescence*, Umweltbundesamt, Dessau-Roßlau, 2020, https://www.umweltbundesamt.de/en/publikationen/influence-of-the-service-life-of-products-in-terms (accessed May 2023).

22. European Parliament, Briefing Note – Right to repair, Brussels, 2022, https://www.europarl.europa.eu/RegData/etudes/BRIE/2022/698869/EPRS_BRI(2022)698869_EN.pdf (accessed May 2023).

23. C. P Balde, V. Forti, V. Gray, R. Kuehr and P. Stegmann, *The Global E-Waste Monitor 2017*, United Nations University (UNU), International Telecommunication Union (ITU) and International Solid Waste Association (ISWA), Bonn/Geneva/Vienna, 2017.

24. ILO, *From Waste to Jobs: Decent Work Challenges and Opportunities in the Management of E-Waste in India*, International Labour Office, Geneva, 2019.

25. K. Daum, J. Stoler and R. J. Grant, *Int. J. Environ. Res. Public Health*, 2017, **14**, 135.

26. M. Oteng-Ababio, E. F. Amankwaa and M. A. Chama, *Habitat Int.*, 2014, **43**, 163.

27. A. Manhart, O. Osibanjo, A. Aderinto and S. Prakash, *Informal E-waste Management in Lagos, Nigeria – Socio-economic Impacts and Feasibility of International Recycling Co-operations*, Basel Convention, Lagos and Freiburg, 2011, http://www.basel.int/Portals/4/Basel%20Convention/docs/eWaste/E-waste_Africa_Project_Nigeria.pdf (accessed June 2023).

28. ANCAT, *Anuário da Reciclagem 2017–2018*; Associação Nacional dos Catadores e Catadoras de Materiais, Brasília, 2019.

29. J. E. B. Migliano, J. Demajorovic and L. H. Xavier, *J. Oper. Supply Chain Manag.*, 2014, **7**, 91.

30. Hindu Business Line, There's much value in store in e-waste, https://www.thehindubusinessline.com/opinion/columns/theres-much-value-in-store-in-e-waste/article24007522.ece (accessed May 2023).

31. M. J. Kishore, *Indian J. Community Med.*, 2010, **35**, 382.

32. M. Heacock, B. Trottier, S. Adhikary, K. A. Asante, N. Basu, M. N. Brune, J. Caravanos, D. Carpenter, D. Cazabon, P. Chakraborty, A. Chen, F. Diaz Barriga, B. Ericson, J. Fobil, B. Haryanto, X. Huo, T. K. Joshi, P. Landrigan, A. Lopez, F. Magalini, P. Navasumrit, A. Pascale, S. Sambandam, U. S. A. Kamil, L. Sly, P. Sly, A. Suk, I. Suraweera, R. Tamin, E. Vicario and W. Suk, *Rev. Environ. Health*, 2018, **33**, 219.

33. The Economic Times, *E-Waste Sector Will Create Half Million Jobs in India by 2025: IFC,* Press Trust of India, Bengal, 2019.

34. X. Chi, M. Streicher-Porte, M. Y. Wang and M. A. Reuter, *Waste Manag.*, 2011, **31**, 731.

35. M. Schluep, T. Terekhova, A. Manhart, E. Muller, D. Rochat and O. Osibanjo, *Where are WEEE in Africa?*, Basel Convention, Châtelaine 2011.

36. H. Schmitz, Green Transformation: Is There a Fast Track?, in *The Politics of Green Transformations*, ed. I. Scoones, M. Leach and P. Newell, Earthscan/Routledge, London, 2015, ch. 11, p. 170.

37. P. Schröder and J. Raes, *Financing an Inclusive Circular Economy: De-risking Investments for Circular Business Models and the SDGs*, Chatham House,

London, 2021, https://www.chathamhouse.org/2021/07/financing-inclusive-circular-economy (accessed May 2023).

38. M. Liboiron, Waste colonialism, Discard Studies, 2018, https://discardstudies.com/2018/11/01/waste-colonialism/ (accessed May 2023).

39. Y. Lembachar, J. Marsden and A. von Schwerdtner, *Thinking Beyond Borders To Achieve Social Justice In A Global Circular Economy*, Circle Economy, London, 2022, https://circulareconomy.europa.eu/platform/sites/default/files/20220602_-_cji_-_iii-evidence_to_action_-_210x297mm.pdf (accessed May 2023).

40. M. Williams, P. Schröder, R. Gower, J. Kendal, M. Retamal, E. Dominish and J. Green, *Bending The Curve: Best Practice Interventions for the Circular Economy in Developing Countries*, Tearfund, London, 2018, https://learn.tearfund.org/-/media/learn/resources/reports/2018-tearfund-bending-the-curve-en.pdf (accessed May 2023).

41. D. N. Perkins, M. N. Brune Drisse, T. Nxele and P. D. Sly, *Annals of Global Health*, 2014, **80**(4), 286. DOI: http://doi.org/10.1016/j.aogh.2014.10.001.

42. Office of the United States Trade Representative, *Annual Review of the State of Intellectual Property (IP) Protection and Enforcement*, Washington, DC, 2022, https://ustr.gov/sites/default/files/IssueAreas/IP/2022%20Special%20301%20Report.pdf (accessed May 2023).

43. W. R. Stahel and R. Clift, in *Taking Stock of Industrial Ecology*, ed. R. Clift and A. Druckman, Springer, Berlin, 2016, pp. 137–158.

44. European Commission, Directorate-General for Environment, *Promoting Remanufacturing, Refurbishment, Repair, and Direct Reuse* – As a contribution to the G7 alliance on resource efficiency 7–8 February 2017, Brussels, Belgium, workshop report, 2017, https://data.europa.eu/doi/10.2779/1810037 (accessed June 2023).

45. A. Asadikia, A. Rajabifard and M. Kalantari, *Sustain. Sci.*, 2022, **17**, 1939.

46. A. G. Fernandes, *Closing the Loop: the Benefits of the Circular Economy for Developing Countries and Emerging Economies*, Tearfund, London, 2016.

47. R. Gower and P. Schröder, *Virtuous Circle: How the Circular Economy can Create Jobs and Save Lives in Low and Middle-income Countries*, Tearfund and Institute of Development Studies, London and Brighton, 2016.

48. A. Lemille, Circular economy 2.0, Huffingdon Post, 2016, www.linkedin.com/pulse/circular-economy-20-alex-lemille (accessed May 2023).

49. P. Schröder, M. Bengtsson, M. Cohen, P. Dewick, J. Hoffstetter and J. Sarkis, *Resour. Conserv. Recycl.*, 2019, **146**, 190.

50. D. C. Wilson, in *The Routledge Handbook of Waste, Resources and the Circular Economy*, ed T. Tudor and C. J.C. Dutra, Routledge, London, 2021, ch. 7, pp. 54–68.

51. WasteAid and Chartered Institute of Wastes Management, *Making Waste Work: A Toolkit*. CIWM, Northampton, 2017, https://www.circularonline.co.uk/wp-content/uploads/2019/11/Making-Waste-Work-Toolkit-v1-161017.pdf (accessed May 2023).

52. P. Schroeder, K. Anggraeni and U. Weber, *J. Industrial Ecology*, 2019, **23**, 77–95.

53. Chartered Institution of Wastes Management (CIWM), *Centenary History of Waste and Waste Managers in London and South East England*, CIWM, Northampton, 2007.

54. J. Barrie, A. L. Latifahaida, M. Albaladejo, I. Barsauskaite, A. Kravchenko, A. Kuch, N. Mulder, M. Murara, A. Oger and P. Schroder, *Trade for an Inclusive Circular Economy: A Framework for Collective Action*, Chatham House, London, 2022, https://www.chathamhouse.org/sites/default/files/2022-06/2022-06-15-inclusive-circular-trade-barrie-et-al.pdf (accessed May 2023).

55. C. K. Barrie, J. P. Bartkowski and M. Haverda, *Social Sciences*, 2019, **8**, 83.

56. US EPA, *Environmental Effects, Backyard Burning*, 2020, https://archive.epa.gov/epawaste/nonhaz/municipal/web/html/env.html (accessed May 2023).

57. P. Krecl, C. H. de Lima, T. C. Dal Bosco, A. C. Targino, E. M. Hashimoto and G. Y. Oukawa, *Sci. Total Environ.*, 2021, **765**(765), 142736.

58. N. Reyna-Bensusan, D. C. Wilson, P. M. Davy, G. W. Fuller, G. G. Fowler and S. R. Smith, *Atmospheric Environment*, 2019, **213**(213), 629.

59. C. A. Velis and E. Cook, *Environmental Science & Technology*, 2021, **55**(11), 7186.

60. D. Moore, Black carbon from burning waste has 'significant climate impact', Circular Online, 2019, https://www.circularonline.co.uk/news/black-carbon-from-burning-waste-has-significant-climate-impact/ (accessed May 2023).

61. J. Jambeck, R. Geyer, C. Wilcox, T. Siegler, M. Perryman, A. Andrady, R. Narayan and K. Law, *Science*, 2015, **347**(6223), 768.

62. SYSTEMIQ, *Radically Reducing Plastic Pollution in Indonesia: A Multistakeholder Action Plan*, National Plastic Action Partnership supported by World Economic Foundation, 2020, https://www.systemiq.earth/resource-category/npap-action-plan/ (accessed June 2023).

63. Sustainable Waste Indonesia, *Waste Data Baseline Report, Banyuwangi Regency*, Jakarta, 2018.

64. J. Danielson P. Chandran, C. Echeverria, J. Luchesi and N. Shekar N, *Leave No Trace*, Denpasar, Vital Ocean, 2020, https://hasirudala.in/wp-content/uploads/2020/12/Leave-No-Trace.pdf (accessed June 2023).

65. Ellen MacArthur Foundation, *Circular Economy in Africa: Examples and Opportunities*, Cowes, 2021, https://emf.thirdlight.com/link/1xpq6ci26lva-52isfm/@/#id=0 (accessed May 2023).

66. C. C. Ihueze and C. C. Okpala, *J. Sci. Eng. Res*, 2018, **5**(4), 8.

67. Zero Waste Living Lab, Kecipir, 2023, https://zerowastelivinglab.enviu.org/our-ventures/kecipir-2/ (accessed May 2023).

68. S. Irianti and P. Prasetyoputra, *J. Health Ecol.*, 2019, **17**(3), 123.

69. W. R. Stahel, *Nature*, 2016, **531**, 435.

70. E. Gunsilius, S. Spies, S. Garcia-Cortes, M. Medina, S. Dias, A. Scheinberg, W. Sabry, N. Abdel-Hady, AL. Florisbela dos Santos and S. Ruiz, *Recovering*

Resources, Creating Opportunities. Integrating the Informal Sector into Solid Waste Management, GIZ, Eschborn, 2011, http://www.giz.de/de/downloads/giz2011-en-recycling-partnerships-informal-sector-final-report.pdf (accessed May 2023).

71. Ellen MacArthur Foundation, *Towards a Circular Economy: Business Rationale for an Accelerated Transition*, Cowes, 2015, https://www.ellenmacarthurfoundation.org/assets/downloads/TCE_Ellen-MacArthur-Foundation_9-Dec-2015.pdf (accessed May 2023).

72. UNEP-ISWA, *Global Waste Management Outlook*, UNEP-ISWA, Paris, 2015.

CHAPTER 4

Challenges Facing SMEs: Political Ideology, Values Prioritisation and the Governance Trap

ANN STEVENSON*

Resource Futures, Create Centre, Smeaton Road, Bristol BS1 6XN, UK
*E-mail: Ann.Stevenson@resourcefutures.co.uk

4.1 Introduction

The circular economy is commonly described as an alternative production and consumption system able to overcome environmental and sustainable development issues and inequalities by greater use of "closed loop" circular business models. With ubiquitous claims of trillions of dollars of benefits for businesses, much of the emphasis in transitioning to a circular economy is on how we can embed innovation in the production side of the system, and, consequently, how businesses are to change. Around the world, Small and Medium-sized Enterprises (SMEs) make up most businesses. In Europe, 99% of established businesses are SMEs, being seen as the "backbone of the EU economy",[1] upon which jobs, innovation and economic growth have become reliant. Even so, only small amounts of research have focused on how established SMEs embedded in existing

Issues in Environmental Science and Technology No. 51
The Circular Economy: Meeting Sustainable Development Goals
Edited by Sadhan Kumar Ghosh and Gev Eduljee
© The Royal Society of Chemistry 2024
Published by the Royal Society of Chemistry, www.rsc.org

linear economy-based supply chains engage with the circular economy and understand their role in it.[2]

In this chapter key findings are presented from research[3] on the perceptions of risks for established manufacturing SMEs in transitioning to a circular economy. The aim is to open out the debate on how existing manufacturing SMEs are to be involved in the circular economy. The context of such debate is the necessity to first recognise and address that the way political ideology, the prioritisation of economic values in societies and governance of transition are embedded in discourse and decision-making practices creates ideological dilemmas for established SMEs.

4.2   The Ideological Dilemma in Circular Economy Decision-making

Every decision involves some form of judgement to be made, a weighing up of rules and facts that may be evaluated consciously or simply as part of the taken-for-granted mundane way of how a decision is expected to be rationally made. However, the rationality or mundaneness of a decision is anything but simple, being influenced by historically established everyday expectations and practices and preferences, and more importantly shared values and ideology that have informed the preconditional boundaries or rules of what is to be seen to be a rational decision.[4] The complexity is that how rules and boundaries are interpreted, and which values are to be prioritised and what is deemed rational in a decision, has subjective dimensions that are also seen as common sense and rational. These subjective evaluations are influenced by a wide range of factors, including the specific situational context of the decision and the decision-maker, the perceived role and agency of the individual, what is socially expected and accepted as everyday practice, and the perceptions of the choices available and the outcomes of a decision.[5,6] The diversity of interpretations of preconditional social influences, whether consciously recognised or not, creates "ideological dilemmas" for individuals when faced with a dilemmatic choice that involves weighing up different values and ideological beliefs,[4] such as when deciding whether to prioritise environmental and sustainable development principles in production and consumption practices. In this context the dilemmatic nature of the decision stems from the complexity of how to measure the pros and cons, the benefits and losses, or, as more commonly referred to, the risks. This is especially difficult when conflicting or contrary values and ideology are part of the decision, which may not be measurable in the same way, and when values, ideology and outcomes may be perceived differently by different people or the same person in a different context. Therefore, making decisions on if or how to engage with the circular economy is highly complex and dilemmatic in nature for established manufacturing SMEs, influenced by subjective evaluations and perceptions of risks.

4.2.1   Perceptions and Risk

All decisions and the actions taken as a result of a decision are influenced by what is perceived as important.[7] The difficulty is that perceptions are never completely fixed, being constantly renegotiated through discursive interactions in specific contexts that call upon historically, culturally, socially and politically defined and reinforced rules.[8] Against this background, business risks are generally spoken about as being easily identified and consistently measured and quantified in some way to enable a *rational* decision to be made that weighs up the negative risks against the positive benefits.[9] But how can businesses consistently measure and quantify and be certain of the risks and benefits associated with how people value or prioritise sustainable development or environmental protection values in purchasing practices? Whereas traditionally risk was treated as a simple calculation of probability and effect, usually related to money, uncertainty is now being accepted more and more as a key component in determining risks and benefits, as highlighted in risk management standards.[10] Due to such uncertainty, judgements are made influenced by a plethora of factors associated with the situational context of the decision-maker. Social relationships, perceptions of agency, knowledge, trust and collective memories are just a few objective and subjective aspects influencing perceptions of risks and benefits.[11,12,13]

4.2.2   The Power of Discourse

The concepts of the circular economy and sustainable development and positioning of risks and benefits in engaging with such concepts are becoming embedded in modern-day discourse. However, how such discourse is interpreted, just as with perceptions of risk and benefits, can be multi-dimensional, with people having contrary and multiple accounts of the same concepts included in the discourse. As highlighted in Chapter 2, there is complexity, debate and uncertainty in how the concepts of the circular economy and sustainable development are understood and what needs to be done. In many cases, rather than coming to a consensus, there is evidence that the discourse of the circular economy and what is to be done is diverging.[14,15] A lack of consensus creates a high degree of uncertainty, confusion and questioning of what is true, enabling people to call upon differing and potentially conflicting arguments to rationalise decisions. This in turn reinforces only those salient aspects deemed important to a decision-maker at a particular moment in time being called upon to decide which truth to select or construct.[16] In this way discourse not only constructs reality and is constituted of existing perceptions of reality, but is always used to achieve an effect during interaction.[17] The constructions and perceptions of the reality of what action to take, who is to take the action and what the risks and benefits are in transitioning to a circular economy are reliant upon the use of familiar and habitual patterns of discourse that enable the sharing of meaning.[18]

The research on perceptions of risk for manufacturing SMEs using discourse analysis showed how there was a wide range of recurrent discourses occurring in published materials and in face-to-face interviews that underpinned seven primary repertoires, as shown in Table 4.1.[3]

Although the SMEs as resource-constrained, SMEs as unknowledgeable and competitive environment repertoires predominantly incorporated consistent discourses by all types of actors, and in a range of settings (*e.g.,* published materials, at events and interviews), contrary maxims that could cause conflict were evident in the others. A deep dive into these repertoires gave insights into the interconnected historical, social, cultural and political influences upon perceptions of risks in transitioning to a circular economy. Established arrangements of entities, structures and rules underpinning power and relationship dynamics, values and political ideology in practice, and trust and truth causal mechanisms[19] go a long way to explaining the engagement of established SMEs.

4.3   Power and Relationship Dynamics

In Chapter 2, a brief overview of the concept of the circular economy, its relationship to sustainability and sustainable development and the complexities involved in defining and transitioning to a circular economy were outlined. In this chapter the circular economy is taken as an umbrella concept bringing together a wide range of ideas and concepts that share the aim of moving away from a traditional linear economy model of production and consumption whilst maintaining economic growth. In this context, the linear economy model is one where nature is constantly exploited to make new materials and products that are used and sold and then thrown away as waste when no longer deemed to have value by individuals and society.[20] However, this linear way of exploiting nature has become a historically hardwired, complicit practice of governments, businesses and all members of society to ensure continued economic growth and progress. As highlighted in Chapter 2, economic growth is seen as fundamental to sustainable development. This drive for economic growth and social progress is underpinned by the desire to bring economic, utility, material and symbolic extrinsic value for humankind. Delivering economic growth and value to humankind by exploiting nature is an unquestioned underpinning of production and consumption business models, whether linear or circular.

Since the 1980s, "business model" has been used as a taken-for-granted term aimed at simplifying the relationships between the wide range of production and consumption actors and explaining what, why and how a business does what it does to survive. There are three interdependent parts of a business model, value proposition, value creation and value capture, each involving a myriad of stakeholders.[21] The value proposition is concerned with what is of value to users of a service or product. Value creation is about how the service or product delivers what is valued by users, whilst value capture is

Table 4.1   Repertoires in perceptions of risks in transitioning to a circular economy. Data from ref. 3.

Repertoire	Dominant discourses	Subordinate discourses
Circular economy as a higher ideal	Interviews • "it's understanding what the circular economy is first of all" • "some of the higher ideals might not make it" Published • "priorities for innovation"	Interviews • "it's just too complex" • "if you've got something really novel" • "well maybe in 20 or 30 years" • "could actually help them" • "nobody's taking ownership"
Circular economy proof of value	Interviews • "need to prove it" • "need to know something is doable" Published • "business barriers" • "real opportunities to be had" • "financial benefit" • "in the absence of public policy intervention" • "would not have happened without access to expertise"	• "legislation" • "there's no incentive" • "certain degree of greenwash" • "OEMs [original equipment manufacturers] are trying to make a difference" • "Blue Planet" • don't plan "further than the next 6 to 12 months" • "small guys usually can't"
SMEs as resource-constrained	Interviews • "everybody's bloody busy" • "day to day pressures" • "always been very lean burn" Published • "time and resources"	• "a guiding light" • "it's not getting any easier" • "industries segmented nowadays"
SMEs as driven by personal motives of owners	• "depends what the SME owner wants"	Published • "new market opportunities" • Waste "reduce, reuse, recycle ... and avoid carbon emissions"
SMEs as unknowledgeable	Interviews • "don't get it all" Published • "lack of awareness"	• "rational decisions ... trade-off ... financial and time costs"
Competitive environment	• "a jungle out there"	
Asymmetric power relations	Interviews • "if the bigger fish say" • "can't influence customers, can't influence suppliers" Published • "large businesses work with... to make savings for themselves"	

about what's in it for the producer of a product or service. Over decades, business models have become optimised working within a linear economy system to create balance of value for all stakeholders. Therefore, to better understand the dilemmatic nature of decisions for established SMEs in proactively adopting circular business models, we first need to understand power and relationship dynamics; the role businesses, and particularly SMEs, have been expected to play in delivering the balance of value in a linear economy; the choices they have; and what has influenced these expectations.

4.3.1 The Role of Businesses

Irrespective of the complexity of the relationship between consumption, production and factors such as population growth, the problem of environmental and social damage and the creation, embedding and reinforcing of a linear economy has been laid at the feet of businesses. Indeed, it would be difficult to argue against the fact that since the age of industrialism, production and consumption practices in the pursuit of economic growth have had detrimental environmental and human impacts. The age of industrialism had its roots in the Industrial Revolution in Britain in the late 18th and early 19th centuries, when technological innovation supported social, political, cultural and economic modernisation of society. As the creators of innovation, manufacturing businesses have historically, and continue to be, celebrated as providers of solutions to the ills, needs and utility, and material and economic wants of societies, particularly creating jobs, supporting higher living standards and delivering economic growth to enable social, political and economic progress, key requirements embedded in conceptualisations of sustainable development. Not only did industrialism provide the means for easier and lower-cost exploitation of energy and nature and to achieve wider social and economic progress, but it heralded in ideological change. A major outcome of this period, which has changed little as we move into the 21st century, was the growth of liberalist, capitalist, free-market economy political ideology that has dominated since the 1970s. Such ideology builds on principles of all natural entities and people being commodities that have an economic value to be determined by global supply and demand mechanisms driven by freedom of choice, wealth creation and limited intervention of the state.[22] Commitments to such ideology are often made explicit in UK government policies, particularly those aimed at businesses.[23]

Working to this ideology, manufacturers' social responsibility is to adopt business models that increase their competitiveness and profits by feeding growing consumerism and freedom of choice of individuals, through the exploitation of nature and people to meet the demands of society.[24] Businesses continue to carry out this responsibility, acting as "a vehicle for economic progress", through innovation aimed at outcompeting other businesses in responding to changing market demands.[25]

However, as this political ideology started to become entrenched in the global economy, and the negative effects of industrialism and pursuit of economic growth became more tangible, environmentalism discourse became a dominant feature in the 1970s. Such discourse was epitomised in the report *The Limits to Growth*[26] and works such as *Blueprint for Survival*.[27] This heralded in an added expectation for manufacturers to take on the role of moral and ethical guardians of the planet and people. However, the *caveat* to this added role, which has remained enshrined in the principles of sustainable development and the circular economy, is to adopt such responsibility without compromising the enactment of current political ideology principles and existing societal expectations of businesses. Namely, continuing to create jobs and economic growth by delivering the price-performance needs', desires' and values' preferences of societies.[28] This added role places expectations for manufacturers to value sustainable development aspirations and values in production as high or higher than the material, utility, symbolic and economic values of stakeholders. This is irrespective of the way political ideology values sustainable development principles, or the way individuals prioritise and enact such values in purchasing practices. An ecological modernisation discourse, reliant on scientific and technological innovation as solutions to reducing costs and increasing performance,[29] underpins the warranting of the placement of this responsibility on business and maintaining the current political ideology in practice. This positioning of technology as the answer to addressing the environmental and social problems of the world without compromising economic growth or the maximisation of value for humankind is embedded in the concept of sustainable development.

If the answer is technological innovation, and SMEs are understood to be sources of high levels of entrepreneurship and innovation, one could expect the significant numbers of established manufacturing SMEs to be leading the charge. However, the evidence of engagement of such businesses in the active adoption of sustainable business models is fragmented and often limited to *niche*, small-scale applications and markets.[30,31] To start to understand why, we need to take account of the history of the role of SMEs and how they fit in to the mechanics of modern-day political ideology in practice and the freedoms of choice they have in changing what they do.

4.3.2   SMEs in a Hierarchy of Power

Before the age of industrialism, all manufacturing businesses in the UK were SMEs serving local needs. With industrialism, the business landscape began to be dominated by monopolistic large organisations with protected product and production rights and large financial assets, predominantly built on mass production innovation.[32,33,34] This transition put many non-specialist SMEs out of business. That was until large-scale globalisation of production took hold in the 1970s and 1980s resulting in sub-contracting and outsourcing beginning to be the norm to maintain low costs, creating significant growth of SME manufacturers in the UK and around the world.[35,36] A key aspect of this transition was that the manufacturing SMEs that flourished in this time manufactured products to a design specified by their customer, or provided "specialist" functions that provided economic benefit to their customers.[37] The study of risks for SMEs[3] demonstrated how this arrangement remains dominant in the discourse of being a manufacturing SME in the UK, such that established SMEs complicitly evolve to serve specialist functions to maximise dependency of customers on their services. What this means is that SMEs are all specialists, locked into a price-performance delivery model to maximise relationship stability and customer dependency, with the ultimate aim of becoming a preferred supplier or strategic partner of customers.[38]

Such dependency-based relationships, although symbiotic, are built asymmetrically,[39] where size matters in terms of who has what type of power and freedoms to determine the price and performance characteristics.[40] The primary difference regarding the nature of dependency is that, for SMEs, dependency is "needs driven" whilst for their customers, dependency is desire-satisfaction driven.[41] In compliance with the principles of liberalist political ideology, what this means is that customers can choose at any moment in time to select a different supplier or to adjust their price-performance preferences.[42] This asymmetry is a normative expectation of production supply chains. The outcome is that established SMEs become more and more socially dependent[40] on their existing customers and markets and constrained to conforming with a hierarchically defined role of delivering specialist services demanded by customers that meet customers' price-performance and value prioritisation preferences to remain competitive and survive.[43] Reinforced by political ideology in practice, SMEs have to earn their right to exist by adopting competitive strategies that ensure that delivering what the customer values exceeds the cost of provision,[43] whether this is lower prices for equivalent benefit or superior performance characteristics. Therefore, the values oriented to and prioritised by existing customers, markets and competitors have a major, if not the key, influence on perceptions of the choices available to established SMEs when making decisions.

These expectations on SMEs have remained unchanged for decades and become an unquestioned established norm. Therefore, the freedoms of SMEs, irrespective of owners' or managers' own values, are bound by how ideology in practice privileges, or otherwise, sustainable development principles and the transition to a circular economy and influences what existing customers and markets value.

4.4 Political Ideology and Values

Political ideology can be defined as a stable system of beliefs of the relationships between the political, social and economic structures of societies that are coherent to all stakeholders that form those societies.[44] As already outlined in Section 4.3, economic growth, maximising utility, material and economic benefits for humankind, supply and demand economics and consumerism are fundamental underpinnings of both the linear economy and circular economy. However, in discourse of the circular economy, the *caveat* is that this is to be achieved "sustainably" through the adoption of innovative circular business models. As explained in Chapter 2, what is meant by acting sustainably is ambiguous, as is what is to be classed as being circular. As such, the value of sustainability in political ideology and the consequent complicit societal enactment of such value frame the rules called upon by SMEs and their customers in decisions on engaging with circular business models.

4.4.1   Sustainability as a Higher Domain of Value

Political ideology in practice is predominantly understood as reinforcing Environmental and Sustainable Development (ESD) values, and thus the circular economy, as being part of a "higher domain of value".[45] In discourse, adopting circular business models is positioned both as the right thing to do for people and the planet, at the same time as being a voluntary ethical or moral obligation on society.[3] This positioning of a relationship to moral and ethical values places an onus on businesses to take account of intrinsic value in decision-making beyond the everyday expected practices and legal obligations regarding extrinsic value. Here, intrinsic value is a characteristic of an entity that cannot be traditionally measured as having economic, material or utility value to humans but has value simply because it exists.[46] However, at the same time, the primacy given to economic growth, economic values and commodification of nature and people positions such obligations as having to be evaluated on an extrinsic cost–benefit basis. This is irrespective of whether intrinsic value can be measured or monetised, or how the legitimisation of the prioritisation of extrinsic value over intrinsic value and what is morally right is embedded in political ideology in practice.[47] The voluntary nature of such obligations, prioritisation of extrinsic values evaluated on a cost–benefit basis and immeasurability of intrinsic value, reinforces the ideological dilemma for established SMEs. These factors work together to create high degrees of uncertainty for SMEs as to whether customers will value ESD values and engage with circular business models if they were to be provided.[3]

4.4.2   Valuing Sustainability

There are embedded understandings that voluntarily engaging with ESD values in production and consumption systems solicits a green premium or negatively affects other extrinsic values, such as quality, convenience or sense of identity.[3] This is against a background of a contrary maxim in circular economy published literature that calls upon economic benefits being achieved through cost savings from using waste as a resource. Evidence exists to support all these positions, again creating dilemmas for decision-makers on which evidence to call on, especially when the same evidence could be interpreted differently and support opposing positions. For example, on the green premium, a fourfold increase in green and ethical purchasing practices between 1999 and 2019 from £11.2 billion to £41.1 billion[48] could be called on to validate how people place a high value on sustainability and are willing to pay a green premium. It would therefore make sense for businesses to look at how they can enter such a market. However, when placed in the context of ethical purchasing growth being equivalent of a rise from 1.7% to 3% of total spending over the same 20 year period,[3] an opposite argument can be construed where the biggest percentage of markets do not preferentially value sustainability in economic practices. On this basis, not moving into the ethical and green product and service market would seem to be a rational decision.

But what if there is no green premium? In that case, manufacturers have to contend with perceptions of sustainable products performing worse than conventional products, namely the "green product performance liability".[49] When a product is associated with ESD values, a trade-off against other performance characteristics is often inferred, which can seriously damage the reputation of a business. It is difficult to ascertain with high levels of confidence why this may be the case, but technological advances in the age of industrialism that improved the performance of historical nature-based products, whether in medicine,[50] farming[51] or hygiene,[52] is likely to have had a long-term effect on perceptions of such products. But what if such products were lower cost?

When it comes to the cost-savings arguments, attempts to reframe waste as a resource may have unintended consequences. As explained in Section 4.2, manufacturing SMEs are predominantly specialist providers of products and services where price-performance characteristics determine the competitive edge of an SME in establishing and maintaining a dependency relationship with their customers. A century of seeing materials and products that were classed as waste as inferior, harmful and unhygienic, and thus a problem, can act to place expectations that products associated with the idea of waste will be of lower quality and must therefore be a lower price, irrespective of the work having to be done to utilise such waste. These negative perceptions have been found in a range of research on remanufactured[53] and refurbished[54] products and the use of recycled materials,[55] being reinforced by practices of downcycling of materials and products.[56] These perceptions are exacerbated by the way people collapse quality, durability and price together such that lower-priced products are inherently perceived as being low quality.[57,58]

Knowledge of these perceptions reinforces the ideological dilemma for established SMEs when carrying out the requisite cost–benefit analysis of whether to proactively engage with circular business models. Decisions would not be dilemmatic if there was certainty that customers (individuals or businesses) preferentially valued ESD values and circular business models in purchasing practices above all other values. However, in evaluating customers' prioritisation of values as part of the expected cost–benefit analysis, the shortcut of willingness to pay is predominantly used as a proxy or "shadow price"[59] for determining the value allocated to ESD values and circular business models by customers. Willingness to pay encompasses accepting a green premium and/or performance losses for equivalent or lower-priced products and services.

4.4.3   The Circular Economy and UN Sustainable Development Goals

It is not only attempts at reframing materials and products classed as waste as a resource that are seen as having unintended consequences. Debate abounds on the unintended consequences of the positioning of the circular economy as a practical means to address a number of the UN Sustainable Development Goals (SDGs), when in discourse and practice action remains focused on maintaining economic growth and feeding growing consumption. Questions are asked as to how a model reliant on technological innovation, better design of products and continued economic growth delivered by businesses as solutions can outweigh the negative impacts of an absolute rise in demand and use of materials as populations grow and people continue to aspire to increase quality of life through materiality and economic growth.[60] By failing to acknowledge such a situation, and because using less material is most likely to result in less output and therefore less economic activity, other potentially more radical economic, dematerialisation and waste reduction ideas and discourses fail to resonate or become sidelined,[61,62] or even ridiculed as idealistic and impractical. There are also arguments that the circular economy is at best a "weak" sustainable consumption model approach that will fail to address global environmental and sustainable issues, given its reliance on the need to incentivise individuals to make the right choice of reduced consumption to stimulate demand in manufacturing.[62] But areas that solicit the most concern in terms of unintended consequences relate to a lack of certainty that the development and adoption of circular business models (1) will result in environmental and social benefits that are essential to the UN SDGs[63] and (2) will not result in a "rebound effect".

The first point is illustrated by the use of biofuels[63] and other nature-based "sustainable" materials that rely on land to grow such materials. This has the potential to result in deforestation, reduced biodiversity, the displacement of people and/or displacing the use of land to grow food, whilst using large quantities of water, pesticides and chemicals. Such moves present ideological dilemmas when weighing up the different UN SDGs. For example, what trade-offs are we willing to make in achieving the targets established for the different goals? To ensure delivery on SDG 2 (Zero hunger) and SDG 6 (Clean water and sanitation) aspirations, should we use fewer, not more, green energy sources or sustainable bio-based materials that take out food-productive land and use water, as part of SDG 12 (Sustainable consumption and production)?

On the second point, there are significant concerns that circular business models in practice will inevitably result in a rebound effect, where subsequent actions of a user or purchaser of a circular product or service will behave in a way that offsets the benefits achieved by engaging with a circular business model.[64] At least four elements of the rebound effect have been demonstrated for circular business models. First, when economic benefits are achieved through improved technological advancement, material efficiency at point of purchase or through in-use energy efficiency or adoption of leasing models, cost savings are used to purchase additional products or services or to use the product more frequently, again stimulating greater demand and consumption.[61] This "Jevons paradox" is not new, having been identified with the roll-out of the steam engine in factories from 1785 onwards, which made the engine's operation affordable to more businesses and used in more remote areas so overall production increased. In more recent times, as demonstrated in a study of leasing and sharing (property, cars and heating system), households were found to use the money they saved on more carbon-intensive services such as more holidays, including in low-income households.[65] A second element of the rebound effect is that recycled, repaired, remanufactured and reusable products and materials do not have a 1:1 displacement relationship to the production of new products, and can result in zero displacement or increased demand. This is because such products compete in different markets and are targeted at different types of customers who may never have had access to such products.[66] Third, many of the features of circular business models that originated as acts of goodwill, such as sharing, reusing and repairing, are argued as fuelling "hyper consumerism"[67] as such acts become commodified, creating increasing demand. This is illustrated by the growth and changing face of peer-to-peer accommodation-sharing platforms, and how they have fuelled the expansion of tourist-based use and demand for more properties to be built as homes become businesses.[68] Lastly, the environmental credentials of a product or service have also been shown to validate decisions to obtain a new "green" product, thus stimulating higher demand and purchasing and generation of obsolete products.[69]

Given such a context of conflicting views, what people perceive to be the truth about the circular economy and sustainable development will have major repercussions on their willingness to engage with circular business models.

4.5 Trust and Truth

In discourse of the circular economy and wider sustainability fields, either there is an implicit assumption that customers preferentially value ESD values in products and services, or rhetoric abounds positioning societal demand for such products. However, in an era of post-modern cynicism, where more and more people are disillusioned with the enactment of moral responsibility of humankind,[70] understandings of the existence of an attitude–behaviour gap in sustainability-oriented purchasing practices has become established in manufacturing.[3] As demonstrated in Section 4.4, it is not disputed that there is evidence of growth in green and ethical purchasing practices. However, increasing research is now demonstrating the existence of a strong "say–do" gap[71] that inhibits growth in such markets. Here, what people say they want and do regarding use of ethical, sustainable or circular products do not consistently translate into behaviour in practice, especially if there is a green premium or extrinsic value is perceived to be negatively affected.[72,73,74]

This is against a background of a wide range of sustainability-oriented products and services already existing.[41]

This cynicism towards what people say was found to extend to claims of demand made by business and public sector customers.[3] In this context, expressions of cynicism manifest themselves in discourse of the existence of greenwash. The concept of greenwash started to become established in the early 1980s[75] and is now used as a shortcut to position organisations as engaging in untruths or superficial and symbolic discursive acts regarding commitments to ESD values.[76] In all cases, cynicism can be understood to be directed towards what people say, such that precedence is given to knowledge and experience of customers' existing prioritisation of values, competitor activities and changes in existing markets in practice.[77] Tacit knowledge and experience of the domination of price-centric purchasing decision-making under the guise of "value for money", and lack of evidence of circular business models replacing existing linear practices beyond *niche* applications and start-up SMEs, will continue to fuel the ideological dilemma of circular economy decision-making for established SMEs. If current systems cannot place an economic value on loss of biodiversity or the impacts of climate change on people's sense of place, or act to place a low economic value on such elements using existing systems whilst economic utility remains privileged in ideology in practice, how can the dilemmatic nature of decisions in engaging with circular business models be overcome? For SMEs, and businesses in general, to fully take on their responsible innovation role, changes are needed to the systems and structures underpinning political ideology and values in practice that prioritise economic values and result in linear modes of production and consumption having lower costs than alternative modes. Such changes require a wider discussion on the governance of transition to a circular economy.

4.5.1   Attribution of Responsibility

In discourse of the circular economy and sustainable development, responsibility for addressing the environmental and social problems we face is a contested issue. As introduced in Section 4.3, manufacturers have been allocated the responsibility to deliver technological solutions that solve societal moral and ethical problems whilst enabling continued consumerism and economic growth, under the guise of "responsible innovation".[78] As such, there are understandings that businesses are the ones responsible for the problems caused, are to be moral and ethical guardians of the planet and are to be responsible for the development of solutions. However, not only can moral or intrinsic values and economic or extrinsic values potentially conflict, but innovators rarely control the costs, acceptance or use of innovation in a free-market economy. Placing the extent of this responsibility on manufacturers whilst avoiding the role and location of consumption is inherently unfair.[78,79] All economic actors and stakeholders have had a role in creating and adding to the global environmental and social problems, have moral and ethical obligations towards society and the planet and have the capacity to contribute to solutions, individually and collectively. The majority of people have historically welcomed and voluntarily involved themselves in modernisation, human development and progress, globalisation, and science and technology innovation, including the establishment of the linear economy that has led to our current problems.[80,81] This "problem of many hands"[82] in the creation of climate change, social injustice, biodiversity loss and resource depletion means that there is major debate on how addressing moral and ethical obligations, paying to fix problems resulting from past actions and protecting the future can be fair and transparent. However, due to the complexity in understanding each actors' contribution, past and present, and in the level of agency different people, societies and organisations have, the attribution of such responsibility is rarely fair, transparent or shared.

4.5.2 The Governance Trap

Even though the global problems we are facing are so daunting, the attribution of responsible innovation in policy can be seen as a symbol of the way that governments shift governance of social issues to individuals, communities and organisations to maintain liberalist, free-market economy political ideology principles in practice.[47] This devolving of governance relies on people truthfully enacting their freedoms in making the right choice for the collective good, as informed by policy messaging. However, understandings of the scale of changes needed means that governments are being looked to more frequently to lead and take strong political action. This builds on inherent understandings that governments have the social power to influence, create and uphold the legal systems that value ESD values in practice at the same time as influencing voluntary moral responsibilities in societies.[83] These differences in expectations on how the embedding of ESD values in practice are to be governed becomes part a "governance trap"[47,79] in transitioning to a circular economy. In this situation, governments' promotion of voluntary approaches to governance, relying on communities and organisations doing the right thing without having the required agency or stable resources, direction or capacities, conflicts with societies' discursive calls for strong political action to address global issues. However, neither governments, businesses nor the people are trusted by each other to take the needed levels of action.

4.6 Conclusions

Established manufacturing SMEs face many daily challenges to their continued existence. Calls to engage in "responsible innovation" and proactively develop circular business models create ideological dilemmas for SMEs that are embedded in needs-based dependency relationships. In these relationships, they rely on adhering to the rules expected of them in delivering price-performance preferences of customers and their perceptions of the truth of how values and ideology are enacted in practice. The needs-based dependency arrangements and lack of social power to influence are historically and socially constituted and reinforced by systems and structures underpinning political ideology and values prioritisation in practice. The primacy of economic values and liberalist, free-market economy principles that reinforce ESD values and what is ethically and morally right as voluntary actions creates an environment of high uncertainty of the value of engaging with the circular economy. Although proponents of the circular economy work to position cost savings resulting from engaging with circular business models and there being demand for more sustainable products, knowledge of the green premium and an attitude–behaviour gap in green purchasing practices fuels this uncertainty.

If governments continue to rely on voluntary enactment of moral values in purchasing practices of individuals, institutions and businesses, including governments themselves, it is likely that few established manufacturing SMEs will proactively engage with circular business models. There has to be coherent strong policy and legislation that economically advantages circularity at all times to grow and stabilise market demand and create a level playing field with the linear system. There needs to be trusted evidence of market demand in practice and a move from giving primacy to price and economic cost savings in purchasing practices. It is critical that there is a move away from responsible innovation discourse towards a shared responsibility discourse to empower all actors and stakeholders to take action. Finally, less reliance is to be placed on discursive indicators of prioritisation of ESD values and voluntary actions of individuals and institutions to avoid commitment to the circular economy being seen as symbolic and idealistic.

But fundamentally, more needs to be done to break the governance trap. Societies need governments to lead.

References

1. European Commission, *An SME Strategy for a Sustainable and Digital Europe*, COM 2020. Communication from the Commission to the European Parliament, the Council and the European Economic and Social Committee and the Committee of the Regions, Brussels, 2020.
2. R. Stewart and M. Niero, *Business Strategy and the Environment*, 2018, **27**, 1005.
3. A. Stevenson, PhD thesis, Cardiff University, 2022.
4. M. Billig, S. Condor, D. Edwards, M. Gane, D. Middleton and A. Radley, *Ideological Dilemmas: A Social Psychology of Everyday Thinking*, SAGE Publications, London, 1988.
5. A. Langley, H. Mintzberg, P. Pitcher, E. Posada, and J. Saint-Macary, *Organization Science*, 1995, **6**(3), 260.
6. Å. Boholm, A. Henning, and A. Krzyworzeka, *Focaal*, 2013, **65**, 97.

7. S. P. Robbins and T. Judge, *Organizational Behaviour*, Pearson/Prentice Hall, London, 13th edn, 2008.
8. J. M. Wittmayer, F. Avelino, F. van Steenbergen and D. Loorbach, *Environmental Innovation and Societal Transitions*, 2017, **24**, 45.
9. S. Maguire and C. Hardy, *Academy of Management Journal*, 2013, **56**(1), 231.
10. ISO, *ISO31000:2018 Risk Management – Guidelines (Preview)*, Geneva, 2018.
11. Å. Boholm and H. Corvellec, *J. Risk Res.*, 2011, **14**(2), 175.
12. Å. Boholm, *Ethnos*, 2003, **68**(2), 159.
13. Å. Boholm, H. Corvellec and M. Karlsson, *J. Risk Res.*, 2012, **15**(1), 1.
14. N. Bocken, I. de Pauw, C. A. Bakker and B. van der Grinten, *J. Ind. Prod. Eng.*, 2016, **33**(5), 308.
15. J. Kirchherr, D. Reike and M. Hekkert, *Resources, Conservation and Recycling*, 2017, **127**, 221.
16. T. R. Shrum, *Climatic Change*, 2021, **165**, 1.
17. J. Potter and M. Wetherell, *Discourse and Social Psychology: Beyond Attitudes and Behaviour*, SAGE Publications, London, 1987.
18. M. Billig, in *Discourse Theory and Practice: A Reader*, ed. M. Wetherell, S. Taylor and S. Yates, SAGE Publications, London, 2005, ch. 15, pp. 210–221.
19. D. Elder-Vass, *Sociologia, Problemas e Práticas*, 2012, **70**, 5.
20. Ellen MacArthur Foundation, *Towards the Circular Economy: Economic and Business Rationale for an Accelerated Transition*, Cowes, 2013.
21. N. Bocken, I. de Pauw, C. A. Bakker and B. van der Grinten, *J. Ind. Prod. Eng.*, 2016, **33**(5), 308.
22. N. Castree, *Geography Compass*, 2010, **4**(12), 1725.
23. Department for Business, Energy & Industrial Strategy, *Industrial Strategy: Building a Britain Fit for the Future*, HM Government, London, 2017.
24. M. Friedman, *The Social Responsibility of Business is to Increase its Profits*, The New York Times Magazine, 1970.
25. D. Henderson, *The Role of Business in the Modern World: Progress, Pressures and Prospects for the Market Economy (Hobart Paper)*, Institute of Economic Affairs, London, 2014.
26. D. H. Meadow, *The Limits to Growth: A Report for the Club of Rome's Project on the Predicament of Mankind*, Universe Books, New York, 1972.
27. E. Goldsmith, R. Allen, M. Allaby, J. Davoll and S. Lawrence, *The Ecologist*, 1972, **2**(1), 48.
28. T. Jackson, *Beyond Consumer Capitalism: Foundations for Sustainable Prosperity*, Centre for the Understanding of Sustainable Prosperity, No 2, 2016.
29. M. Jänicke, *J. Clean. Prod.*, 2008, **16**(5), 557.
30. J. Kirchherr and R. van Santen, *Resources, Conservation and Recycling*, 2019, **151**, 104480.
31. S. Ritzén, and G. Ö. Sandström, *Procedia CIRP*, 2017, **64**, 7.
32. P. L. Payne, *The Economic History Review*, 1967, **20**(3), 519.
33. R. McIntyre, *The Role of Small and Medium Enterprises in Transition: Growth and Entrepreneurship Research for Action (Research for Action No. 49)*, World Institute for Development Economics Research, Helsinki, 2001.

34. T. Sturgeon, *Industrial and Corporate Change*, 2002, **11**(3), 451.
35. L. Abramovsky, R. Griffith and M. Sako, *SSRN Electronic Journal*, 2004, 1463419. DOI:10.2139/ssrn.1463419.
36. M. Kitson and J. Michie, *The Deindustrial Revolution: The Rise and Fall of UK Manufacturing 1870–2010*, Centre for Business Research (CBR) research programme on enterprise and innovation No 459, 2014.
37. J. Bolton, *Small Firms: Report of the Commission of Inquiry on Small Firms*, Cmnd 4811, HMSO, London, 1971.
38. J. Gosling, L. Purvis and M. M. Naim, *Int. J. Prod. Econ.*, 2010, **128**(1), 11.
39. B. Huo, B. B. Flynn and X. Zhao, *J Supply Chain Management*, 2017, **53**(2), 52.
40. B. H. Raven, *Analysis of Social Issues and Public Policy*, 2008, **8**(1), 1.
41. E. Tory-Higgins, in *Social Psychology: Handbook of Basic Principles*, ed. A. W. Kruglanski and E. Tory Higgins, Guildford Publications, Guildford, 2007, ch. 19, pp. 454–472.
42. S. Sandhu, L. K. Ozanne, C. Smallman and R. Cullen, *Business Strategy and the Environment*, 2010, **19**(6), 356.
43. M. E. Porter, *Competitive Advantage: Creating and Sustaining Superior Performance*, The Free Press, New York, 1985.
44. A. Novikau, *The Evolution of the Concept of Environmental Discourses: Is Environmental Ideologies a Useful Concept?* Western Political Science Association Annual Meeting, San Diego, 2016.
45. D. Ludwig, *Ecosystems*, 2000, **3**(1), 31.
46. M. J. Zimmerman, B. Bradley and E. N. Zalta, *Intrinsic vs. Extrinsic Value*, Stanford Encyclopedia of Philosophy Archive, 2019.
47. N. Pidgeon and C. Butler, *Environmental Politics*, 2009, **18**(5), 670.
48. Co-op, *Twenty Years of Ethical Consumerism*, Ethical Consumer, 2019.
49. B. Usrey, D. Palihawadana, C. Saridakis and A. Theotokis, *J. Advert.*, 2020, **49**(2), 125.
50. H. Chung, Britain's industrial revolution, 2016, https://industrialrevolution britain.weebly.com/medicine.html (accessed June 2023).
51. Farming First, Celebrating science and innovation in agriculture, https://farmingfirst.org/science-and-innovation#home (accessed May 2023).
52. S. Hoy, *Chasing Dirt: The American Pursuit of Cleanliness,* Oxford University Press, New York, 1997.
53. W. Gu, D. Chhajed, N. C. Petruzzi and B. Yalabik, *Int. J. Prod. Econ.*, 2015, **162**, 55.
54. E. Van Weelden, R. Mugge and C. A. Bakker, *J Clean. Prod.*, 2016, **113**, 743.
55. L. Veelaert, E. Bois, I. Du Moons, P. Pelsmacker, S. De Hubo and K. Ragaert, *Sustainability*, 2020, **12**, 27.
56. M. de Wit, J. Verstraeten-Jochemsen, J. Hoogzaad and B. Kubbinga, *The Circularity Gap Report: Closing the Circularity Gap in a 9% World*, Circle Economy, Amsterdam, 2019.
57. European Commission, *Behavioural Study on Consumers' Engagement in the Circular Economy*, European Commission. Brussels, 2018.
58. H. Deval, S.P. Mantel, F. R. Kardes and S. S. Posavac, *J. Consum. Res.*, 2013, **39**(6), 1185.

59. J. O'Neill, *Economic and Political Weekly*, 2001, **36**(21), 1865.
60. J. M. Allwood, T. G. Gutowski, A. C. Serrenho, A. C. H. Skelton and E. Worrell, *Phil. Trans. R. Soc. A*, 2017, **375**(2095), 20160361.
61. G. Kallis, *Phil. Trans. R. Soc. A*, 2017, **375**(2095), 20160383.
62. K. Hobson, *Environment and Planning C: Government and Policy*, 2013, **31**(6), 1082.
63. A. Murray, K. Skene and K. Haynes, *J. Bus. Ethics*, 2017, **140**(3), 369.
64. F. Figge and A. S. Thorpe, *Ecological Economics*, 2019, **163**, 61.
65. S. Junnila, J. Ottelin and L. Leinikka, *Sustainability (Switzerland)*, 2018, **10**(11), 4077.
66. T. Zink and R. Geyer, *J. Ind. Ecol.*, 2017, **21**(3), 593.
67. K. Hobson, *Climatic Change*, 2019, **163**(1), 99.
68. K. Hobson and N. Lynch, *Futures*, 2016, **82**, 15.
69. F. Valenzuela and S. Böhm, *Ephemera: Theory & Politics in Organization*, 2017, **17**(1), 23.
70. S. A. Stanley, *The French Enlightenment and the Emergence of Modern Cynicism*, Cambridge University Press, New York, 2012.
71. Ipsos Mori, *Sustainability and the Say–Do Gap: Understanding the Consumer Mindset*, Webinar, 2020.
72. J. Bray, N. Johns and D. Kilburn, *J. Bus. Ethics*, 2011, **98**(4), 597.
73. M. L. Johnstone and L. P. Tan, *J. Bus. Ethics*, 2015, **132**(2), 311.
74. H. J. Park and L. M. Lin, *J. Bus. Res.*, 2020, **117**, 623.
75. J. A. Ottman, in *The New Rules of Green Marketing: Strategies, Tools, and Inspiration for Sustainable Branding*, Greenleaf Publishing, Sheffield, 2014.
76. Y. S. Chen and C. H. Chang, *J. Bus. Ethics*, 2013, **114**(3), 489.
77. J. Jansson, J. Nilsson, F. Modig and G. Hed Vall, *Business Strategy and the Environment*, 2017, **26**(1), 69.
78. I. van de Poel and M. Sand, *Synthese*, 2018, **198**, 4769.
79. P. Newell, H. Bulkeley, K. Turner, C. Shaw, S. Caney, E. Shove and N. Pidgeon, *Climate Change*, 2015, **6**, 535.
80. A. Giddens, *The Modern Law Review*, 1999, **62**(1), 10.
81. I. van de Poel and J. N. Fahlquist, in *Essentials of Risk Theory*, ed. S. Roeser, R. Hillerbrand, P. Sandin and M. Peterson, Springer Briefs in Philosophy, Springer Netherlands, 2012, ch. 5, pp. 107–143.
82. B. Sena, *Int. Soc. Sci. J.*, 2014, **65**(215–216), 79.
83. M. Alznauer, *Inquiry*, 2008, **51**(4), 365.

CHAPTER 5

The Role of Design in Conserving Product Value in the Circular Economy

DEBORAH ANDREWS*[a] AND BETH WHITEHEAD[b]

[a]London South Bank University, School of Engineering,
103 Borough Road, London, SE1 0AA, UK; [b]Operational Intelligence,
DC107 The Clarence Centre, 1 St George's Circus, London, SE1 0AP, UK
*E-mail: deborah.andrews@lsbu.ac.uk

5.1 Design

5.1.1 The History and Role of Design

The Oxford English Dictionary defines *design* as a noun and a verb. In this context, design is described as 'the purpose, planning or intention that exists or is thought to exist behind an action, fact or material object';[1] in the broadest sense, therefore, 'design' is integral to human activity. Evidence of design dates from prehistoric times when artefacts such as cutting, digging, and hunting tools were fashioned to fulfil a particular function.

Since then, design has been – and still is – practised across societies, and in many contexts and related activities, outputs have simultaneously developed alongside social, economic, and environmental change. While some design activity remains informal and intuitive, design has evolved, become increasingly sophisticated, and developed into a profession. Design activities have also become more specialised and distinct disciplines have emerged, examples of

Issues in Environmental Science and Technology No. 51
The Circular Economy: Meeting Sustainable Development Goals
Edited by Sadhan Kumar Ghosh and Gev Eduljee
© The Royal Society of Chemistry 2024
Published by the Royal Society of Chemistry, www.rsc.org

which include fashion, textiles, graphics, communication, furniture, interiors, film, exhibition, architecture, service, user experience (UX), product, and industrial design. Design is also integral to engineering, which is defined as 'the branch of science and technology concerned with the design, building and use of engines machines and structures' in the Oxford English Dictionary.[1] The evolution of engineering from an intuitive and informal activity to a profession mirrors that of design. As engineering has become more complex for example, more specialist disciplines have emerged. The history of the profession is now discussed.

5.1.2   The Concurrent Development of Engineering, Design, and Industry

Engineering was practised in ancient civilisations, but 'modern' and formal engineering and the engineering profession began during the first and second Industrial Revolutions in the 18th and 19th centuries, respectively. Engineering evolved concurrently with the various Industrial Revolutions, each of which simultaneously initiated and drove changing technologies. For example, the invention of the steam engine and electrical systems encouraged the first and second Industrial Revolutions, while the third (computing-based) Industrial Revolution began in the 1950s and the fourth Industrial Revolution that integrates cyber-physical technologies is now underway. During the first and second Industrial Revolutions, the introduction of new manufacturing processes (including mechanisation), energy systems, and materials facilitated a shift from craft-based and batch production to one of higher volume and mass, as did changes to working practices such as the division of labour. Production rates and output were further accelerated during the 1950s and 1960s following the introduction of industrial robots and Computer-aided Design (CAD) and Computer-aided Manufacture (CAM); since then their utilisation has become increasingly widespread and integral to many industries. Digital manufacturing technologies progressed to include processes such as additive manufacturing (3D printing), and a shift from centralised to decentralised local and domestic production, facilitated by increasing connectivity and the emergence of cyber-physical systems via the Internet of Things (IoT).

Formal design practice and the design profession also emerged and evolved from the first and second Industrial Revolutions. Initially design practice was inherent to architecture, craft, and engineering activities; however, the introduction of new manufacturing processes, technologies, and materials encouraged the development of distinct applied arts (*i.e.*, design). The importance of design as a driver of innovation was formally recognised when the Society for the Encouragement of Arts, Manufactures and Commerce (now known as The Royal Society for Arts (RSA)) was founded in 1754 in the UK to stimulate industry. The need for specialist design training became apparent at this time, which led to the establishment of design schools (known as Normal schools) near to manufacturing centres in 1836. Their mission was to simultaneously prepare students for employment in and support the industry, and to develop 'good' design.[2] Since then, the RSA has continued to promote and consolidate links between the arts, manufacturing, and society.

The design process comprises a series of stages, which were formalised by the UK Design Council in 2005. The Double Diamond Design Method for Innovation[3] shown in Figure 5.1 defines the stages as discover, define, develop, and deliver. The method is not linear and involves considerable iteration to ensure that the outcome meets all constraints. Like engineering, design has pushed and pulled development of technologies in response to the demands and requirements of industry and users, and design has become increasingly important as a mediator between technology and user in line with rising sophistication and complexity.

5.1.3   Design, Society, and Economics

Design and engineering had major impacts on society. For example, at the beginning of the Industrial Revolution western societies were predominantly agrarian but rural labourers were encouraged with the promise of higher incomes to seek employment in industrial facilities and to move close to mines, mills, and factories. Because of this, urban population increased from 9% in 1800 to 62% in 1900. Improved diets,[4] antibiotics, and other developments in medical care, equipment, and products have also increased global average life expectancy from less than 30 in 1800 to 73 years in 2019. This has contributed to global population growth from around 1 billion in 1800 to 8 billion in 2022, and increased demand on resources.

Design and engineering have also been critical to economics and in 1776 Adam Smith stated that, in addition to agriculture, gold, and silver, the 'wealth of nations' was grounded in 'national production', that is, manufacture.[5] 'National production' was critical to and developed because of the demand–supply–income cycle,[6] and, as predicted by Smith, national and personal wealth increased. In the UK, for example, between 1700 and 1871 Gross

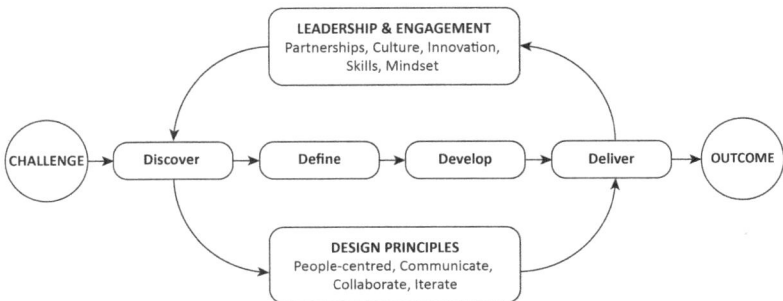

Figure 5.1   The framework for innovation: Double Diamond design methodology.

Domestic Product (GDP) increased 53% per capita,[7] and between the beginning and end of the 20th century an even greater increase of 400% per capita was recorded.[8]

Design encouraged demand, consumption, and industrial activity, through aesthetic and technical interventions. For example, in 1932 Bernard London,[9] an American economist, stated that increasing demand for goods would stimulate the economy following the 1929 Wall Street financial crash and Great Depression. He proposed planned obsolescence, that is, that products should be designed to break to encourage replacement.[10] This phenomenon had already proved successful for the lighting industry, members of which formed a cartel and purposefully reduced the life of tungsten bulbs from 2500 hours to 1000 hours to increase their profits. Similarly in 1927 Harley Earl was employed to lead the newly established Art and Colour Section at General Motors to increase profits by increasing sales of new vehicles; this was achieved by changing colour and minor design details to encourage product replacement without the need for major changes to engineering and tooling.[11] At a time when there was little if any concern about resource supply and environmental degradation, the predominant perception of design was as a key driver of consumption and economics. While the industrial designer Brooks Stevens defined his role as creating the 'desire to own something a little newer and a little better, a little sooner than necessary',[12] another renowned designer – Raymond Loewy – apparently stated that 'Industrial design keeps the customer happy, his client in the black and the designer busy' and that 'The most beautiful curve is a rising sales graph'.[13] These strategies certainly encouraged and continue to encourage demand and sales, but they also generate waste and adverse environmental and social impacts.

5.1.4   Sustainable Design

Not all designers and engineers agreed with the aforementioned strategies, however, and in 1938 Richard Buckminster Fuller – a 'comprehensive anticipatory design scientist'[14] – advocated resource efficiency by 'doing more with less', which he termed ephemeralisation.[15] Similarly, during the 1960s, the designer Victor Papanek adopted a radical environmental and ethnographically sensitive approach to design. However, he was castigated by his peers and forced to resign from the Industrial Designers Society of America (IDSA) when he stated that they produced 'shoddy, stylised work that wasted natural resources, aggravated environmental crises, and ignored their social and moral responsibilities'[16] and that 'There are professions more harmful than industrial design, but only a few of them'. Many designers were responsible for planned obsolescence and had created 'whole species of permanent garbage to clutter up the landscape, and by choosing materials and processes that pollute the air we breathe, designers have become a dangerous breed'.[17]

Growing concern about the environment, social justice, and human needs led the United Nations to establish the World Commission on Environment and Development (WCED) in 1983, and in 1987 the group published a highly influential report, *Our Common Future*. It includes one of the most frequently cited definitions of sustainable development as 'development that meets the needs of the present without compromising the ability of future generations to meet their own needs'.[18] This has significant implications for design and engineering practice, and output regarding, for example, use of resources, conflict minerals, waste, emissions, and pollution. Some designers and professional bodies responded to these concerns, as a result of which Buckminster Fuller and Papanek's ideas gradually spread and in 1991 the UK Design Council published 'Green Design'[19] as part of the Design Issues series. Since then, Design for Sustainability (DfS) has evolved considerably as a discipline that can be integrated into other design and engineering specialisms. As shown in Figure 5.2, initially product/industrial design was solely object and technology focused and supported the environmental and economic dimensions of sustainability; it has evolved to be more systems-focused, and to include socio-ethical factors in addition to addressing environmental and economic concerns, and therefore to support more radical change and transition to potentially more sustainable systems overall.[20] The inclusion of social dimensions also suggests that there is potential to align with more of the UN Sustainable Development Goals (SDGs) (see Section 5.3).

5.1.5   The Introduction of the Linear Economy and Design

Early mechanical products were easy to repair and upgrade because of the way in which the constituent parts were manufactured and assembled; for example, mechanical components such as gears, pistons, and belts found in steam engines or looms were exposed and could be removed and replaced. Similarly, mechanical fixings such as nails, screws, and bolts permitted relatively easy disassembly while welded, soldered, and joints that used animal-based glues could be reheated and separated. This in conjunction with the relatively high cost of producing goods and limited expendable income, encouraged thrift and careful use of resources until the general level of expendable income rose and relatively inexpensive (*e.g.*, plastic) goods were introduced. Changes in purchasing power and, as previously discussed, built-in obsolescence and aggressive marketing encouraged product replacement before necessary, increasing waste; this was exacerbated by changes in materials selection, assembly, and joining processes that limited or prevented easy repair. The second-hand market and 'make-do-and-mend' culture had thrived when expendable income was limited and during wars when resources were scarce or required for manufacture of defence goods. In addition to rising incomes, and more available and cheaper resources design strategies encouraged the linear economy which also increased disposal of unwanted and/or damaged goods and decreased interest in and market for second-life goods.

In a linear model of consumption, we take resources from the ground, make them into a product, use the product, then dispose of it. In this take–make–waste approach, the product and the materials it contains are considered

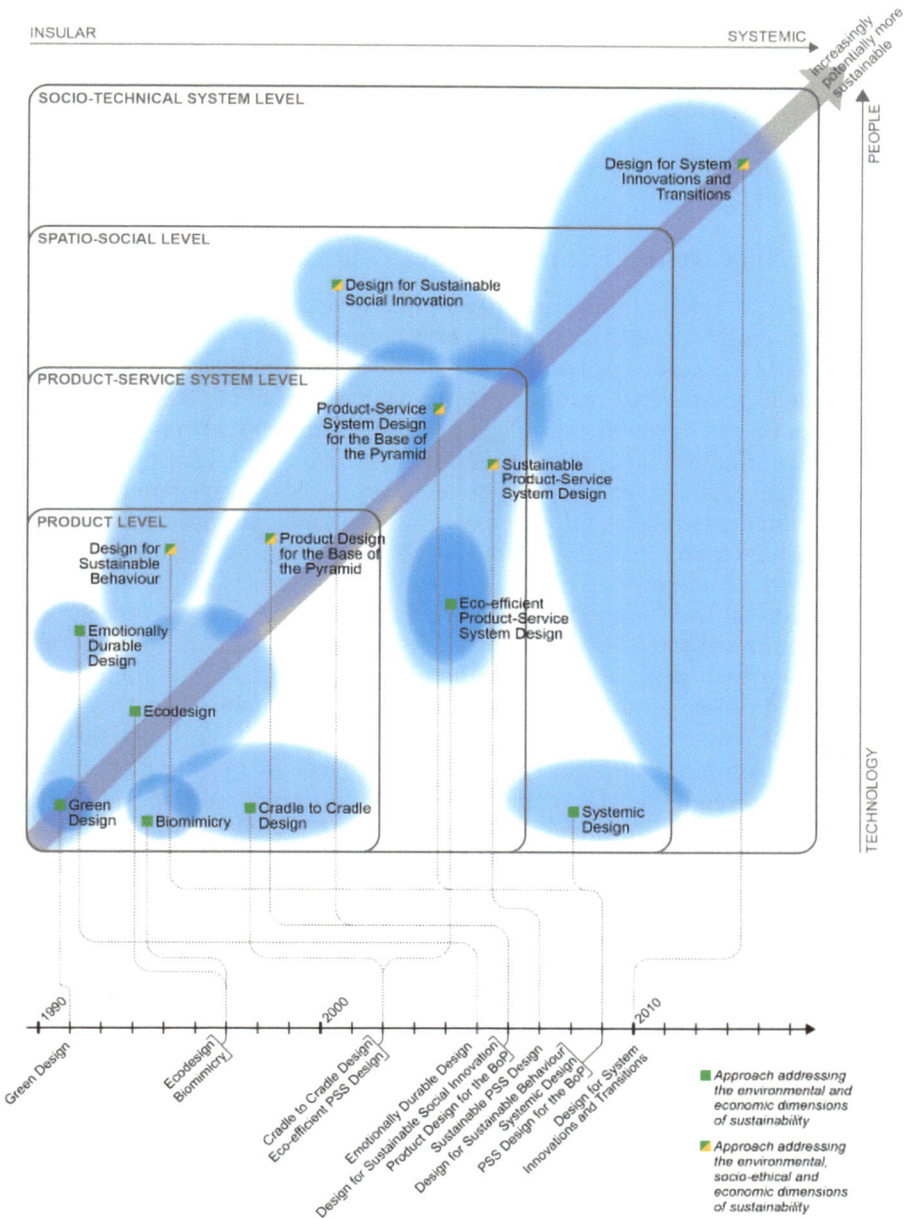

Figure 5.2 Evolution of design for sustainability: from product design for system innovations and transitions. Reproduced from ref. 20, https://doi. org/10.1016/j.destud.2016.09.002, under the terms of the CC BY 4.0 license, https://creativecommons.org/licenses/by/4.0/.

waste at the end of its life. This is a problem not only because of the pollution created from waste disposal, but also from the continued extraction of virgin raw materials, which have a finite supply. A model of consumption in which products and their components and materials have no value at the end of their first life is inherently flawed when planetary boundaries put limits on the resources available to us and the amount of pollution it can remove.

5.1.6   The Circular Economy and Design

In contrast, a circular model of consumption takes a whole-systems approach based on a fully renewables-based energy supply and in which no product or material ever becomes waste. The circular economy echoes natural systems where dead/redundant materials decompose and become nutrients for the subsequent year's growth and/or generations. There are two main loops in the circular economy:[21] one includes technical 'nutrients' and the other biological 'nutrients'. In the second loop, systems are designed to regenerate soil, by feeding nutrients released through composting and anaerobic digestion into the land on which renewable materials like cotton and wood are grown and fed back into the economy.[22]

In the technical loop materials are circulated for as long as is technically and economically feasible, and products are designed to allow for reuse, repair, refurbishment, and remanufacturing; at end of life the materials are recycled, reclaimed, and reused. Closing the materials loop ensures that we do not lose resources from the system; rather they flow continuously within the loop at a product, component, or material level. By minimising or ideally removing the concept of waste altogether, pollution from disposal and raw material extraction can at least be minimised and ideally eliminated. A circular economy includes physical artifacts/goods and performance/functional services in which products are leased and/or shared. This practice supports dematerialisation and resource efficiency.[23] Figure 5.3 illustrates the technical loop, the principal focus of this chapter.

An example is the Rolls Royce business model in which the company does not sell aircraft engines but sells Power by the Hour and therefore maintains and controls all their products from design to, and beyond, end of life.[24]

5.1.7   Design and the Value Chain

According to the World Business Council for Sustainable Development (WBCSD), the value chain is 'the full life cycle of a product or process, including material sourcing, production, consumption, and disposal/recycling processes'.[25] It differs from the supply chain because, in addition to economic factors, it considers ethical, social, and environmental factors that are added internally and externally throughout the lifecycle.[26] Furthermore, a sustainable value chain is one that 'enables both business and society to better understand and address the environmental challenges associated with the

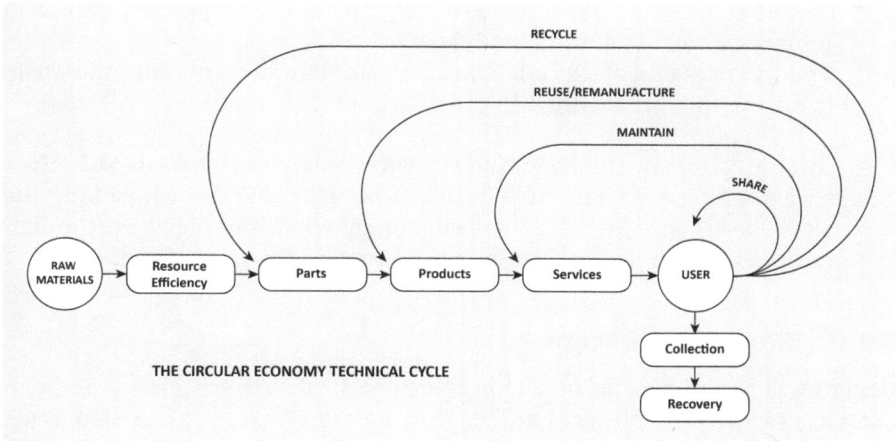

Figure 5.3   The technical loop of the circular economy.

life cycle of products and services'.[26] It is evident that informal and formal design decisions and related activities have always influenced impacts throughout the value chain. Impacts may be environmental, social and/or economic; therefore, design can affect the overall sustainability or otherwise of a product.

Such is the power of design that it is said to determine up to 80% of the environmental impact of a product over its lifetime.[27] Although frequently cited,[28–31] there does not appear to be any empirical evidence to support this exact percentage. However, the figure may originate from an article that describes green product design as 'a proactive approach to environmental protection that addresses life-cycle environmental concerns in the product design stage. Decisions made during that stage profoundly influence the entire life cycle of the product and determine 80 to 90 percent of its total life-cycle costs'.[32] Whether 80% is accurate or otherwise, the influence of design on environmental impact and overall sustainability is indisputable.

To facilitate the move to a circular economy, designers need to consider positive and negative impacts at each stage of the product life cycle (*i.e.*, down- and upstream). Factors include the following:

- **Design:** physical durability, cognitive and physical functionality, and emotions.
- **Resource efficiency:** materials selection and efficiency – their criticality; social and environmental impacts; cost; extraction/processing methods required; and ease of recovery from the product at end of life.
- **Operational energy** inputs and efficiency.
- **Manufacturing methods:** efficiency and waste reduction/elimination.
- **Ability to disassemble** products economically for life extension and end-of-life treatment.

- **Potential to extend life** through reuse, repair, component upgrade, and remanufacture, including modularity.
- **Treatment at end of life:** enable maximum recycling and minimal waste to ensure that little if anything goes to landfill.

In addition to influencing physical factors such as materials selection, design also influences users who decide how, where, and for how long the product is used. These factors can be encouraged or discouraged by stimulating emotion, which is another key role for design in the value chain.

5.1.8   Emotional Design

Design has significant influence on the quality of user experience and can decrease or increase physical and cognitive ease of use.[33] It can also make products inclusive and enable easy use across generations and genders such as OXO Good Grips products. These were developed for use by people with limited dexterity and grip because of, for example, arthritis. Some products for users with different abilities appear alienating or cumbersome and different from mainstream products; however, Good Grips products have a distinct visual brand identity that was created to appeal to and be used by everyone, and therefore the products do not look out of place alongside other mainstream products. While product aesthetics give brands visual identity, research shows that users also find aesthetically pleasing 'beautiful' products and interfaces easier to learn how to use than less visually appealing products or interfaces[34] and, consequently, 'attractive things' may even be perceived as working better.[35] Making interfaces more comprehensible and guiding users through processes are increasingly important as products and their functions become more complex. Visual beauty is overt and generates emotional response; however, other less overt design features such as surface finish also appeal to other senses, in this case touch, which, alongside sound and smell, can also stimulate emotional responses. A user is more likely to develop an emotional bond with a product that they find pleasing even if they are unable to explain why. This emotional bond encourages the user to keep the product, extending its lifetime and helping to reduce waste.

5.2   The Role of Design in the EU Circular Economy Action Plan and UK Circular Economy Package

Concern about climate change and other environmental factors, increasing waste, and supply chain security encouraged the European Commission to devise and publish various guidelines and policies to increase the design and manufacture of more resource-efficient consumer[27,36] and capital goods.[37] The initial focus was energy efficiency, which was rated, and goods and properties were labelled from A* to E accordingly. In 2020 the European Union (EU) extended this incentive to include embodied energy and physical resources in the Circular Economy Action Plan for a cleaner and more competitive Europe;[27] the UK also published a similar Circular Economy Package in 2020. Some of the initial guidelines for better practice in the Action Plan, for example, the right to repair (with emphasis on consumer goods), are already being developed into policy and it is expected that in time all 35 recommendations will become mandatory. Several key product groups are prioritised in the Action Plan because there is significant potential to increase circularity and reduce negative impact throughout the value chain. These are textiles, furniture, high-impact intermediary products (steel, cement, chemicals) and electronics and ICT equipment. Some proposed actions are beyond the role of design and will depend on legislative action; these include introducing a ban on the destruction of unsold durable goods, restricting single-use plastics and products, countering premature obsolescence, and rewarding products based on their different sustainability performance, including by linking high-performance levels to incentives. However, design can directly influence many other proposed actions. These include increasing product energy and resource efficiency by improving durability, reusability, upgradability, and reparability, and enabling remanufacturing and high-quality recycling; increasing recycled content in products, while ensuring their performance and safety; reducing carbon and environmental footprints and addressing the presence of hazardous chemicals in products. A combination of design and legislation will drive other actions including the incentivisation of 'product as a service'; other models where producers keep the ownership of the product or the responsibility for its performance throughout its lifecycle; and mobilising the potential of digitalisation of product information, including solutions such as digital passports, tagging and watermarks. We now consider the challenges presented by and opportunities for change through design and the value chain for the four priority areas.

5.2.1 Textiles

The global textile industry is well established and creates yarns and fabrics that are essential to different sectors including clothing, footwear, and household textiles. Products from these sectors are essential to well-being and in addition to creating privacy, protecting wearers from injury, maintaining comfortable environmental and body temperature, for example, they also offer wearers and users a sense of cultural and social identity.[38] Currently 175 million tonnes of primary raw materials are used to produce textiles annually for the EU, 40% of which become clothing, 30% household textiles, and 30% footwear.[39,40]

Textiles can be composed of natural and/or synthetic fibres and the global fibre market is split into four categories, of which synthetic fibres, and in particular polyester, account for the greatest proportion. The global split of production in 2021 was as follows:[41]

- **Synthetic fibres** made from crude oil (64%) – polyester, polyamide (nylon), polypropylene, acrylics, elastane.

- **Plant (natural cellulose) fibres** (28%) – cotton and others like jute, flax, and hemp.
- **Manmade cellulosic fibres** (6.4%) extracted from plant-based materials (such as wood and bamboo), processed into pulp and extruded (6.4%) – viscose, acetate, lyocell, modal, cupro, and rayon.
- **Animal fibres** (1.6%) – wool (sheep, goats, rabbits, alpaca), down and feathers, silk (from the cocoons of insect larvae), leather, and other fibres.

Polyester accounts for 54 percentage points of the 64% synthetic fibre production, driven in large part by fast fashion. Many qualities have led to its dominance, for example it is relatively low cost; easy to dye, and weave and blend with other fibres; lightweight; strong; and wrinkle-free.[42]

5.2.1.1   Problems Faced by the Textiles Industry

Fibre mixing is commonplace and while 100% natural textiles may be composed of (for example) cotton, wool, and silk, 100% synthetics may be composed of mixed acrylic, nylon, and elastane yarns/fibres. Natural–synthetic hybrid mixes (such as polyester cotton and mixes of three and more types of fibre) are also common as are 'performance' textiles, which may be coated for fireproofing, waterproofing and/or to block out light.

From 1975 to 2020, global fibre production almost tripled from 34 to 109 million tonnes. In 2021 around 8.5% of production was from recycled fibres, of which less than 1 percentage point derived from textiles. In the case of polyester for example, most of this content was from open-loop recycling of bottles rather than textiles. Low recycling rates of polyester are partly due to a lack of infrastructure for textile–textile recycling, but also because of the degradation in quality over successive cycles of recycling, and combination with other materials such as cotton (polycottons), which makes recycling impossible. This pattern is typical to other textiles. Many textiles end up in landfill or incineration in countries outside the EU and are often discarded rather than donated.[43] Furthermore, cheaper clothes mean that more are being bought, worn for shorter periods, and discarded in less responsible ways.

Natural virgin fibres require a lot of land and water for cultivation and energy for materials production. Textile production is heavily reliant on chemicals and utilises 25% of those produced globally per year.[44] Water consumption is also very high, and it takes 100–150 litres water to produce 1 kg of fibre depending on type, while the fashion industry alone uses about 93 billion m^3 water per year.[45] The textile industry is also responsible for 20% of water pollution from dyeing and finishing and 10% of global greenhouse gas emissions per year,[43] while the washing and laundering of synthetics release microplastics and cause water pollution.

These criteria make identification and separation of constituent yarns/fibres for reuse and recycling very challenging at best, and impossible at worst, resulting in very high adverse impacts. At present less than 1% are recycled into new products; some textiles are exported from the EU, and some of the non-reusable fraction is downcycled into industrial rags, upholstery filling, and insulation; however, 87% of textile products are sent to landfill or incinerated at end of life.[43] Consequently, textiles have the fourth highest impact on the environment and climate change after food, housing, and mobility in the EU.[39]

5.2.1.2   Design and Change in the Textiles Value Chain

As previously explained, textiles fulfil several important functions; however, it is evident that current practices in the sector are not sustainable. Design needs to mitigate or eliminate the adverse impacts of natural and synthetic fibres used by changing technologies and behaviours. Technical change includes use of materials that can be easily recycled (including single rather than composite yarns), and use of recycled as opposed to virgin materials, which will be expedited by an increase in the rate of recycling. Clothing and household products also need to be designed for durability, longevity, and repair as and when appropriate. However, these changes must be supported by change in user behaviour, some examples of which are already taking place. Examples include clothing hire and purchase of 'pre-loved' items. Nonetheless, one of the biggest challenges is 'fast' (and low-cost) fashion, which is also encouraged by online business and social media. Design could encourage 'slow fashion', although this will be a challenge because the fashion industry is built around two seasonal design collections every year. Consumption could be reduced by increasing the perceived, emotional, and actual value of items through use of textiles that wash well, and retain shape and colour, for example.

Several initiatives have been established in response to the status quo to encourage better practices across the value chain. One initiative[46] recognises that 'design continues to play a significant role in the damaging impacts on natural systems and human populations throughout the textile supply chain' and is undertaking research to address this by developing circular systems for fashion and textiles that are already in circulation. Example projects include use of post-consumer waste, the development of associated materials' production technologies, and a sustainable design toolbox. A second initiative is focusing on future fashion and household textile challenges by turning waste bio-feedstocks into functional and regenerative textiles designed for circularity; by developing better tools to facilitate circular supply chains; and by improving links between resource flow and human well-being and changing consumer experience.[47]

5.2.2   Construction and the Built Environment

The human-made buildings, roads, infrastructure, service networks, and public spaces that support our home and working lives are collectively known as the built environment, which is underpinned by the construction industry.

In the UK in 2022, the construction industry accounted for 6% of employment, 6% of the total Gross Value Added (GVA)[48] and 10% of the total EU GVA.[49] The industry is a driver for social and economic growth and is one of the most important industries to the EU economy.

The construction industry also adds a heavy burden to the environment. It is the world's largest consumer of raw materials,[50] accounting for 50% of all extracted material,[27] and one of the biggest sources of waste in Europe.[51] In the EU in 2020, construction accounted for 37.5% of the total waste stream, whilst mining and quarrying accounted for 23.4%.[52] In the EU the industry creates 9.4% of total domestic carbon footprint, and buildings consume 40% of total energy consumption, whilst the cement, steel, aluminium, and plastics industries – used extensively in the industry – account for 15% of total EU carbon emissions.[49] Globally, cement alone accounts for 8% of total CO_2 emissions. In the UK this equates to 5–12% of all greenhouse gas emissions originating from materials extraction, construction product manufacturing, construction, and renovation.[27] (Construction and the built environment is also discussed in Chapter 9.)

The main materials used in construction are as follows:

- metals such as steel, aluminium, and copper (structural elements, pipes, and ductwork)
- cement (mortars and concrete)
- ceramics products (bricks, tiles, conduits, and electrical insulators)
- glass for windows
- wool insulation
- glass fibres (composite products such as cladding, and insulation)
- wood (including composites such as glulam and cross-laminated timber (CLT))
- plastics.

The choice of material depends on the function of the element. For example, concrete works well in compression, is strong, durable, relatively cheap, and fire-resistant. However, it does not perform well in tension and is therefore used with aluminium reinforcement bars or mesh. Steel is fast to erect on site, generally leads to lighter superstructures and therefore a reduced substructure (foundations), and can be used to create longer column-free spans; however, it requires fire protection, which concrete does not. Often composite metal decking with a slimmer concrete slab is used to reduce the depth of material use, and the need for temporary formwork.

5.2.2.1   Problems Faced in Construction and the Built Environment

The industry still relies heavily on a linear (*i.e.*, take–make–dispose) model of consumption. It consumes huge volumes of virgin raw materials for construction and energy for operation, and accounts for one of the biggest waste streams globally. To reach the EU 2030 goal of reducing CO_2 emissions by 55% below 1990 levels[53] and 2050 targets for reaching carbon net zero, the industry must reduce its global emissions by 2.1 billion tonnes[54] – a target the industry is currently not on track to achieve.[55]

There are several barriers to the slow implementation of a circular economy that would reduce these impacts. Globally, investment in building energy efficiency is low, alongside a moderate move to renewable sources of energy, and there is a lack of mandatory energy codes in many countries.[55] Policy is needed to incentivise circularity, as will be set out in the European Commission's Strategy for a Sustainable Built Environment,[56] for example the setting of required recycled content for construction products,[27] EU material recovery targets, and less prescriptive waste regulations.[57]

Contractual barriers between different stakeholders in the building lifetime create a silo effect, making it hard for the industry to collaborate and share knowledge on circular designs and solutions. In addition, teams that could improve outcomes later in the project – such as commissioning and demolition engineers – are often not included early enough in the design process to be able to influence outcomes and support a paradigm shift, and lessons learned are often not fed into future designs. Furthermore, the industry has substantial inertia, meaning the positive effects of any systemic and behavioural changes take longer to appear.

Underdeveloped infrastructures make material recovery hard. Steel can be recycled almost indefinitely without degradation to its performance. Globally 85% of steel is recycled.[58] However, the highest embodied impact is retained when it is reused. To support reuse the industry needs initiatives such as take-back schemes, secondary markets, mechanisms to guarantee quality (for example methods of demolition that retain the structural integrity of elements), and initiatives that can help eliminate the 15% of construction materials that leave site as waste without having been used.[54] Although many other construction products are recyclable – such as cement and plastics – reuse and recycling rates are much lower (with huge EU variance) due to poor recycling infrastructure and disassembly outcomes. Concrete can contain recycled aggregates such as pulverised fuel ash (PFA) and ground granulated blast-furnace slag (GGBS); however, in the UK in 2020, they accounted for only 28%[59] of total aggregates used.

5.2.2.2   Design and Change in the Construction and Built Environment Value Chain

The Ellen MacArthur Foundation describes three circular economy strategies that provide the biggest opportunities for reducing emissions from the built environment. The first is to make better use of existing buildings,[54] by building nothing and reusing and renovating existing buildings, or increasing the utilisation of already underused spaces.

The second strategy is to design new spaces in ways that eliminate waste[54] and allow for better outcomes from the first strategy. Although typical design lifetimes for buildings are over 60 years, buildings are often demolished after 30 when they are no longer deemed desirable or fit for function.[58] To improve circular economy outcomes, new buildings should be designed for longevity using durable and maintainable materials that are repairable, and with spatial consideration of the needs of different users to increase building utilisation through concurrent use of spaces by multiple tenants. Modular designs should be used to allow for future adaptability and reconfiguration, which can meet the needs of new occupants as well as any future implications of climate change such as change in space heating and cooling requirements. Design complexity, building spans and building dimensions should be minimised, and high-strength materials (resulting in smaller elemental masses) used to avoid overspecification and reduce material use. Simplified designs are also easier to construct, reducing risks of error and wastage during construction. Wherever possible, prefabricated elements should be used to reduce raw materials consumed and waste generated. By lengthening lifetimes and reducing construction volumes this effectively circulates products and materials to retain their embodied impact.[54]

The final strategy is to reuse and recycle materials.[54] Currently only 20–30% of construction and demolition waste is reused or recycled.[58] Designs need to take a whole-life approach that increases recycled content and reduces the use of virgin materials, critical raw materials (*e.g.*, recycled aluminium reduces the consumption of primary bauxite from which it is extracted), and hazardous materials. Designs should allow for disassembly, using methods such as reversible mechanical fixings, and limiting composite materials that cannot be reused in their initial state or separated. On practical completion, buildings should be handed over with repair, maintenance, adaptability, and disassembly manuals in digital logbooks,[27] and material passports[60] to ensure improved circularity of materials and retention of embodied impacts.[54] Finally, it is imperative that the construction process and building operation transition to renewable sources of energy and eliminate the use of fossil fuels.

5.2.3   Furniture

Traditionally most furniture was made from wood because it was an abundant resource and, unlike metals, it could be harvested, shaped, and joined relatively easily. Accurately cut joints also rely on friction fit and do not require glue. Traditional glues were made from animal parts, which was very sustainable because the parts used – skin, bone, hooves, and horn – were unfit for human consumption. Although glues were not used for some highly skilled furniture and building manufacture (such as those found in Japan), they were often used by batch and mass manufacturers, particularly for lower-cost items.[61] Metal furniture and components were introduced concurrently with the development and evolution of metal manufacturing processes for the construction and transport industries. Early examples from the 1830s include cast iron tables and benches that were used in gardens and parks. Tube bending processes were originally developed for bicycle handlebars, but the technology was imaginatively transferred to furniture production by members of the Bauhaus (Modernist German design school and movement) during the 1920s; the new processes and materials allowed for new lighter-weight forms such as the Model B3 Club Chair designed by Marcel Breuer in 1925–26 and the MR10 cantilevered chair designed by Mies van der Rohe in 1927. Both chairs can be easily disassembled for repair, component reuse, and recycling.

While solid wood furniture can also be repaired and disassembled relatively easily if joined with animal-based glues, it is subject to shrinkage and swelling in certain humid conditions if it is not cured properly. Cutting planks from tree trunks can also generate considerable waste, although this is frequently reduced to particles or fibres and used to manufacture chip, strand, and medium-density fibreboard (MDF). Temperate hardwoods such as oak, beech, walnut, poplar, cherry, and ash are used to make furniture,[62] while softwoods from managed plantations (pine and spruce) are also used in the furniture and construction sectors.

Although wood is regarded as a renewable resource and some woods (eucalyptus, pine, and spruce) are relatively fast growing, other temperate woods are much slower growing and therefore regeneration is slow. Many forests are managed and certified as sustainable by the Forest Stewardship Council but almost 100 types of trees are listed as (critically) endangered, threatened, vulnerable or in need of conservation.[63] This is a point of concern because the supply chain lacks transparency, and it is difficult to identify the exact origins of timber without extensive due diligence; therefore, wood used could be from unsustainable sources. Many imported products are not covered by EU timber regulations. At present over 50% of furniture and components are imported from outside the EU, 42% of which are from Chinese forests and 50% by volume come from countries with illegal logging/trade issues.[62] These issues are compounded by the fact that import partners do not necessarily represent the origin of the timber[62] and retailers are not doing enough to understand or validate their supply chains, all of which damage local ecosystems and societies and reduce potential carbon sequestration.

Examples of engineered timber by-products include chip (particle) board, MDF, high-density fibreboard (HDF) or hardboard, laminated wood where the grain of successive glued layers is parallel, and plywood where layers/veneers of wood are glued together with the grain of each layer at right angles to the next, providing strength and stiffness in both directions. The manufacture of by-products from timber processing reduces waste, which is positive. However, these materials are bonded with synthetic resins and are heated and pressed into boards. Resins used include polyester, vinyl ester, epoxy, and polyurethane resins, which are thermoset and converted from liquid to solid through polymerisation (cross-linking). Consequently, they cannot be returned to their former liquid state, which makes economic and/or simple recycling impossible. Melamine – another thermoset plastic – can be made in sheets; it is available in a wide range of colours and surface patterns, and it is frequently used to face chip and fibre boards. Being water- and heat-proof it protects the engineered timber and prolongs product life. However, it is glued to the timber, which, in conjunction with being a thermoset material, compounds the challenge of recycling. In addition to adhesives, inexpensive flatpack and mass-produced furniture is often joined using knock-down fittings. Some of these devices can be unfastened to allow for component separation, but many are made from plastic and designed to click fit one way, which prohibits disassembly. Therefore, it is not surprising that in 2018 the North London Waste Authority found that 22 million furniture items were thrown away each year, 11% of which ended up in household rubbish.[64] Some of these discarded items would have been incinerated in specialised plants to generate energy (electricity) from waste and therefore have some benefit. Other items would have been sent to landfill, which is both a waste of resources and a slow process because use of polymer-based resins and adhesives means that decomposition is slow. Finally, formaldehyde can 'off gas' and leak into the atmosphere but levels tend to be so low that any adverse effects are rare, and most will occur prior to delivery of boards or furniture to homes.

It is evident that current furniture design and manufacture methods are simultaneously environmentally and socially advantageous and disadvantageous. For example, use of solid timber can increase durability, and perceived and actual product value, but unless it comes from an approved supply chain, it contributes to the destruction of forests. Melamine-faced chip and fibre board furniture can be more economical and available to more consumers, but it is less durable, which limits life in service, it cannot be recycled, and it contributes to a growing waste stream. Once again design decisions will have a considerable impact on environmental, social, and economic factors throughout the value chain and designers need to consider and balance all factors to develop products that simultaneously meet users' requirements and minimise adverse impacts.

Design features could increase product life. Examples include adjustable furniture that 'grows' with children, supply of repair and customisation kits to update and refresh tired products, and design for disassembly to encourage and enable easy reuse of components and recycling at end of life. Designers also need to identify and consider less energy-intensive manufacturing processes and use of low-impact and alternative materials. For example, by-products such as walnut shells reduce the need for virgin materials, significantly reduce formaldehyde emissions, and improve water resistance in fibre and particle board at the same time as reducing stress on forests and creating jobs and a second income stream for farmers.[65]

5.2.4   Electronics and ICT

5.2.4.1   *Electricity*

The Industrial Revolution saw the transition from an economy based on hand production to one of industrial and machinery-based production. This was due to technological advances that included new energy sources such as coal, steam engines, and electricity.[66] The development of electricity-based technology began in the 1750s when Benjamin Franklin noticed that power could be generated from sparks emitted from lightning strikes.[67,68] From the 1880s, other pioneering scientists subsequently extended the knowledge, and developed components, products, and systems including electric batteries, motors,[68] and incandescent light bulbs; coal-fired, hydro and wind generation plants;[66,68] and distribution networks. The number of suppliers increased in line with the growing demand for electricity and reliable transmission, and in the UK by 1937 all local grids were combined and run synchronically with the same 132 kV voltage and 50 Hz frequency without failure.[69] By 1950 the network could no longer meet demand and work started on a 175 kV supergrid, which has subsequently been extended, with most of the UK now connected to the grid and/or with access to an off-grid electricity supply.

The growth of mass power generation and distribution encouraged rapid developments in the design and production of commercial and domestic electrical equipment. Manufacture was also accelerated by the invention of Bakelite – another synthetic thermoset plastic made from phenol (and later urea) formaldehyde by the Belgian chemist Leo Baekeland in the early 1900s. The materials were excellent insulators and ideal for electric lighting fixtures, plugs and sockets, as well as phonographs, kettles, irons, washing machines, and vacuum cleaners, which became popular because they enabled entertainment in the home and alleviated some of the drudgery of housework. Electricity also initiated changes in machinery and manufacturing processes, although early products were assembled in the same way as non-electrical goods, that is, plastic components tended to be bolted/screwed together with those made from the same and other materials. This meant that product repair, replacement, and reuse of parts was straightforward, and the purchase and use of second-hand goods was commonplace. Electrical and other metal components were separated and recycled, and waste was usually limited to the thermoset plastic housings. Telephones and wireless radio sets benefitted from plastic casings and cabinets; they also proved very popular and changed communication permanently. The earlier invention of cinema had already transformed entertainment, and this was further transformed by the introduction of television in the 1930s, which, like telephony and radio[70] were enabled by the discovery of the electron and gradual development of electronics.

5.2.4.2   Electronics

The evolution of modern-day electronics dates from 1897 when the British physicist Sir Joseph John Thomson discovered the electron. Various subsequent discoveries led to the development of the thermionic valve, which converted alternating current (AC) radio signals to weak direct currents detectable by telephone receivers and radios[70,71] and were more reliable than early 'cat's whisker' radios[72] composed of fine wires in contact with a crystal of semiconductor material, which could not be amplified.[70] A semiconductor (such as germanium and silicon) conducts electricity to a degree between that of conductors (such as copper and aluminium) and insulators (such as glass and diamond); trace amounts of other elements can be added to alter its electrical properties. These materials proved incredibly important as the predecessor of modern-day solid-state components. Further discoveries led to the amplification of radio signals and advances in radio broadcasting, long-distance telephony,[70] and television.[68] Transistors, invented in 1948,[70,71] were made of semiconductors, which were originally made from germanium[73] and subsequently silicon.[70] Research found it was more abundant, cheaper, and better able to perform at higher temperatures. Transistors soon replaced vacuum tubes and valves because they were very reliable[73] and because it was possible to create tiny transistors in silicon, all of which contributed to the development of modern computing.

As with other technologies, computing was (and is) simultaneously led and driven by scientific discoveries, technical and conceptual innovation, and development. Modern computing began in the 1820s in the UK when Charles Babbage developed his mechanical calculation machines – the Difference Engine and the Analytical Engine – with support from Ada Lovelace. She was the first person to recognise that a universal computer could do anything provided that it was given the right data and instructions; therefore, she is often referred to as 'the first programmer'.[74] Development continued slowly through the 19th and early 20th centuries with the introduction of electro-mechanical machines until 1937, when the British mathematician Alan Turing made a significant conceptual breakthrough and published a paper describing an imaginary machine that performed simple mathematical tasks by following precise logical steps.[75] At the time, reuse of scavenged components was commonplace and processing speed was increased through use of thermionic valves from telephone exchanges, which in 1943 facilitated development of the world's first completely programmable, electronic, digital computer.

Other notable innovations include use of binary language to produce the first electronic calculator in 1938 and the first high-level programming language around 1944. During and since the 1950s, developments in programming, manufacturing, and other technologies have further transformed the computing sector; examples include keyboard input capability, transistor-based technologies, and Integrated Circuits (ICs), which were invented to address the growing need for lightweight electronics. These components are assemblies of interconnected miniaturised devices such as transistors, diodes, capacitors, and resistors embedded into a thin layer of silicon. Miniaturisation continued and a chip that could accommodate several components in the 1950s was able to accommodate 1000 components by the 1970s. In 1971 Intel developed the first microprocessors,[76] which integrate the arithmetic, logic, and control circuitry necessary to perform the functions of a computer's Central Processing Unit (CPU) and link the various distinct parts of a computer system. Early computers were large stand-alone mainframe products; however, these various advances led to streamlined computing in the 1970s and then to microcomputers that could perform more than just data processing or scientific calculations, paving the way for computing as we know it today. The resulting demand for microprocessors led to high-volume production and reduced costs, allowing for entry into the household appliance market, which previously could not afford such technologies. For the first time, personal computers became a financially viable option and by the mid-1980s cheap computerisation of home appliances such as microwaves and thermostats was more widespread, whilst industry benefited from automated factory lines and shops, and retail from point-of-sales systems.

In addition to miniaturisation and lower costs, the ubiquitous use of computing technologies was encouraged by the development of networks that linked computers for data exchange. The first were internal but in 1984 external networks for business and academia were introduced. During the 1980s, the British engineer and computer scientist Sir Tim Berners-Lee also developed a new digital information and communication language and network, which subsequently evolved to become the World Wide Web in 1989. Such is the popularity of this technology that over 4.95 billion people and 62.5% of the global population were 'connected' via the internet in 2022. Of these individuals, 97% owned a smartphone, 64% a laptop or desktop PC, and 34% a tablet.[77]

5.2.4.3   Materials and Manufacturing Processes Used for Electronics and ICT

Early computers were room-sized 'mainframe' entities, and, although large, their functionality was very limited. The term mainframe now refers to large non-movable machines that support thousands of applications and input/output devices simultaneously to serve thousands of users. Expansion of digital communication and computing capability has transformed societies around the world by enabling access to information, education, health, commerce, and entertainment by individuals for whom access would otherwise be difficult or impossible. However, it has also increased demand for resources, which will continue as more digital technology is embedded into products (including clothing), automation increases in industry, and the internet expands.

Electronic equipment can contain more than 50 different materials including ferrous and non-ferrous metals, precious metals (PMs), platinum group metals (PGMs), rare earth elements (REEs) – which are all finite resources – plastics, and ceramics. In addition to the IT sector, demand for these resources is increasing in sectors of strategic significance including renewable energy, e-mobility, defence, and aerospace.[78] The EU has identified a list of Critical Raw Materials (CRMs), which are of high economic and technical significance to the EU, and have high-risk supply chains[79] because of the limited amount of material that remains in the earth's crust and current low recycling rates. Furthermore, their specific properties mean that substitution is either very difficult or impossible at present; some are located in politically volatile areas and/or in countries with poor human rights records, and they are often concentrated in one main location. Consequently in 2008, as part of the Raw Materials Initiative the European Commission committed to compiling and regularly updating a list of Critical Raw Materials. In 2011 the first list

Table 5.1   2020 Critical raw materials.[a] Data from ref. 80.

Antimony	**Lithium**
Baryte	Magnesium
Bauxite	Natural graphite
Beryllium	Natural rubber
Bismuth	Niobium
Borate	Platinum group metals[c]
Cobalt	Phosphate rock
Coking coal	Phosphorus
Fluorspar	Scandium
Gallium	Silicon metal
Germanium	**Strontium**
Hafnium	Tantalum
Heavy rare earth elements[a]	**Titanium**
Indium	Tungsten
Light rare earth elements[b]	Vanadium

[a] Materials in bold are new in the 2020 CRM list.
[b] Dysprosium, erbium, europium, gadolinium, holmium, lutetium, terbium, thulium, ytterbium, yttrium.
[c] Cerium, lanthanum, neodymium, praseodymium, samarium.

contained 14 CRMs. This grew to 20 in 2014, 27 in 2017, and 30 in 2020. The 2020 list of CRMs to the EU is shown[80] in Table 5.1.

5.2.4.4   Design in the Electronics and ICT Value Chain

Electrical components vary in size according to type, function, and the product in which they are used, and dimensions range from millimetres to metres. The case of electronic components is rather different because they are comparatively small and material value per item is low. For example, a single silicon wafer may be 0.13 μm thick while surface-mounted capacitors for use with printed circuit boards may be 0.4 mm × 0.2 mm (although some are larger at 7.4 mm × 5.1 mm). Capacitor composition depends on type and application, but they comprise metals (such as aluminium, tantalum, niobium, palladium, or silver), oxides of metals, and insulators (waxed paper, mica, ceramic, or plastics such as polypropylene (PP) or polyethylene terephthalate (PET)); the surface of the metals/oxides may be etched and sintered, and the various material layers wound together.[81] Another example – printed circuit boards (PCBs) or printed wiring boards (PWBs) – are also made up of several materials. These include a substrate (base) layer, which may be rigid (and made from glass fibre and epoxy composites) or flexible polyimide (a thermoplastic elastomer). The second layer is copper, and the third – solder mask – and fourth – silkscreen ink – layers are usually epoxy.[82] The various components either sit on top of (surface mounted) or poke through holes in the boards and all are soldered into position to ensure connectivity. Traditional solder – a fusible metal alloy – is made of tin and lead; lead is a hazardous substance and consequently use in consumer electronics has been restricted in the EU since 2006, as a result of which use in the global market has also been reduced.[83] Lead-free solder is now widely used and may include tin, copper, silver, bismuth, indium, zinc, antimony, and traces of other metals.[84]

Electrical and electronic components were developed to perform specific functions and changes to design have been implemented to improve their functionality and performance as well as to reduce physical size. This has increased the portability of consumer products such as mobile phones, tablets, and laptop computers, which have become ubiquitous. Electrical and electronic components are also embedded in numerous household products and vehicles and there is a growing interest in the creation of wearable electronics that may be embedded into products – for example, cycle lights on or in helmets – and clothing. The scale, scope, and capability of digital manufacturing technologies has also grown as have connectivity and data transmission between products and manufacturers to enable monitoring of the frequency of product use and performance, for example.

The design and manufacture of electrical and electronic equipment can be separated into two distinct categories: external housing/casings and internal components. The external casings are the interface between user and technology and should be designed to facilitate accessibility, easy interaction, and use. In addition to size and manufacturing processes, the ease and difficulty in disassembly and the value of the embodied materials determine the economic viability of reuse and recycling at end of life. Early electrical products were screwed and/or bolted together using 'off-the-shelf' mechanical fixings that were exposed and easy to reach; recent safety regulations mean that many fixings are now concealed and require special tools to access and open them. Some manufacturers also produce unique fixings to prevent access to the internal components and opening products negates the guarantee.[85] This could be for safety reasons, but it is also to prevent interventions by anyone other than Original Equipment Manufacturers (OEMs). While miniaturisation benefits product users, it has increased use of adhesives, rather than mechanical fixings, to save space, which prevents easy separation of components and materials and deters repair and recycling. This means that the products are not fit for a circular economy, and at end of life (in use), unless they are donated or sold via a secondary market, they contribute to the growing e-waste stream, for example by being either stockpiled or added to domestic waste by owners. Designs of these products and components could be changed to facilitate repair and upgrade by users and/or non-OEM businesses, and recycling and reclamation of materials at end of life. Several companies have already proved that it is possible to design for circularity and develop electronic consumer products that meet the above criteria; the most notable examples are Framework (laptops)[86] and Fairphone (mobile phones),[87] which also avoid use of conflict and unethically produced materials and are exemplars of good practice.

The second design category is internal components, which, as described previously, is extremely challenging, although there is evidence of research into components with lower impacts than those on the market. Examples include materials substitution such as use of paper[88] and biopolymer substrates in PCBs and biodegradable capacitors;[89] 3D-printed PCBs where the boards and components are printed as one entity;[90] nano-sized circuits to reduce embodied materials; and use of inks loaded with conductive materials.[91] Some of these proposals will reduce energy use and waste from manufacture and others will facilitate component separation for recycling. However, none of these proposals has been commercialised yet and there is a real need to design, develop, and test whole systems to ensure that a positive impact in one area does not create a negative impact in another area. Unfortunately, the possibility of developing more circular electronic components at present is remote because of the physical properties and behaviour of the materials involved, the performance of which is determined at atomic level. This is very problematic because only 17% of e-waste is formally collected and sustainably recycled; the rest is either sold on informal secondary markets and/or exported and recycled using unregulated, hazardous, and toxic processes.[92] The current recycling infrastructure is underdeveloped, and materials reclamation and reuse are either poor or non-existent and the only recycled metals are ferrous, aluminium, copper, and gold (although research into reclamation of other critical raw materials and metals is ongoing); consequently, many millions of tonnes of potentially useful resources are wasted, and while many cause environmental damage as they downgrade and leach into water supplies for example, the mining of many materials to replace those that are lost is also environmentally and socially damaging. Furthermore, as demand for resources grows, the lack of closed loops for materials is a potential threat to supply, which will affect the transition to green technologies and could lead to conflict.

5.3   Design, Circularity and the SDGs

In an ideal world waste should be minimised and any that arises should be processed in the least environmentally and socially damaging way possible. Similarly, when there is no alternative to the use of virgin materials, any negative environmental or social impacts associated with mining and extraction processes should be minimised or eliminated and all supply chains should be ethical. The extent to which design and designers can influence these factors is unknown. Change will come about only with a collaborative approach that allows knowledge transfer between every stakeholder in the value chain, and which is supported by policy and legislation. Although the extent to which design can influence choice of materials based on source location and supply chain diligence is unknown at present, in Table 5.2 we briefly consider the role of design in relation to the UN SDGs in an ideal situation.

It is clear there is huge potential to inform decision-makers throughout the value chain about the benefits of these and other practices that will also have a positive impact on the SDGs as indicated in Table 5.2.

Table 5.2 The potential positive impact of design on the UN sustainable development goals.

Goal		Materials/ component selection – due diligence along supply chain	Materials selection – durability – physical	Materials selection – durability – emotional design	Manufacturing technology – safe and clean, proper disposal of waste	Place of manufacture – energy sources	Place of manufacture – respect for human rights, health and safety	Place of manufacture – job creation and employment, fair wages	Disassembly, repair, remanufacture and recycling	Formal, legal recycling	Usability and inclusion
1	No poverty	✓					✓	✓			
2	Zero hunger						✓	✓			
3	Good health and wellbeing						✓			✓	✓
4	Quality education							✓			
5	Gender equality							✓			
6	Clean water and sanitation	✓			✓		✓		✓		
7	Affordable and clean energy					✓		✓			
8	Decent work and economic growth		✓				✓	✓	✓		
9	Industry, innovation, and infrastructure		✓	✓				✓	✓		
10	Reduced inequalities										✓
11	Sustainable cities and communities				✓		✓		✓	✓	✓
12	Responsible consumption and production		✓	✓							
13	Climate action	✓	✓	✓	✓	✓			✓	✓	
14	Life below water				✓	✓				✓	
15	Life on land				✓	✓				✓	
16	Peace, justice, and strong institutions		✓	✓	✓			✓	✓		✓
17	Partnerships for the goals	✓	✓	✓	✓	✓	✓	✓	✓	✓	✓

5.4   Conclusions

This chapter opened with a definition of design and brief history of design and engineering, which revealed how design has always had a significant impact throughout the value chain. Several priority areas – textiles, construction and the built environment, furniture, electronics, and ICT – were specifically identified in the EU Circular Economy Action Plan and their history and evolution was discussed in order to understand the challenges and opportunities for change through design now and in the future.

Electrical and electronic equipment is unlike the other materials and product groups discussed in that textiles, buildings and civil infrastructure, and furniture all originated before the first Industrial Revolution. At that time, they were manufactured from natural and renewable resources and fabricated using processes that enabled easy disassembly, repair, and remanufacture. At end of life, organic materials were composted, timber was sometimes used as fuel, and metals were recycled. In other words, the pre-industrial world was based on circular systems, which could be replicated now by using the same or similar materials and manufacturing processes. This is not the case for many materials, manufacturing processes, and technologies that developed during and after the Industrial Revolution, which accelerated the development of many new materials that were engineered to fulfil specific functions. Examples include plastics, polymers, and composites, which may perform very well during use phase, but were not developed with any consideration of what happens to them after use. The chemical composition of thermoset plastics prevents degradation like organic/biomaterials in air, on land or in water. Those that can be reprocessed – for example, thermoplastics – lose their inherent properties within the process and rather than being recycled and (re)made into the same product, they are generally downcycled or used with virgin material because of changes in their performance. At present many composites comprise glass or carbon fibres and a thermoset matrix; therefore, at best, they are very difficult, and at worst impossible, to recycle successfully and/or economically; although they may be ground up as filler, many are incinerated or dumped. Pre-industrial structures and products were not consciously developed to be circular; rather they happened to fit circular systems because of the nature of available materials, manufacturing, and assembly processes.

Although some post-industrial inventions created during the 20th and 21st century were developed to prohibit intervention by anyone other than OEMs, the majority were not developed to prevent recycling or disassembly or degradation back to elemental level, rather development focused on functionality and performance during and up until the end of the use phase because materials were abundant and there was space for local landfill sites, for example. The ever-increasing demand on resources, in particular Critical Raw Materials that are essential to digital and clean technologies, potential threats to supply chain security, and rising costs and waste, mean that current strategies are untenable.

We conclude that design has generated positive and negative impacts and that it also has the potential to conserve value in the circular economy in the priority sectors identified by the EU and, therefore, other sectors. However, design must encourage behavioural as well as technological change, which will prove more challenging in some areas than others. The extent to which societies, industries, and services around the world have become reliant on digital technologies and associated services means that the technology is here to stay but, for the foreseeable future, electronic components will remain difficult, if not impossible, to recycle, and reclamation and reuse limited; a problem compounded by underdevelopment of recycling technologies and infrastructure.

References

1. Oxford English Dictionary, *OED Online*, Oxford University Press, Oxford, 2019.
2. H. T. Wood, in *A History of the Royal Society of Arts*, Cambridge University Press, Cambridge, 2011.
3. Design Council, Framework for innovation: Design Council's evolved Double Diamond, https://www.designcouncil.org.uk/our-work/skills-learning/tools-frameworks/framework-for-innovation-design-councils-evolved-double-diamond/ (accessed March 2023).
4. R. Huston and R. A. Dastrup, *People, Places, and Cultures*, Tulsa Community College, Tulsa, 2020.
5. A. Smith, *The Wealth of Nations*, Wordsworth Editions, Ware, 2012.
6. Y. Goossens, A. Mäkipää, P. Schepelmann, I. van de Sand, M. Kuhndt and M. Herrndorf, *Policy Department Economic and Scientific Policy. Alternative Progress Indicators to Gross Domestic Product (GDP) as a Means Towards Sustainable Development. IP/A/ENVI/ST/2007-10. PE 385.672*, Brussels, 2007.
7. S. Broadberry, B. Campbell, A. Klein, M. Overton and B. van Leeuwen, in *Quantifying Long Run Economic Development*, The University of Warwick in Venice, Palazzo Pesaro Papafava, Venice, 2011.
8. J. Hicks and G. Allen, *A Century of Change: Trends in UK Statistics since 1900*. House of Commons library research paper 99/111, London, 1999.
9. B. London, *Ending the Depression through Planned Obsolescence*, New York, 1932, https://babel.hathitrust.org/cgi/pt?id=wu.89097035273&view=1up&seq=1 (accessed May 2023).
10. M. Krajewski, *IEEE Spectr.*, 2014, **51**, 56.
11. S. Clarke, *J. Des. Hist.*, 1999, **12**, 65.
12. G. Adamson, *Industrial Strength Design: How Brooks Stevens Shaped your World*, The MIT Press, Cambridge, MA, 2005.
13. Estate of Raymond Loewy, About Raymond Loewy, Quotes by Raymond Loewy, https://www.raymondloewy.com/about/quotes/ (accessed February 2023).

14. Buckminster Fuller Institute, R. Buckminster Fuller, 1895–1983, https://www.bfi.org/about-fuller/biography/ (accessed February 2023).

15. R. B. Fuller, *Nine Chains to the Moon*, Cape, London, 1973.

16. V. Papanek, *Design for the Real World: Human Ecology and Social Change*, Thames & Hudson, London, 2nd edn, 1985.

17. A. Rawsthorn, *Victor Papanek, an early champion of good sense*. New York Times, 15 May 2011, www.nytimes.com/2011/05/16/arts/16iht-design16.html?_r=1&adxnnl=1&pagewanted=all&adxnnlx=1417093227-SA24SiC/Gd3tz/L3DX8h8w (accessed May 2023).

18. World Commission on Environment and Development (WCED), *Report of the World Commission on Environment and Development. Our Common Future*, United Nations, New York, 1987.

19. P. Burall, *Green Design*, Design Council, London, 1991.

20. F. Ceschin and I. Gaziulusoy, *Des. Stud.*, 2016, **47**, 118.

21. Ellen MacArthur Foundation, *The Butterfly Diagram: Visualising the Circular Economy*, https://ellenmacarthurfoundation.org/circular-economy-diagram (accessed March 2023).

22. Ellen MacArthur Foundation, *Circulate Products and Materials*, https://ellenmacarthurfoundation.org/circulate-products-and-materials (accessed February 2023).

23. Product-Life Institute, Welcome to the product-life institute, http://www.product-life.org/ (accessed February 2023).

24. Rolls-Royce, Rolls-Royce celebrates 50th anniversary of power-by-the-hour, https://www.rolls-royce.com/media/press-releases-archive/yr-2012/121030-the-hour.aspx (accessed February 2023).

25. WBCSD, *Collaboration, Innovation, Transformation. Ideas and Inspiration to Accelerate Sustainable Growth – A Value Chain Approach*, https://docs.wbcsd.org/2011/12/CollaborationInnovationTransformation.pdf (accessed February 2023).

26. University of Cambridge, Cambridge Institute for Sustainability Leadership, *What is a value chain? Definitions and characteristics*, https://www.cisl.cam.ac.uk/education/graduate-study/pgcerts/value-chain-defs#:~:text=World%20Business%20Council%20for%20Sustainable%20Development%20(WBCSD)&text=A%20value%20chain%20refers%20to,and%20disposal%2Frecycling%20processes.%E2%80%9D (accessed February 2023).

27. European Commission, *Circular Economy Action Plan. For a Cleaner and More Competitive Europe*, COM/2020/98 final, Brussels, 2020, https://ec.europa.eu/environment/circular-economy/pdf/new_circular_economy_action_plan.pdf (accessed February 2023).

28. Ellen MacArthur Foundation, Recycling and the circular economy: what's the difference?, https://ellenmacarthurfoundation.org/articles/recycling-and-the-circular-economy-whats-the-difference (accessed February 2023).

29. B. Murray, *Embedding Environmental Sustainability in Product Design*, WRAP, Banbury, 2013.
30. European Commission, Directorate-General for Energy and Directorate-General for Enterprise and Industry, *Ecodesign Your Future: How Ecodesign can Help the Environment by Making Products Smarter*, DOI: 10.2769/38512, Brussels, 2014.
31. Flanders DC and Circular Flanders, Close the loop. A guide towards a circular fashion industry, https://www.close-the-loop.be/en/phase/5/design (accessed February 2023).
32. T. E. Graedel, P. R. Comrie and J. C. Sekutowski, *AT&T Technical Journal*, 1995, **74**, 17.
33. D. A. Norman and S. W. Draper, *User Centred System Design: New Perspectives on Human–Computer Interaction*, L. Erlbaum Associates, Hillsdale, NJ, 1986.
34. N. Tractinsky, A. S. Katz and D. Ikar, *Interact. Comput.*, 2000, **13**, 127.
35. D. Norman, *Interactions*, 2002, **9**, 36.
36. European Commission, First circular economy action plan. Implementation of the first circular economy action plan, COM/2019/190 final, Brussels, 2019, https://ec.europa.eu/environment/circular-economy/first_circular_economy_action_plan.html (accessed March 2023).
37. European Union, *Directive 2009/125/EC of the European Parliament and of the Council of 21 October 2009 Establishing a Framework for the Setting of Ecodesign Requirements for Energy-related Products*, Official Journal of the European Union, 2009.
38. N. Akdemir, *Gaziantep University Journal of Social Sciences*, 2018, **17**, 1371.
39. T. Duhoux, K. le Blévennec, S. Manshoven, F. Grossi, M. Arnold and L. Fogh Mortensen, *ETC/CE Report 2/2022: Textiles and the Environment: the Role of Design in Europe's Circular Economy*, European Topic Centre Circular Economy and Resource Use, Boeretang, 2022.
40. European Environment Agency, Briefing: Textiles and the environment: the role of design in Europe's circular economy, Copenhagen, 10 February 2022, https://www.eea.europa.eu/publications/textiles-and-the-environment-the (accessed March 2023).
41. S. Opperskalski, A. Franz, A. Patanè, S. Siew and E. Tan, *Preferred Fiber & Materials Market Report*, Textile Exchange, Lamesa, 2022.
42. Common Objective, Fibre briefing: polyester, London, 22 October 2021, https://www.commonobjective.co/article/fibre-briefing-polyester (accessed February 2023).
43. European Parliament, *The Impact of Textile Production and Waste on the Environment*, DG Communication, Brussels, 26 April 2022.
44. O. Edwards, Toxic textiles: the chemicals in our clothing, PCIAW, Milton Keynes, 8 November 2022, https://pciaw.org/toxic-textiles-the-chemicals-in-our-clothing/ (accessed February 2023).

45. P. Senthil Kumar and K. Grace Pavithra, in *Water in Textiles and Fashion*, ed. S. S. Muthu, Elsevier, London, 2019, pp. 21–40.

46. University of the Arts, Centre for Circular Design, https://www.arts.ac.uk/research/research-centres/centre-for-circular-design (accessed February 2023).

47. Royal College of Art, Textiles circularity centre, https://www.rca.ac.uk/research-innovation/research-centres/materials-science-research-centre/textiles-circularity-centre/ (accessed February 2023).

48. G. Hutton, *Industries in the UK*, House of Commons library research briefing, UK Parliament, 6 December 2022.

49. European Commission, Revised construction products regulation – factsheet, Brussels, 30 March 2022, https://ec.europa.eu/docsroom/documents/49314 (accessed February 2023).

50. K. Breene, *Can the Circular Economy Transform the World's Number One Consumer of Raw Materials?* World Economic Forum, Geneva, 2016.

51. European Commission, *Closing the Loop – An EU Action Plan for the Circular Economy*, COM(2015) 614 final, Brussels, 2015.

52. *Eurostat, Waste statistics. Waste generation 2020*, Luxembourg, 2023.

53. European Commission, *Stepping up Europe's 2030 Climate Ambition. Investing in a Climate-neutral Future for the Benefit of our People*, COM(2020) 562 final, Brussels, 2020.

54. Ellen MacArthur Foundation, *Completing the Picture. How the Circular Economy Tackles Climate Change*, 2021 reprint, Cowes, 2021.

55. UNEP, *2022 Global Status Report for Buildings and Construction: Towards a Zero Emission, Efficient and Resilient Buildings and Construction Sector*, New York, 2022.

56. European Parliament, Legislative train schedule. Strategy for a sustainable built environment in 'A European green deal', Brussels, February 2023, https://www.europarl.europa.eu/legislative-train/theme-a-european-green-deal/file-strategy-for-a-sustainable-built-environment (accessed February 2023).

57. D. Acharya, R. Boyd and O. Finch, *From Principles to Practices: First Steps Towards a Circular Built Environment*, Arup and Ellen MacArthur Foundation, Cowes, 2018.

58. Ellen MacArthur Foundation, Building a world free from waste and pollution, https://ellenmacarthurfoundation.org/articles/building-a-world-free-from-waste-and-pollution (accessed February 2023).

59. Mineral Products Association, *The Contribution of Recycled and Secondary Materials to Total Aggregates Supply in Great Britain – 2020 Estimates*, London, 2022.

60. Arup and Ellen MacArthur Foundation, d.Hub. Circular buildings toolkit, https://ce-toolkit.dhub.arup.com/ (accessed February 2023).

61. S. Ebnesajjad, *Handbook of Adhesives and Surface Preparation*, Elsevier, London, 2011.

62. C. Drewe and T. Barker, *Are you Sitting Comfortably? Sustainable Timber Sourcing and the UK Furniture Industry*, WWF-UK, Working, 2016.

63. IUCN, The IUCN red list of threatened species. Version 2022-2, IUCN, Cambridge, 2023, https://www.iucnredlist.org/ (accessed February 2023).
64. North London Waste Authority, 22 million damaged furniture items and 11,000 bust bicycles thrown away each year, London, 8 October 2018, https://www.nlwa.gov.uk/news/22-million-damaged-furniture-items-and-11000-bust-bicycles-thrown-away-each-year (accessed February 2023).
65. H. Pirayesh, H. Khanjanzadeh and A. Salari, *Compos. B. Eng.*, 2013, **45**, 858.
66. S. Pain, *Nature*, 2017, **551**, S134.
67. National Grid, The history of energy in the UK, London, 2023, https://www.nationalgrid.com/stories/energy-explained/history-of-energy-UK (accessed February 2023).
68. R. Forrester, *History of electricity*, *SSRN Electronic Journal*, 30 November 2016, DOI:10.2139/ssrn.2876929.
69. National grid, History of electricity in Britain, London, 2023, https://www.nationalgrid.com/about-us/what-we-do/our-history/history-electricity-britain (accessed February 2023).
70. J. Collett, in *The History of Electronics. Companion Encyclopaedia of Science in the Twentieth Century*, ed. J. Krige and D. Pestre, Taylor & Francis, London, 2013, pp. 253–274.
71. S. K. Routray, in *2004 IEEE Conference on the History of Electronics (CHE2004)*, Bletchley Park, 28–30 June 2004.
72. G. Gandhi, V. Aggarwal and L. O. Chua, *IEEE Circuits and Systems Magazine*, 2013, **13**, 8.
73. L. Łukasiak and A. Jakubowski, *Journal of Telecommunications and Information Technology*, 2010, **1**, 3.
74. L. Carlucci Aiello, *Artif. Intell.*, 2016, **235**, 58.
75. A. M. Turing, *Proceedings of the London Mathematical Society*, 1937, **s2–42**, 230.
76. S. Furber, *Proceedings of the Royal Society A: Mathematical, Physical and Engineering Sciences*, 2017, **473**, 20160893.
77. Kepios Pte. Ltd, *Digital 2022. Global Overview Report. The Essential Guide to the World's Connected Behaviour*, We Are Social and Hootsuite, Singapore, 2022.
78. S. Bobba, S. Carrara, J. Huisman, F. Mathieux and C. Pavel, *Critical Raw Materials for Strategic Technologies and Sectors in the EU. A Foresight Study*, EU Joint Research Centre, Luxembourg, 2020.
79. European Commission, Critical raw materials, Brussels, 2022, https://single-market-economy.ec.europa.eu/sectors/raw-materials/areas-specific-interest/critical-raw-materials_en (accessed February 2023).
80. European Commission, *Critical Raw Materials Resilience: Charting a Path Towards Greater Security and Sustainability*, COM(2020) 474 final, Brussels, 2020.
81. R. Wiens, Digi-key electronics academic component reference guide: capacitors, Digi-key, Thief River Falls, https://www.digikey.co.uk/en/pdf/d/digikey/acrg/capacitors (accessed February 2023).

82. EMSG, The basics of PCB design and composition, EMSG, York, PA, https://emsginc.com/resources/basics-pcb-design-composition/ (accessed February 2023).

83. European Chemicals Agency, Substances restricted under REACH, ECHA, Helsinki, 2016, https://echa.europa.eu/substances-restricted-under-reach/-/dislist/details/0b0236e1807e30a6 (accessed February 2023).

84. S. Das, Lead free solder and composition. Pb-free solder, Electronics and You, 13 December 2019, https://www.electronicsandyou.com/blog/lead-free-pb-free-solder-and-composition.html (accessed February 2023).

85. W. Gordon, The most common ways manufacturers prevent you from repairing your devices, IFIXIT, 17 April 2019, https://www.ifixit.com/News/15617/the-most-common-ways-manufacturers-prevent-you-from-repairing-your-devices (accessed February 2023).

86. Framework, Fix consumer electronics, Framework Inc., San Francisco, https://frame.work/gb/en/about (accessed February 2023).

87. A. Fischer, S. Pascucci and W. Dolfsma, *J. Clean. Prod.*, 2022, **364**, 132487.

88. Cadence, Paper circuit boards: The future of PCB design, Cadence Design Systems, Dublin, 15 December 2015, https://resources.pcb.cadence.com/blog/2020-paper-circuit-boards-the-future-of-pcb-design (accessed February 2023).

89. A. Staat, R. Mende, R. Schumann, K. Harre and R. Bauer, in *2016 39th International Spring Seminar on Electronics Technology (ISSE)*, IEEE, Pilsen, 18–22 May 2016.

90. Stratasys Inc., *Transform manufacturing by 3D printing end-use parts*, Stratasys Inc., Eden Prairie, 2023, https://www.stratasys.com/uk/resources/whitepapers/end-use-parts/ (accessed May 2023).

91. ReFREAM, Conductive printing technology, European Commission STARTS Programme, 2019, https://re-fream.eu/portfolio/conductive-printing-technology/ (accessed February 2023).

92. V. Forti, C. P. Baldé, R. Kuehr and G. Bel, *Global E-waste Monitor 2020. Quantities, Flows, and the Circular Economy Potential*, United Nations University, International Telecommunication Union and International Solid Waste Association, Bonn/Geneva/Rotterdam, 2020.

CHAPTER 6

The Circular Economy, Responsible Consumption and the Consumer

HYEONG-JIN CHOI AND SEUNG-WHEE RHEE*

Department of Environmental Engineering, College of Engineering, Kyonggi University, Republic of Korea
*E-mail: swrhee@kyonggi.ac.kr

6.1 Introduction

Concomitant with mass production in a capitalist economic society, the purchasing and consumption of goods and services has increased, and the concept of the consumer as the locus of consumption has emerged.[1] In many countries, the consumer is defined as "those who use the goods and services provided by business entities for their daily lives as consumers or for their production activities" in legislation such as the Consumer Protection Act.[2] This includes those at the end of the chain who use any goods or services supplied, or who use the furnished goods or services for agricultural (including the livestock industry) and fishery activities.

Consumed goods and services are discharged in the form of waste at their end-of-life stage. Global solid waste generation was 2.01 billion tonnes in 2016 and is expected to increase to 3.40 billion tonnes by 2050.[3] Massive generation of wastes causes global environmental problems such as open dumping, soil contamination by improper management, global warming and a

Issues in Environmental Science and Technology No. 51
The Circular Economy: Meeting Sustainable Development Goals
Edited by Sadhan Kumar Ghosh and Gev Eduljee
© The Royal Society of Chemistry 2024
Published by the Royal Society of Chemistry, www.rsc.org

resource crisis. Sustainable Development Goal (SDG) 12 (Ensure sustainable consumption and production patterns – Responsible production and consumption) articulates these concerns (see Chapter 2 for details).

In order to prevent waste pollution and hence to also address SDG 12, practices and management based on the 3Rs (Reduce, Reuse, Recycle) along with the concept of the supply chain can be a key solution. The term "upstream" in the supply chain refers to the earlier stages of production, such as raw material extraction and component manufacturing, while "downstream" in the supply chain refers to the later stages of production, such as distribution and retail. These terms are used to distinguish the different stages of production and distribution of goods or services. Environmentally sound consumption practices can potentially be made by most stakeholders at all stages of the supply chain, both upstream and downstream, with all stakeholders working together with a strong focus on reducing waste. For the reduction of waste, the role and responsibility of consumers is very important for waste discharged from households. Specifically, the reduction of packaging waste and single-use products is very important because of their short lifetimes. Manufactures should also consider redesigning products to reduce packaging and to change materials to producing green products. To reuse solid wastes, some countries have enforced the system of deposit refund to collect glass bottles and polyethylene terephthalate (PET) beverage containers. In the deposit-refund system, consumers are incentivised to return these products to be reimbursed the deposit fee. In the field of recycling management, source separation by consumers can be very important to collect recyclable wastes from households. In the Republic of Korea (henceforth abbreviated to Korea), six types of wastes (plastics, paper, cans (steel and aluminium), Expanded Polystyrene (EPS), textile and glass) from households are separately discharged by consumers without paying a discharge fee.[4]

Responsible consumer behaviour is important to build a circular economy and simultaneously to achieve SDG 12. This chapter describes the role and the responsibilities of consumers in establishing a circular economy society in Korea. Based on the 3Rs concept and strategies for environmentally sustainable consumption, key success factors introduced in regulations and systems supporting responsible consumption and reduction in waste generation are discussed from the consumer's point of view. The term "discharge" used in this chapter is synonymous with "discard" used in other jurisdictions, the former representing the transfer of waste from source to sink.

6.2   Roles and Responsibilities of the Consumer

Consumers are often unable to ensure fair and sustainable consumption due to disparities in economic conditions, educational levels and bargaining power. To address these disparities, the concept of consumer rights was born. The concept of consumer rights has been expanded since John F. Kennedy's speech to the US Congress in 1962, in which he spoke of the four rights (the right to safety, to choose freely, to be informed and to be heard). In 1985, the General Assembly of the United Nations adopted eight consumer rights, with Consumers International defining the eight rights as follows:[5]

1. **The right to safety:** The right to be protected against products, production processes and services that are hazardous to health or life.
2. **The right to be informed:** The right to be given the facts needed to make an informed choice, and to be protected against dishonest or misleading advertising and labelling.
3. **The right to choose:** The right to be able to select from a range of products and services, offered at competitive prices with an assurance of satisfactory quality.
4. **The right to be heard:** The right to be heard, to have consumer interests represented in the making and execution of government policy, and in the development of products and services.
5. **The right to satisfaction of basic needs:** The right to have access to basic, essential goods and services: adequate food, clothing, shelter, healthcare, education, public utilities, water and sanitation.
6. **The right to redress:** The right to receive a fair settlement of just claims, including compensation for misrepresentation, shoddy goods or unsatisfactory service.
7. **The right to consumer education:** The right to acquire knowledge and skills needed to make informed, confident choices about goods and services, while being aware of basic consumer rights and responsibilities and how to act on them.
8. **The right to a healthy and sustainable environment:** The right to live and work in an environment that is non-threatening to the well-being of present and future generations.

In April 2022 the European Parliament adopted a topical resolution on the right to repair for electronics to reduce costs for consumers and promote the development of a circular economy. The concept of "the right to repair" is not specifically defined, but addresses the following issues:[6]

1. The right to free repair of defective products during the guarantee period.
2. The right to repair after the guarantee period.
3. The right for consumers to self-repair the products.

Through these consumer rights, consumers are in a position to respond to business operators as actors in the market economy. In other words, consumers have become influential in the market economy, escaping from weak or passive positions. Consumers have these rights as well as roles and responsibilities,[7] details of which are shown in Table 6.1.

The role of consumers is to be a chooser. Consumer choice is not only a matter of which goods and services to buy, but also whether and how to

Table 6.1   Roles and responsibilities of consumers.

Type		Contents
Role	Chooser	Consumers choose the types of goods and services, whether they consume them, and how they consume them
	Communicator	Consumers communicate with stakeholders through consumption behaviour
	Activist	Consumers contribute to improving the share of all consumers in a moral context
Responsibility	Exercise fundamental rights	Consumers shall make the right choice of goods/services and exercise justly their fundamental rights through recognising themselves as the main constituent of the free-market economy
	Acquire knowledge and information	Consumers shall endeavour to acquire knowledge and information necessary for promoting their own rights and interests
	Resource-saving and environment-friendly consumption	Consumers shall engage in resource-saving and environment-friendly consumption independently and rationally, and thereby play a positive role in the improvement of their lives as consumers

consume them. Consumers must be informed to make more environmentally sound choices.[8] Consumers play the role of communicators. Consumers communicate with stakeholders through consumption behaviour. Consumers can point out the lack of the goods or services rather than just paying for them. Manufacturers can reflect matters communicated with consumers in the production stage. Communication between the government and consumers is also important in the public sector, such as in formulating circular economy policies. In order to implement efficient circular economy policies, it is necessary to communicate with consumers at all stages, from purchase to discharge. As a consumer, an activist may use their consumption behaviour as a means of communication with stakeholders such as manufacturers, the government and other consumers. For example, an activist may choose to boycott certain products or companies that engage in practices that are harmful to the environment or to marginalised communities.

Consumers have a responsibility to exercise their fundamental rights. Consumers should be aware of rights such as the right to safety and the right to be informed, and exercise them when deprived of their basic rights. Consumers also have a responsibility to acquire knowledge and information. Consumers should get complete information about the quantity, price, quality, *etc.*, of goods or services without blindly trusting the seller.[9] Consumers can avoid unfair practices by acquiring knowledge and information to improve their own safety and rights. For instance, consumers can consume energy-efficient products and green-certified goods by obtaining information on products and manufacturers. And consumers have a responsibility to conserve resources and consume environmentally friendly goods. On the other hand, consumers must preserve the environment to pass on to the next generation through eco-friendly consumption.

By fulfilling the roles and responsibilities of consumers, individuals can contribute to responsible consumption practices that promote sustainability and environmental stewardship. The concept of responsible consumption refers to the idea that individuals and organisations should consider the environmental, social and economic impacts of their consumption patterns.[10] Responsible consumption also involves consideration of the long-term impacts of our actions, taking steps to minimise negative impacts and maximise positive ones.[11] This can include choosing products that are made from sustainable materials, produced in an environmentally friendly way, and designed for reuse or recycling, as well as reducing overall consumption and waste. Responsible consumption can also involve supporting businesses and practices that are socially and environmentally responsible, such as those that promote fair labour practices and minimise their environmental impact – reducing negative impacts on natural resources, climate change, and social and economic inequalities.

From the perspective of SDG 12 and of the circular economy, consumers should participate in the zero-waste movement to reduce the usage of plastic packaging products and single-use products. Furthermore, consumers should reduce household waste when considering the eight rights and the right to repair. However, it is not easy to reduce waste in most households because of the significant increase in single-occupancy households and doorstep delivery of goods and food. Hence, in Korea the mandated Volume-based Rate System (VBRS) has been enforced to reduce waste generation. Consumers can also consume products that utilise renewable resources, properly discharge consumed goods and voluntarily refrain from using single-use products.

6.3 Strategies at each Stage of the Circular Economy for Responsible Consumption

In order to implement the circular economy, sustainable consumption must be prioritised, for which transparent disclosure of environmental information and efforts to fulfil consumer responsibilities are required. To explain sustainable consumption and consumer responsibility, the consumption stage is subdivided into purchase for consumption, discharge to reduce and recycle waste, and collection for recycling of waste. This section introduces strategies and examples of consumer responsibility at each stage, focusing on the Republic of Korea.

6.3.1 Purchase for Consumption

In the purchasing stage of goods and services for consumption, the following three measures have been established to measure responsible consumption:[12]

1. Legal restrictions on consumption of products that fall out of the loop in the circular economy.
2. Encouraging consumers to make responsible purchases through economic measures such as a deposit system and taxes.
3. Encouraging consumers to voluntarily reduce consumption of environmentally unsound products or to purchase products that are easy to recycle.

For the first measure in the consumption stage, restrictions on purchases by government policies are being promoted mainly for plastic packaging materials and single-use products.[13] In Korea, the free provision and use of disposable bags is prohibited, as detailed in Table 6.2.

Korea plans to ban the use of disposable plastic bags in bakeries and general retail from 2022 and expand it to all industries by 2030. Policies banning single-use products are being promoted worldwide. In the EU, Directive 2019/904 restricts the placing on the market of nine single-use plastic products, including cutlery, drinking straws, drink stirrers and food and beverage containers made of EPS.[14] Japan announced its Resource Circulation Strategy for plastics in 2019. Under the strategy, Japan plans to reduce single-use plastics by 25% by 2030. China banned free provision of plastics bags by retailers in 2008 and plans to limit single-use plastic products such as shopping bags, straws and utensils by the end of 2025 through "Opinions on further strengthening the control of plastic pollution".[15]

For the second measure, Korea is implementing a deposit-refund system for single-use cups and empty bottles. This system promotes collection and reuse of the spent containers by establishing the deposit-refund system on recyclable containers, which customers can redeem when they are returned.[16] A deposit-refund system for empty bottles in Korea has

Table 6.2 Policy on banning the use and free provision of disposable bags in Korea.

Year	Ban of use	Ban of free provision
2022	Bakery, general retail (small and medium-sized supermarkets, convenience stores, *etc.*)	Restaurant and tavern business (excluding bakeries)
2025	Wholesale and retail business (exceeding 33 m^2) Restaurant and tavern business (including food packaging/delivery)	Wholesale and retail business (less than 33 m^2) (small shops, shops in traditional markets, *etc.*)
2030	All industries (including wholesale/retail business and other service businesses under 33 m^2)	–

been implemented since 1985, the deposit being 70 to 350 won per bottle (about 0.06 to 0.29 USD per bottle) depending on the size of bottle. The deposit-refund system for single-use cups has been demonstrated, initially in Sejong City and Jeju Island, since December 2022, and is planned to be expanded nationwide in the future.[17] The deposit for single-use cups is 300 won per cup (about 0.25 USD per cup). However, there are many barriers to the expansion of the deposit-refund system for single-use cups in Korea:

1. Who will pay the deposit fee and what is the optimal level of the fee?
2. Where and what size are the applicable stores to which the product is returned?
3. What is the return system for single-use cups?

The demonstration project aims to overcome these barriers and revise the deposit-return system for single-use cups to successfully implement nationwide.

Owing to issues such as microplastics and climate change, economic measures for plastics are being implemented worldwide. Freiburg in Germany has been running The FreiburgCup since 2016, which charges a 1 EUR (about 1.1 USD) deposit for reusable cups,[18] encouraging consumers to take responsibility and adopt correct discharge behaviours by paying a deposit. As a result, the system can increase the collection rate of waste, and the waste can be collected in a relatively homogeneous and clean condition, building a circular economy through effective recycling. In the EU, a plastic tax of 0.8 EUR per kg has been imposed from 2021 on plastic packaging waste that is not recycled.[19] In Japan, a deposit of 3–5 yen (about 0.02–0.05 USD) has been imposed from 2020 as a mandatory fee for each plastic bag. These economic measures help consumers to practise their responsibilities through rational and efficient consumption.

A final measure is to encourage consumers to voluntarily reduce their consumption of environmentally unsound products that are not easy to recycle. Environmental information, including greenhouse gas generation, energy consumption and recyclability, should be delivered to consumers in order to persuade consumers to make rational purchases. In Korea, environmental information is provided by an energy-efficiency labelling system and a material and structure assessment system, as shown in Figure 6.1. The energy-efficiency labelling system is a mandatory reporting system for manufacturers and importers so that consumers can easily purchase highly efficient electrical and electronic products. By improving energy efficiency in the consumption stage of the life cycle, the system can reduce raw materials and waste required for energy production. The material and structure assessment system encourages manufacturers to voluntarily produce eco-friendly products by inducing consumers to purchase eco-friendly products.[20] In the system, the recyclability grade, which indicates the ease of recycling, is divided into four grades from poor recyclability to excellent recyclability. The lowest

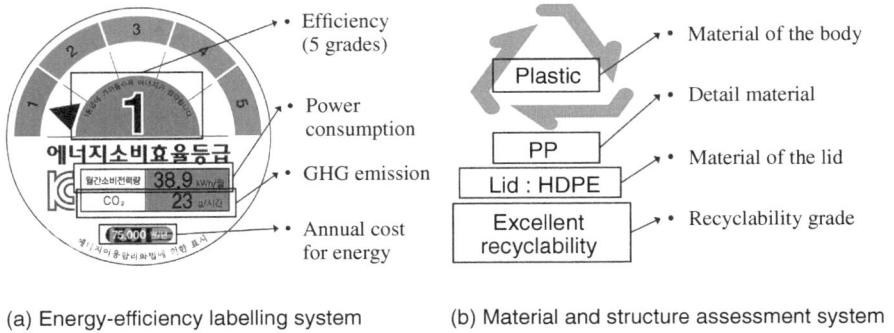

(a) Energy-efficiency labelling system   (b) Material and structure assessment system

Figure 6.1   Information to encourage responsible consumption.

level, "poor recyclability", requires mandatory labelling on the packaging. The label can encourage consumers to separate waste easily and to prefer reusable and recyclable products.

The three measures for responsible consumption either directly or indirectly reduce consumption or encourage the consumption of eco-friendly goods and services. Reducing unsound practices is a high-priority method in the hierarchy of waste management due to its effectiveness in reducing waste and greenhouse gas emissions. In addition, inducing consumption of eco-friendly goods or services is a driving force for consumer goods within the circular economy to stay in the economic system for as long as possible.

6.3.2   Discharge to Reduce and Reuse Waste

The development of urbanisation, income increase and consumption acceleration have resulted in environmental pollution as well as large-scale waste generation. In many countries, such as Korea, the USA, the Philippines and the Republic of Kazakhstan, sanitary landfills and incinerators are necessary to dispose of large amounts of waste, but these facilities cause great social conflict within local communities, such as the Not in My Backyard (NIMBY) syndrome. Therefore, it is not possible to build facilities to dispose of waste. To overcome these waste problems, the paradigm of waste management has begun to shift from "how to treat" to "how to reduce".

Korea implemented the VBRS in 1995 as a policy to reduce waste generation and maximise recycling.[22] This system, in which the following three waste management principles are applied, aims to establish consumption and discharge lifestyles that can reduce waste by imposing a waste-generation charge on waste generators (consumers):

1. Polluter Pays Principle (3Ps): The person who caused the pollution or its generation should bear the costs.
2. User Pays Principle: This calls upon the user of a natural resource to bear the cost of running down natural capital.[21]

3. Prevention Principle: Prevention takes precedence over disposal, which shifts the direction of waste policy from a supply-oriented approach to a demand-control approach.[22]

The system introduces the concept of imposing a charge for waste to the general population by applying the "Polluter Pays Principle", which obliges the person who discharges waste to pay the treatment cost according to the quantity of wastes. It also provides incentives for households to separate their recyclables from other solid waste by offering a free recycling collection service.[23]

The target wastes and operation of the VBRS are shown in Table 6.3. The target wastes of the VBRS are Municipal Solid Waste (MSW), bulky waste and food waste. Recyclable resources such as single-material plastics, cans, Styrofoam, scrap metal and waste paper are excluded from the VBRS as they are subject to separate discharge/collection. In addition, hazardous wastes from households, such as waste fluorescent lamps, waste medicines and waste batteries, are excluded.

Table 6.3 Operation of the VBRS by type of waste in Korea.

Type	Target waste	Operation
Municipal Solid Waste (MSW)	• Municipal solid waste generated from workplaces and households • Excluding briquette ash, bulky waste, and recyclables	• Discharge of waste by consumers using standard waste plastic bag sold by local governments • Discharge waste is collected by the local government
Bulky waste	• Home appliances such as refrigerators, washing machines and TVs • Furniture such as desks and sofas • Household items such as bedding, bicycles, *etc.*	• Recyclable bulky waste: Report bulky waste that can be recycled, such as home appliances, to the recycling centre operated by the local government (local governments collect it at no cost) • Other bulky waste: Discharge with stickers sold by local governments or report to private waste collection companies
Food waste	Food waste (not including foreign substances such as animal bones, toothpicks and tissue paper)	Consumers discharge of waste in one of the following ways: • Discharge of standard waste plastic bag • Discharge with stickers attached • Discharge using Radio-Frequency Identification (RFID) system

In the VBRS, waste is discharged using a Standard Waste Bag (SWB) or sticker according to the size and capacity of the waste. The SWB is used mainly for MSW and food waste. The SWBs used in households are sold as prescribed by local government ordinances. Generally, the capacity of SWBs for MSW and food waste is 5–75 L and 2–20 L, respectively. Transportation is the major cost in the supply chain from waste generation to final treatment. The reduction in waste quantity and volume provides economic advantages for waste transportation.[24] In the VBRS, SWBs are charged by volume, so consumers manually compress their waste at the discharge stage. In addition, consumers separate recyclable wastes to reduce the volume of residual waste. Discharged waste using a SWB is collected by local governments but waste not placed in SWBs is not collected by local governments, and fines may be imposed on the discharger.

The SWBs cannot be used for bulky waste such as refrigerators, washing machines, furniture and large-size household items. Recyclable bulky waste can be reported to the recycling centre operated by the local government and collected free of charge. Other bulky waste such as bedding and bicycles can be discharged with stickers sold by local governments or collected by private waste collectors.

Since the introduction of the VBRS in 1995, the amount of MSW in Korea reduced dramatically from 2.10 kg *per capita* per day in 1991 to 1.08 kg *per capita* per day in 1996. In 2010, the amount of MSW in Korea was less than 1.0 kg *per capita* per day. The generation of MSW from 2000 to 2020 increased by only about 1.4% in terms of compound annual growth rate, demonstrating that the introduction of the VBRS has effectively and successfully reduced MSW nationwide. In addition, in terms of treatment of MSW, the recycling rate increased significantly from 23.7% in 1995 to 59.5% in 2020. By separating and discharging MSW under the VBRS, the quality of separated MSW has improved and recycling has become easier.[25]

The sticker attachment method can be used for food waste as well as bulky waste in some local authorities. In particular, those who discharge a large amount of food waste (more than 20 L), such as from restaurants and apartment complexes, can attach a sticker to the large-capacity food waste collection container. The sticker can be used for wastes up to 120 L and must be attached to the handle of the collection container. However, if a large amount of food waste is discharged at a place where various consumers gather such as an apartment, the fee is charged collectively by dividing the total amount by the number of people. This may result in lower costs for consumers who generate a lot of waste and may violate the principles applied in the VBRS. Since 2010, Korea has applied a Radio Frequency Identification (RFID)-based food waste management system that can collect and manage individual food waste discharge information. The RFID-based system automatically measures the amount of food waste and charges a fee based on the weight of discharged food waste. The RFID-based system was trialled in pilot projects for 18 local governments in 2010. Moreover, in 161 out of 226 local governments, a RFID-based system for collecting food waste was established in 2020.[26] The total number of RFID-based food waste collection units was 99 402 in 2020.[27]

The discharge of food waste using the RFID-based system is shown in Figure 6.2. Consumers discharging food waste can open the RFID-based food waste collection equipment to discharge the food waste by contacting the RFID card. The RFID-based food waste collection equipment automatically measures the amount of discharged waste and informs consumers of the discharge fee at the end of each month. Information on food waste is sent to the Korea Environment Corporation and used as a basis for policy design. Local government is responsible for installation and operation of the RFID food waste collection equipment, and database management based on the IT system for food waste is done by the integrated central system in the Korea Environment Corporation (KECO).[28]

Even though the VBRS on MSW was implemented in 1995, food waste in Korea was not separated from MSW: food waste was put into the SWB and discharged along with MSW. Because the reduction of food waste was activated by the Korean government's management strategy from 1993, the amount of food waste subsequently decreased.[29] The amount of food waste generated decreased from 0.45 kg *per capita* per day in 1993 to 0.28 kg *per capita* per day in 1997. Through the implementation of the VBRS, consumers have learned sound practices to reduce food waste. However, it was hard to recycle the

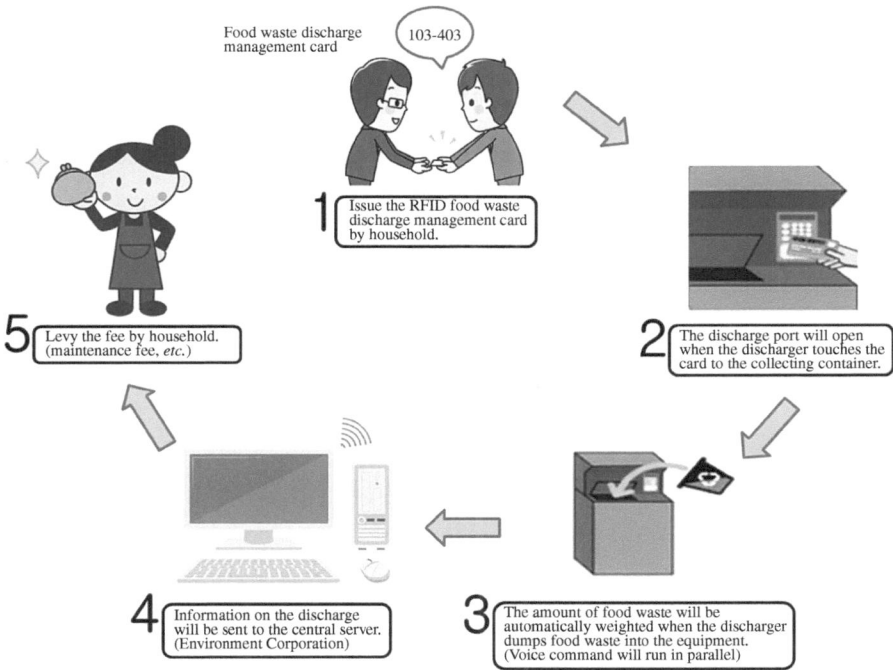

Figure 6.2   Discharge of food waste using RFID system.

discharged food waste in the SWB because it also contained other foreign substances such as plastic. Until 1997, the recycling rate of food waste in Korea was very low, at less than 10%. In order to expand the recycling of food waste, all local governments in Korea amended their ordinances in 1998 to implement separate collection of food waste, with the cost of collection, transportation and recycling paid by local governments.[30] As a result, the recycling rate of food waste increased significantly from 9.8% in 1997 to 93.3% in 2005. Table 6.4 summarises the changes brought about by the VBRS.

It is against the Polluter Pays Principle for local governments to bear the cost of food waste treatment. In February 2010, the government announced the "Comprehensive Measures to Reduce Food Waste", which contains the VBRS on food waste. According to this measure, the VBRS on food waste was fully implemented in June 2013 after a 2 year pilot project. By separating and collecting food waste by VBRS, the quality of food waste has improved, and recycling has become easier, enabling consumers to continue reducing food waste and discharging it separately responsibly and economically under the VBRS.

Recyclable wastes excluded from the VBRS are separately discharged and collected. In Korea, a system of separate discharge of waste has been implemented since 1991 in accordance with the Act on the Promotion of Saving and Recycling of Resources. The purpose of the act is to contribute to the preservation of the environment and sound development of the national economy by facilitating the use of recycled resources with means of controlling the generation of wastes and facilitating recycling.[31] The act stipulates that the waste generator can separate and store the waste by type, material and condition so that it can be recycled more effectively. In order for consumers to properly separate discharged waste in Korea, the central and local governments established guidelines for recyclable wastes, and manufacturers are obliged to label their products with a separate discharge mark.

Table 6.4 Generation and treatment of food waste in Korea.

Year	Generation (tonnes per day)	Recycling (tonnes per day) [%]	Incineration (tonnes per day) [%]	Landfill (tonnes per day) [%]	Waste generation per capita (kg per capita per day)
1993	19 764	N/A[1]	N/A	N/A	0.45
1995	15 075	316 [2.1]	372 [2.5]	14 387 [95.4]	0.34
1997	13 063	1275 [9.8]	815 [6.2]	10 973 [84.0]	0.28
2000	11 434	5161 [45.1]	1088 [9.5]	5185 [45.4]	0.25
2005	12 977	12 104 [93.3]	516 [4.0]	356 [2.7]	0.27
2010	13 671	13 055 [95.5]	422 [3.1]	194 [1.4]	0.27
2015	15 340	13 861 [90.4]	1089 [7.1]	390 [2.5]	0.30
2020	15 369[2]	13 809 [90.7]	1140 [7.4]	421 [2.7]	0.30[2]

[1] Not available.
[2] Intermediate disposal of waste such as anaerobic digestion is not included (138 tonnes per day).

Consumers should separate and discharge by removing other contents and foreign substances and reducing the volume as much as possible. In addition, it is necessary to prevent the separated waste from getting wet in the rain or being contaminated by other substances.[32] In Korea, separately discharged wastes are divided into 11 types and 18 sub-items. However, if separation and collection of wastes is not possible due to regional characteristics, some items may be discharged together. For example, in rural areas where waste generation is low, paper types can be classified into cartons and non-cartons (newspaper, booklets, corrugated boxes, *etc.*). Types and separately discharged items for recyclable wastes in the guideline are shown in Table 6.5.

In order for consumers to identify the types and the materials of waste and to discharge them properly, the separate discharge mark must be labelled according to certain standards.[33] Table 6.6 is a compilation of the relevant representations. The separate discharge mark includes the type of waste (inside the triangular symbol) and the material (bottom of the symbol). The colour of the triangle matches the type of waste, making it easy for identification. For example, in the case of plastic with other materials, the phrase "Plastic" inside the triangle figure and the phrase "Other" at the bottom of symbol are displayed. If other materials are coated or bonded to the

Table 6.5 Types and separately discharged items in Korea.

Type	Separately discharged items
Paper	Paper pack (sterilisation pack)
	Newspaper
	Booklets, notes, flyers
	Paper cup
	Other paper
Corrugated cardboard	Corrugated boxes, *etc.*
Glass bottle	Glass bottle, others
Metal cans	Beverage cans, food cans
	Other cans (butane gas, pesticide containers, *etc.*)
Colourless PET bottle	Colourless PET bottle
Foamed synthetic resin	Styrofoam
Synthetic resin containers, trays	Containers and packing materials such as PET, PVC, PE, PP, PS, PSP
	PET containers and trays, excluding colourless PET bottles
Vinyl	Plastic packaging material, disposable plastic bag
Clothing and fabrics	Cotton clothing, other clothing
	Vegetable fibres (cotton, hemp, *etc.*), animal fibres (wool, *etc.*), synthetic fibres (polyester, nylon, acrylic, polyurethane, *etc.*) and other synthetic fibres
Batteries	Mercury batteries, silver oxide batteries, nickel/cadmium batteries, lithium primary batteries, manganese batteries, alkaline manganese batteries, nickel hydrogen batteries
Fluorescent lamp	Lighting products containing mercury

PE, polyethylene; PP, polypropylene; PS, polystyrene; PVC, polyvinyl chloride.

Table 6.6 Standard for the separate discharge mark in Korea.

Type	Text	Colour	Material	Example of labelling
Plastic	Transparent PET	Yellow	Bio, Discharge by VBRS[1]	
	Plastic	Blue	PET, HDPE, LDPE, PP,	
	Vinyl	Purple	PS, Others, Bio-PET, Bio-HDPE, Bio-LDPE, Bio-PP, Bio-PS, Discharge by VBRS[2]	**OTHER**
Metal	Cans	Grey	Iron, aluminium	**IRON**
Paper	Paper	Black	–	
	Carton pack	Green	–, Discharge by VBRS	
	Sterile pack	Blue-green	–, Discharge by VBRS	
Glass	Glass	Orange	–	
Others	–	Red	–	

[1] Discharge by VBRS refers to the case where components other than the body are coated or bonded.

[2] HDPE, high-density polyethylene; LDPE, low-density polyethylene; PP, polypropylene; PS, polystyrene.

components of products and packaging materials, consumers cannot separate the wastes by themselves. Since coated or bonded waste is difficult to recycle, the separate discharge mark for the wastes is indicated by a red triangle with a slash mark through it.

Through the implementation of the separate discharge system, MSW management in Korea has changed in a very positive direction from an environmental point of view.[29] In 1989, before the separate discharge system was implemented, the amount of waste generated was very high at 1.84 kg *per capita* per day. When the separate discharge system was implemented, the amount of waste generated decreased significantly to 0.99 kg *per capita* per day by 2015. Recently, the ratio of single-person households in Korea has increased significantly from 27.2% in 2015 to 31.7% in 2020.[34] Consumption practices changed as a result, and the amount of MSW generation gradually increased from 0.99 kg *per capita* per day in 2015 to 1.12 kg *per capita* per day by 2019. In particular, the amount of MSW generation significantly increased to 1.19 kg *per capita* per day in 2020. It is speculated that waste generation has increased due to the increase in personal protective equipment (PPE) waste caused by the COVID-19 pandemic and the use of packaging materials such as delivery and courier.[35] In addition, the separate discharge mark in some products can support an increase in the efficiency of recycling. In 1989, the recycling rate of MSW was 1.2%, and most of it was disposed of in landfills. The recycling rate in 2020 is 61.5%, and the amount of waste disposed of is significantly reduced. Figure 6.3 summarises the achievements from 1989 to 2020.

Implementing the VBRS and the separate discharge system for responsible consumption by consumers is very effective in reducing and reusing waste. However, responsible consumption is not easily achieved. Consumers should understand each system and make efforts to fulfil their responsibilities. Education and advertisement must be promoted to encourage consumers to fulfil their responsibilities. In addition, the realisation of consumer responsibility requires the establishment of infrastructure such as collection services as well as consumer education. Details on factors of realising consumer responsibility are covered in Section 6.4.

6.3.3   Collection for Recycling of Waste

Even if consumers support responsible consumption, they cannot fulfil their responsibilities if social infrastructure such as waste collection is not established. Among the wastes generated by consumers, electrical and electronic waste (e-waste) requires special management because it contains not only

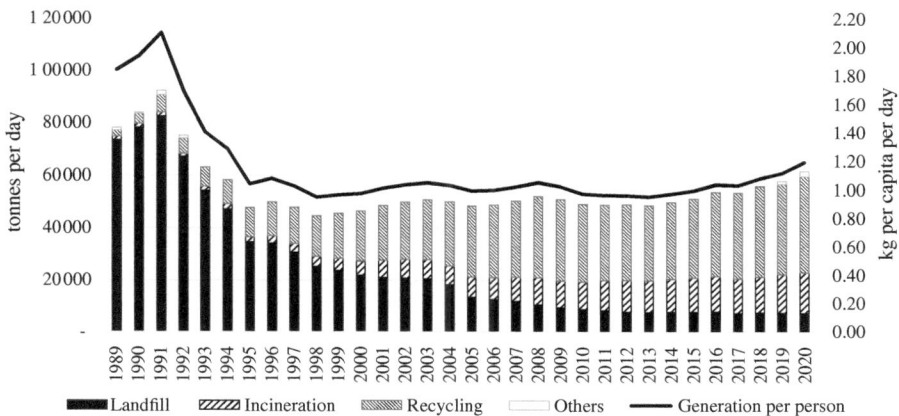

Figure 6.3   Generation and treatment of MSW in Korea.

hazardous substances such as heavy metals and refrigerants, but also resources such as gold, silver, copper and lithium.[36] In this section, a collection system for recycling is introduced, focusing on e-wastes that are not easy to discharge and collect.

E-waste in Korea began to be managed by the Extended Producer Responsibility (EPR) system in 2003; since 2008 it has been separated from the EPR and managed by the Eco-Assurance System (Eco-AS). The purpose of Eco-AS for e-waste management is to minimise environmental loads through systematic management of the entire life cycle of electrical products and electronic devices, from design and production to disposal, in order to reduce wastes and promote recycling activities.[37] The collection of e-waste in Eco-AS is operated by municipal collections, a take-back system and door-to-door systems. Through the operation of these collection systems, the amount of recycled e-waste in Korea has significantly increased from 58 000 tonnes in 2003 to 377 000 tonnes in 2020.[38,39] Municipal collection is a method without any fee in which consumers can discharge e-waste to a collection box at a specific date and site managed by a local government and Producer Responsibility Organisation (PRO). The take-back system is a method in which waste electronic products are returned to producers or retailers when consumers purchase new products. In particular for large electronic products, manufacturers or retailers visit the home for installation. The take-back system involves the collection of old electronic products free of charge by manufacturers or retailers who visit the home.

Discharging large-size e-waste from home to a recycling centre or collection point is very difficult. The door-to-door system is a method through which consumers request the collection of e-waste from producers or a PRO.[40,41] The procedure of door-to-door system is shown in Figure 6.4. When consumers discharge of e-waste, they request the collection of e-waste to institutions operated by the PRO. Collection reservations can be requested through various channels such as online or by phone. The producer or retailer visits the consumer's residence to collect the e-waste free of charge. This system is utilised mostly by consumers because it helps consumers to avoid buying new products. The amount of e-waste collected by the door-to-door system increased from 1.22 million units (49 000 tonnes) in 2016 to 3.68 million units (98 000 tonnes) in 2020.[39]

Korea is a peninsula with 3348 islands.[42] Among the islands, there are 54 inhabited islands, where 49 916 people reside. Residents of small islands cannot fulfil their consumer responsibilities because they do not have e-waste collection services. In addition, e-waste left on the island flows into the ocean, causing serious marine pollution. To solve these problems, Korea started the "E-waste in island" shipping project in 2014. Initially, the project targeted residents of 11 islands; it expanded to 34 islands in 2020.[39] The results of the project are summarised in Figure 6.5.

A total of 3254 tons of e-waste was collected from 2014 to 2020.[39] By collecting e-waste discharged from the island, residents can be relieved of the inconvenience of disposal and implement the responsibility of consumers. As a

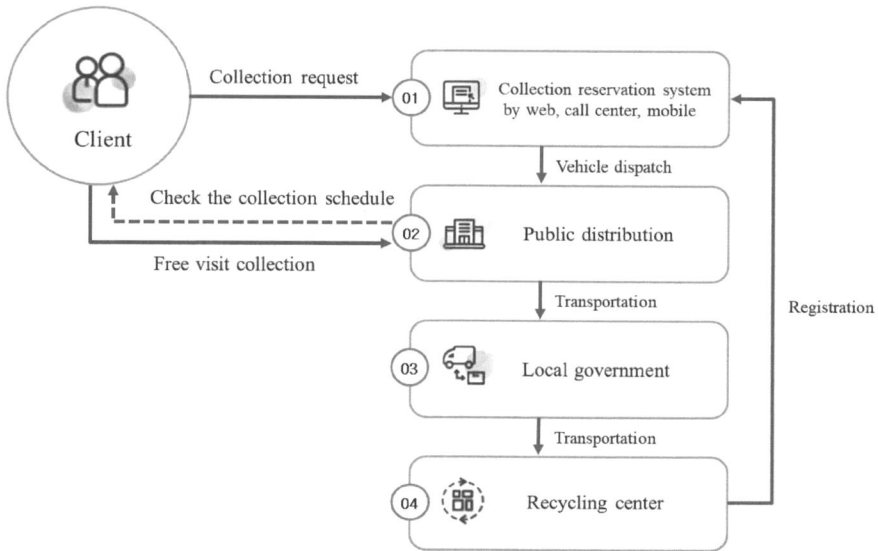

Figure 6.4   The process of a door-to-door collection system.

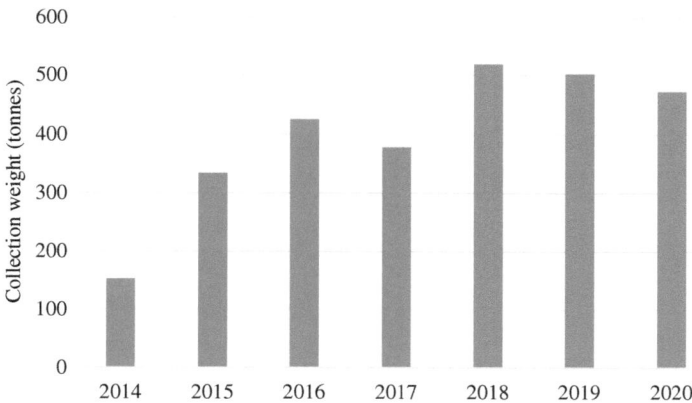

Figure 6.5   The result of e-waste in island shipping project.

result, consumers in island regions can build a clean island that prevents environmental pollution from e-waste.

6.4 Key Success Factors of Responsible Consumption in a Circular Economy

Responsible consumption in the circular economy requires not only fulfilling the responsibility of consumers, but also the cooperation of various stakeholders such as central governments, local governments, manufacturers and sellers of goods and services. Each stakeholder is an executor in each process of the circular economy, and a cycle is formed when the stakeholders fulfil their responsibilities. The formation of a circular cycle serves as a driving force for consumers to continuously implement responsible consumption. In this section, six key success factors for realising responsible consumption are introduced, as shown in Figure 6.6.

Six key factors were examined in order to realise responsible consumption in the circular economy. It is important for consumers to recognise the meaning of circular economy and healthy consumption behaviour as consumers. On the other hand, governments must set clear policy targets for the circular economy and allocate responsibilities to each stakeholder. Manufacturers must provide information on goods and services transparently, and consumers must be aware of their rights and responsibilities. In addition, all stakeholders must communicate to promote responsible consumption, and infrastructure must be established for consumers to fulfil their responsibilities. These key factors can create an environment in which consumers can act responsibly and create environmentally sound practices.

Responsible consumption requires setting clear policy targets. Establishing clear policy targets is the most important prerequisite for effective policy implementation.[43] Clear policy targets mean that not only the "goal to achieve" but also the "strategy for promotion" must be specific and clear. Policy targets must be quantitatively measurable so that the effectiveness of the policy can be assessed. Evaluation of responsible consumption policy in the circular economy can utilise indicators such as *per capita* usage of single-use products, tonnage of separate discharge and tonnage of waste generation.

Stakeholder responsibilities must be clearly allocated to ensure responsible consumption. A clear mechanism must be established to identify all

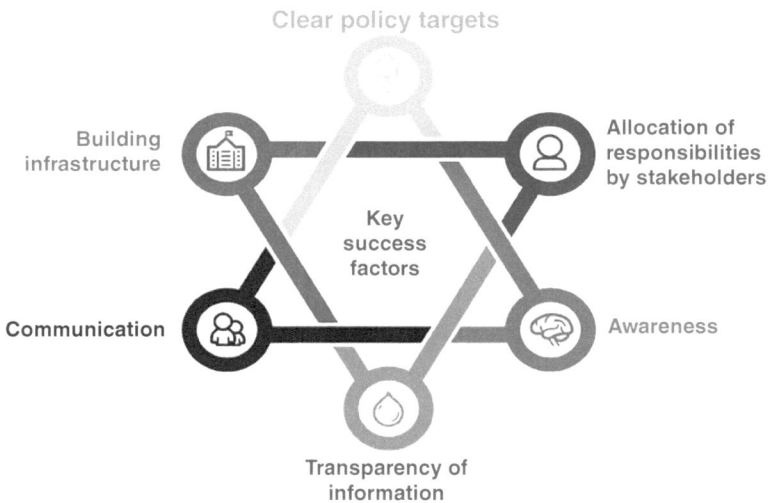

Figure 6.6   Key success factors of responsible consumption.

stakeholders who have legal obligations to protect the rights of consumers and fulfil their responsibilities. Stakeholders include consumers, manufacturers, retailers, central governments, local governments and waste collection operators. Governments should establish a master plan for consumer policies and operate an education and support system for consumers. Consumers must fulfil the responsibilities noted in Section 6.1. Manufacturers and retailers must faithfully and accurately provide information about goods to consumers and comply with the quality standards set by the government to fulfil the rights of consumers. In addition, waste collection operators must ensure that dust and leachate are not generated when collecting and transporting waste.

Consumers should know their rights and be aware of the adverse effects of illegal or inappropriate consumption on human health and the environment. Consumer awareness can be raised through educational programmes. Consumer education can increase awareness of products made from renewable resources. This is a driving force for manufacturers to actively utilise recyclable materials. In addition, consumer education includes product reuse and correct separation of household waste; as a result, it has a great impact on the upstream and downstream industries of the consumption stage in the circular economy.[44] In the past, consumer education was conducted in analogue ways such as promotional materials and brochures. Recently, digital media such as video media and online platforms have been used to improve educational accessibility. In addition, consumer education through digital media is being tailored by identifying targets such as age and education level. These educational programmes should be designed to provide user-friendly content so that consumers can easily access the information for responsible consumption.

Information about products, businesses and waste management is a basic requirement for consumers to make decisions. Therefore, information must be delivered transparently and openly to consumers. However, some businesses are confusing consumers in their choices by introducing products that pretend to be eco-friendly, namely greenwashing, defined as:[45]

> [T]he practice of falsely promoting an organisation's environmental efforts or spending more resources to promote the organisation as green than are spent to engage in environmentally sound practices. Thus, greenwashing is the dissemination of false or deceptive information regarding an organisation's environmental strategies, goals, motivations, and actions.

Since greenwashing occurs extensively, it is not efficient for the government to intervene and control the market on all issues. Therefore, to secure transparency of information, it is necessary to monitor greenwashing businesses by acquiring knowledge and information not only from the government but also from consumers.

With respect to achieving a circular economy, consumers cannot achieve it alone because it includes the entire process from manufacturing to recycling. Therefore, each stakeholder in the circular economy must communicate to achieve responsible consumption. Communication tools can vary depending on the method of stakeholder participation.[46] Table 6.7 lists the tools available for communication.

The communication tools can be expanded in the order of information disclosure (step 1), consultation (step 2), participation (step 3) and collaboration (step 4) according to the degree of stakeholder participation. Finally, memoranda of understanding and partnerships can be concluded between consumer organisations and manufacturers to achieve responsible consumption. Communication tools help consumers to control environmentally unscrupulous businesses, motivate them to operate in an environmentally sound manner and achieve a circular economy through rational exchange of information.

In order for consumers to fulfil their responsibilities, tangible and intangible infrastructure must be provided. Tangible infrastructure refers to the physical, material structures and facilities that are necessary for the functioning of a society or economy. Examples of tangible infrastructure include bins, waste collection equipment, waste recycling facilities and product analysis equipment. Intangible infrastructure refers to the non-physical system, processes and institutions such as consumer policies, consumer education systems, waste collection systems and consumer networks. Tangible and intangible infrastructure can be utilised in each stage of the circular economy. In the stage of purchasing for consumption, the government must have facilities to test and inspect the quality and safety of goods or services to ensure consumers' right to safety. Manufacturers and retailers must also build infrastructure to provide information about their goods and services. In the discharge stage for reuse and reduction of waste, the government should install waste bins so that consumers can easily discharge their waste and provide educational infrastructure for correct behaviours. In the collection stage for waste recycling, bulky waste collection services and in-island waste shipping projects should be considered and implemented. Through the EPR system

Table 6.7   Communication tools for responsible consumption.

Type	Tools			
	Indirect	Slightly indirect	Slightly direct	Direct
Disclosure of information	Exhibition	Public hearing	On-site disclosure and on-site investigation	Collecting opinions on the Internet
Consultation	Tea party	Fieldwork and surveys	Suggestion box	Lecture, seminar
Participation	Advisory committee	Workshop	–	Conference
Collaboration	Collaborative workshop	–	Forum	Memorandum of understanding, partnership

and Eco-AS, manufacturers and retailers must build infrastructure to collect waste. It is desirable to integrate these tangible and intangible infrastructures so that consumers can easily access them. In Korea, a consumer support system, including consumer education and damage relief services, is operated that includes information on all stages from purchase to discharge.

6.5   Conclusions

In a circular economy and to achieve the goals of SDG 12, consumers take responsibility for their consumption as an important role between upstream and downstream industries in the supply chain. Consumers have the responsibility to exercise their rights, to acquire knowledge and information, to conserve resources and to consume environmentally friendly products. The stages in which consumers fulfil their responsibilities are divided into purchasing for consumption, discharge for reuse and reduction of waste, and the collection of waste for recycling. Various strategies can be used at each stage to fulfil consumer responsibilities.

There are three strategies for responsible consumption in the purchasing stage for consumption: legal restrictions on consumption of environmentally unsound products; economic measures, such as deposit systems and taxes; and encouraging appropriate consumption practices. These strategies are very effective in reducing and recycling waste, and ultimately in reducing greenhouse gases.

At the discharge stage for waste reduction and reuse, the Volume-based Rate System (VBRS) and separate discharge systems can be implemented. In Korea, the VBRS, which imposes a fee according to the amount of waste generated, is applied to MSW, bulky waste and food waste, but not industrial waste. Consumers endeavour to reduce and reuse waste and compress waste at the discharge stage in order to avoid waste fees under the VBRS. Recyclable waste that is not subject to the VBRS is separately discharged by type and material. Korean consumers have fulfilled their consumer responsibilities and have achieved remarkable results in reducing waste through these two systems. The amount of MSW generated in Korea has significantly decreased from 1989 to 2015, but has increased from 2015 to 2020 due to changes in consumption practices associated with a rise in single-person households, and with the COVID-19 pandemic.

Owing to geographic and physical factors, in some special cases consumers themselves may not be able to discharge waste properly and bespoke collection systems must be in place to fulfil consumer responsibilities. In Korea, collection of e-waste is carried out by municipal collection, take-back systems and door-to-door systems whereby manufacturers and retailers visit homes to collect e-waste. In addition, the Korean government supports consumer responsibility by actioning e-waste in-island shipping projects. By applying these collection systems, the collection of e-waste has increased significantly, contributing to cleaner islands.

The circular economy is a system created not only by consumers but also by other stakeholders. Therefore, all stakeholders including consumers must fulfil their respective responsibilities. In order to realise responsible consumption, various factors must be considered. There are six key success factors to make responsible consumption: clear policy targets, allocation of responsibilities by stakeholders, awareness, transparency of information, communication and building infrastructure. The key success factors can create an environment in which consumers can act responsibly and create environmentally sound practices.

In order to action SDG 12 and to establish a circular economy society, it should be emphasised that the roles and responsibilities of consumers among stakeholders are very important in reducing and recycling wastes at the stage of consumption and discharge, supported by awareness programmes such as education programmes through digital media, communication tools and partnerships. It is hoped that all stakeholders work together to reduce, reuse and recycle waste, building strong relationships through both domestic and international cooperation.

References

1. S. W. Rhee and D. K. Min, *J. Korean Soc. Waste Manage*, 2000, **17**, 537.
2. H. J. Song, *Uniform Definition of the Concept of the Consumer and the Business for Cross-border Consumer Contracts: Focus on International Contract Laws, EU Directives and the Laws of the European Countries*, Dong-A Law Review, 2017.
3. S. Kaza, L. Yao, P. Bhada-Tata and F. Van Woerden, *What a Waste 2.0: A Global Snapshot of Solid Waste Management to 2050*, World Bank Publications, New York, 2018.
4. S. W. Rhee, S. K. Choi and D. K. Min, *J. Korean Soc. Waste Manag.*, 2000, **17**, 489.
5. Consumers International, How are consumer rights defined? World consumer day: history and purpose, London, 2009, http://www.consumers international.org (accessed Nov 2022).
6. K. Šajn, Right to repair, European Parliament Research Service, Brussels, 2022, https://www.europarl.europa.eu/RegData/etudes/BRIE/2022/698869/ EPRS_BRI(2022)698869_EN.pdf (accessed Nov 2022).
7. Y. Gabriel and T. Lang, *The Unmanageable Consumer*, Sage, Thousand Oaks, 2nd end, 2006.
8. S. C. Kim and D. P. Jeong, *A Study on the Social Responsibilities and Roles of Consumers*, Korea Consumer Agency, Seoul, 2012.
9. U. Schrader, *Int. J. Innov. Sustain. Dev.*, 2007, **2**, 79.
10. E. Ulusoy, *J. Bus. Res.*, 2016, **69**, 284.
11. J. Morelli, *J. Environ. Sustain.*, 2011, **1**, 1.
12. H. J. Choi, Y. Choi and S. W. Rhee, *J. Korean Soc. Waste Manag.*, 2018, **35**, 709.
13. Ministry of Environment, *Korea, Establishment of a Korean Circular Economy – Implementation Plan for Carbon Neutrality*, Korean language, Seoul, 2021.

14. European Parliament, Directive 2019/904 of the European Parliament and of the Council of 5 June 2019 on the reduction of the impact of certain plastic products on the environment, *Official Journal of the European Union*, 12.62019, L 115/1, Brussels, 2019.

15. National Development and Reform Commission, *Opinions of the National Development and Reform Commission and the Ministry of Ecology and Environment on Further Strengthening Plastic Pollution*, 2020, https://perma.cc/A56P-KWPW (accessed Nov 2022).

16. Korea Environment Corporation (KECO), Operation and management of resource circulation system: beverage container deposit system, Incheon, 2022, https://www.keco.or.kr/en/core/operation_Beverage/contentsid/3086/index.do (accessed Nov 2022).

17. COSMO (Container Deposit System Management Organization), Deposit system for single-use plastic cups, 2022, https://www.cosmo.or.kr/home/sub.do?menuNo=43 (accessed Nov 2022).

18. Freiburgcup, ASF und Stadt Freiburg sagen Danke!, 2022, https://freiburg-cup.de/de/was.php (accessed Nov 2022).

19. European Parliament, *Commission Implementing Decision (EU) 2021/958 of 31 May 2021 laying down the format for reporting data and information on fishing gear placed on the market and waste fishing gear collected in Member States and the format for the quality check report in accordance with Articles 13(1)(d) and 13(2) of Directive (EU) 2019/904 of the European Parliament and of the Council*, 15.6.2021, L 211/51, Brussels, 2021.

20. Korea Packaging Recycling Cooperative (KPRC), Evaluation of material and structure, 2020, https://kprc21482.cafe24.com/html/eng/main/sub.htm?MN=03&PN=01 (accessed Nov 2022).

21. United Nations, *Glossary of Environment Statistics, No. 67*, United Nations Publications, New York, 1997.

22. D. K. Min and S. W. Rhee, in *Municipal Solid Waste Management in Asia and the Pacific Islands*, ed. A. Pariatamby and M. Tanaka, Springer, Singapore, 2014, pp. 173–194.

23. Y. J. Kim, *2011 Modularization of Korea's Development Experience: Volume-based Waste Fee System in Korea*, Ministry of Strategy and Finance, Seoul, Korea, 2012.

24. Y. Choi, H. J. Choi and S. W. Rhee, *Resources Recycling*, 2018, **27**, 3.

25. J. H. Seo and K. H. Jung, *J Korean Association for Policy Studies*, 2007, **16**, 147.

26. Ministry of Environment, Korea, Reducing food waste! The great action to save the earth, Seoul, 2014, http://www.me.go.kr/home/web/board/read.do?menuId=10181&boardMasterId=54&boardCategoryId=&boardId=357384 (accessed Nov 2022).

27. Korea Environment Corporation (KECO), *Present status of Project*, Incheon, 2022, https://www.keco.or.kr/en/core/waste_rfid/contentsid/1984/index.do (accessed Nov 2022).

28. Korea Environment Corporation (KECO), *Operation system*, Incheon, 2022, https://www.keco.or.kr/en/core/waste_rfid/contentsid/1984/index.do (accessed Nov 2022).

29. Ministry of Environment, Korea, *Statistics of National Waste Generation and Treatment 2020*, Korean language, Seoul, 2022.
30. S. I. Kang, *5-Year Plan for Recycling Food Waste*, National Archives of Korea, Seoul, 2007, https://www.archives.go.kr/next/search/listSubjectDescription.do?id=007056&pageFlag=&sitePage= (accessed Dec 2022).
31. Korea Legislation Research Institute, *Act on the Promotion of Saving and Recycling of Resources*, Act No. 17326, 2020a, https://elaw.klri.re.kr/kor_service/lawView.do?hseq=54858&lang=ENG (accessed Dec 2022).
32. Korea Legislation Research Institute, *Guidelines for Separate Discharge of Recyclable Resources*, Ministry of Environment Ordinance No. 1462, 2020, https://www.law.go.kr/LSW//admRulLsInfoP.do?chrClsCd=&admRulSeq=2100000192190#AJAX (accessed July 2023).
33. Ministry of Environment, Korea, *Guidebook on Separated Discharge Marking*, Seoul, 2021.
34. Korean Statistical Information Service (KOSIS), Percentage of single-person households (city/province/city/county/district), Seoul, 2022, https://kosis.kr/statHtml/statHtml.do?orgId=101&tblId=DT_1YL21161&conn_path=I2 (accessed Dec 2022).
35. S. W. Rhee, *Waste Manag. Res.*, 2020, **38**, 820.
36. S. W. Rhee, *Procedia Environmental Sciences*, 2016, **31**, 707.
37. Korea Environment Corporation (KECO), Eco-assurance system, Seoul, 2022, https://www.keco.or.kr/en/core/operation_eco/contentsid/1978/index.do (accessed Nov 2022).
38. Korea Environment Corporation (KECO), *Extended Producer Responsibility Enforced 13 Years*, (Assessment on result of operation), Internal document, 2017.
39. Korea Electronics Recycling Cooperative (KERC), *2020 KERC Operation Report*, 2021, 17 Daehak 4-ro, Suwon, 16226, Republic of Korea.
40. J. Park, C. Ahn, K. Lee, W. Choi, H. T. Song, S. O. Choi and S. W. Han, *Resources, Conservation and Recycling*, 2019, **144**, 90.
41. Korea Electronics Recycling Cooperative (KERC), Main contents of business, 2022, https://15990903.or.kr/portal/cnts/selectContents.do?cntnts_id=A1000002 (accessed Dec 2022).
42. Korean Statistical Information Service (KOSIS), Population and area of inhabited islands, Seoul, 2022, https://kosis.kr/statHtml/statHtml.do?tblId=DT_75001_A001025&orgId=750&language=kor&conn_path=&vw_cd=&list_id= (accessed Dec 2022).
43. A. Hindmoor, *Parliamentary Affairs*, 2005, **58**, 272.
44. S. Ishak and N. F. M. Zabil, *Asian Social Science*, 2012, **8**, 108.
45. K. Becker-Olsen and S. Potucek, in *Encyclopedia of Corporate Social Responsibility*, ed. S. O. Idowu, N. Capaldi, L. Zu and A. Das Gupta, Springer, Berlin/Heidelberg, 2013, pp. 1318–1323.
46. H. J. Kim, *Regional Governance and Roles by Subjects*, Gwangju University, 2021, https://www.city.go.kr/data/pdf/05.03.pdf (accessed Dec 2022).

CHAPTER 7

Plastics: Sustainable Development Goals and Circular Solutions

FABIULA D. B. DE SOUSA*

Technology Development Center, Universidade Federal de Pelotas, Rua Gomes Carneiro 1, 96010-610 Pelotas – RS, Brazil
*E-mail: fabiuladesousa@gmail.com

7.1 Introduction

Plastics are ubiquitous. They are part of our lives, and impossible to ban or to replace in some sectors of modern society. Plastics make our lives easier and safer. For example, some of their properties, such as low density, make them useful as packaging material.

From a particular point of view, there is no more outstanding example of the benefit of plastic to society than during the COVID-19 pandemic. Since the first cases of pneumonia were reported in Wuhan, China, in December 2019, life in society has completely changed, with the World Health Organization (WHO) declaring COVID-19 as a pandemic in early 2020.[1] Hundreds of researchers worldwide mobilized in the search for new vaccines; meanwhile, populations lived with social distancing and lockdowns. To prevent the transmission of the SARS-CoV-2 virus, the use of personal protective equipment (PPE), especially face masks, was mandatory in many places: "Polymer materials have become the first line of defence against the virus and its further spread".[2]

Issues in Environmental Science and Technology No. 51
The Circular Economy: Meeting Sustainable Development Goals
Edited by Sadhan Kumar Ghosh and Gev Eduljee
© The Royal Society of Chemistry 2024
Published by the Royal Society of Chemistry, www.rsc.org

In thousands of hospitals, people fought for their lives in intensive care units, with the help of various devices that were made of plastic or have plastic parts in their structure such as respirators, heart monitors, and blood pressure monitors; flow meters, controllers, and sensors for saline solutions, medicines, and food; oximeters; thermometers; and others no less important such as tubes, suction and urinary probes, catheters, and diapers.[3,4] When patients died, their bodies were packed in at least one waterproof bag – made of plastic.

Just as plastic affects our lives in society, it also affects the development and achievement of the 2030 Agenda for Sustainable Development. In 2015, at the Rio+20 Conference, the 2030 Agenda for Sustainable Development was discussed, resulting in an action plan containing 17 Sustainable Development Goals (SDGs), mainly aimed at eradicating poverty.[5] The SDGs are[6] no poverty (SDG 1); zero hunger (SDG 2); good health and wellbeing (SDG 3); quality education (SDG 4); gender equality (SDG 5); clean water and sanitation (SDG 6); affordable and clean energy (SDG 7); decent work and economic growth (SDG 8); industry, innovation, and infrastructure (SDG 9); reduced inequality (SDG 10); sustainable cities and communities (SDG 11); responsible consumption and production (SDG 12); climate action (SDG 13); life below water (SDG 14); life on land (SDG 15); peace, justice, and strong institutions (SDG 16); and partnerships to achieve the goals (SDG 17). Further details are provided in Chapter 2.

The global production of plastics has increased over the years, as have their consumption and applications. Global plastics production reached 368 million tonnes in 2019.[7] For instance, in China, the use of some polymers such as polyvinyl chloride (PVC), polystyrene (PS), polyethylene (PE), polypropylene (PP), and acrylonitrile-butadiene-styrene (ABS) increased considerably from 1 kg *per capita* in 1978 to 46 kg *per capita* in 2017.[8]

Considering the total production of plastics, around 49% are used to produce single-use items.[9] Around 40% of the consumption of all plastic material produced globally is accounted for by the packaging industry.[10] However, single-use and packaging items are chemically stable materials with a short service life, resulting in plastics creating a major environmental problem owing to incorrect waste management practices prevalent in many places worldwide. As an example, a plastic bottle made of high-density polyethylene (HDPE) with an approximate wall thickness of 500 μm will take 250 years to lose the first 50% of the polymer mass (half-life) when buried on land, or 58 years in a marine environment.[11]

It would thus appear that plastics and the SDGs are not aligned. The question posed in this chapter is as follows: how can we comply with the 2030 Agenda for Sustainable Development given that the use of plastics is so pervasive in modern society?

This chapter discusses the different roles plastic plays in relation to the SDGs, identifying their benefits and adverse effects. Commencing with a search performed in the Scopus database to obtain the relevant literature, a simplified bibliometric analysis and mapping were performed, with an emphasis on the authors' keywords to obtain an overview of the role of plastics in the SDGs.

7.2 Methodology

The methodology follows de Sousa,[12] as described below. To obtain the bibliographic data inputs, a Scopus search was conducted on 15 December 2022 using the keywords ("sustainable development goal*" OR "SDG*") AND ("plastic*" OR "polymer*"). Reviews and articles in English were considered.

Refinements to the search were first performed. The research area of biochemistry, genetics, and molecular biology have some terms similar to "SDG" but with entirely different meanings unrelated to the subject of the present work.[12] Therefore, this area was excluded. The Scopus search was also limited to articles and reviews in English from 2015 to 2022 since 2015 was the year in which the 2030 Agenda for Sustainable Development was discussed at the Rio+20 Conference.

A scopus.csv file containing the data was then recorded. The data contained in the file were analysed and mapped using Bibliometrix (an R software package) and VOSviewer version 1.6.18 software programs.

7.3 Results and Discussion

The Scopus search obtained by using the keywords ("sustainable development goal*" OR "SDG*") AND ("plastic*" OR "polymer*") resulted in 253 publications: 201 articles and 52 reviews. The number of publications per year is shown in Figure 7.1.

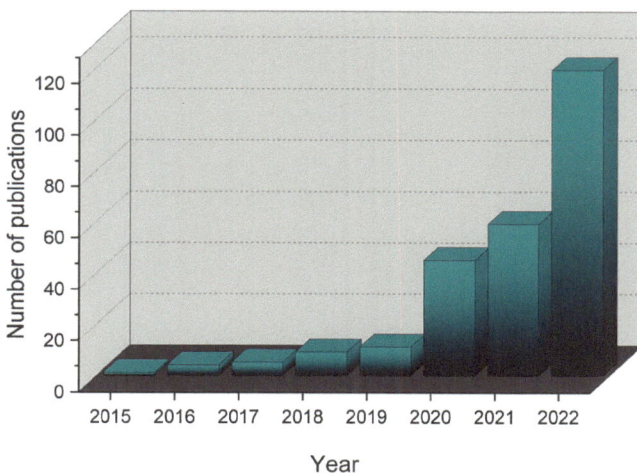

Figure 7.1 Number of publications per year.

The subject is interdisciplinary. The research areas with the highest number of publications were environmental science (118 publications), chemistry (56), energy (37), engineering (37), and medicine (35).

The search shows an increase in the number of publications over the period, with an annual growth rate of 97.7%, according to Bibliometrix. This increase has been more significant from 2020, the year in which COVID-19 was considered a pandemic by the WHO.[1] It is known that managing contaminated residues generated during the global pandemic was a challenge, and remains so. Achieving the SDGs was affected by the COVID-19 pandemic, especially SDG 6 (Clean water and sanitation) and SDG 14 (Life below water), which were deeply affected due to the improper disposal of PPE made of plastic, especially the disposal of face masks.

7.3.1 Bibliometric Analysis and Mapping

Since the authors' keywords are considered as the signal of the study's central questions,[13] an analysis of the authors' keywords in terms of their relevance was the focus of the present work. An in-depth study of the authors' keywords can elicit a panoramic view of a given research field, and can be used as the basis for extending the frontiers of knowledge by informing new data and information.[14]

The word cloud containing the 30 most frequent authors' keywords is presented in Figure 7.2. The size of the letters is proportional to the frequency of the keyword.

The most frequent authors' keywords, apart from the keyword "sustainable development goals" used in the Scopus search (the keyword is not present in the word cloud), are (number of occurrences in parentheses): circular economy (25), sustainability (20), COVID-19 (11), plastic pollution (10), life cycle

Figure 7.2   Word cloud containing the 30 most frequent authors' keywords.

assessment (9), waste management (9), environment (8), microplastics (8), plastic waste (8), and sustainable development (8).

By careful analysis of the word cloud, the judgement of the literature on plastics concerning the achievement of the SDGs is clear: due to the presence of the keywords "marine litter" and "marine pollution", plastic is a polluter, mainly of water bodies. Based on the literature, plastics are therefore the most substantial obstacle to the achievement of SDG 14.[12] Also in relation to SDG 14, the word cloud depicts the connection with the COVID-19 pandemic in the generation of plastic residues and the subsequent increase in water pollution due to the incorrect disposal of PPE made of plastic, mainly disposable face masks. The COVID-19 pandemic has resulted in such a significant increase in plastic pollution that it has been likened to a "plastic pollution pandemic". It has been argued that the plastic pandemic is a more significant threat to humankind than the COVID-19 pandemic,[15] with the socio-economic–ecological crisis caused by the pandemic delaying progress toward the SDGs[16] (see also Chapter 2).

The most frequent SDGs correlated with plastics are SDG 7 (Affordable and clean energy) and SDG 9 (Industry innovation and infrastructure). Concerning these SDGs, plastic plays a positive role.[12] Regarding SDG 7, the main pros of plastic are the support for obtaining energy from renewable sources, even considering the high costs involved. In the case of both SDGs, the pros are the research and development of new materials, and the cons are the difficulties inherent in research and development processes.[12] On the other hand, plastic has a negative role regarding SDG 6.

The word cloud also indicates some strategies to mitigate the problems caused by plastics, such as "waste management", "circular economy", "life cycle assessment", and "recycling". According to some studies,[17] the correct management of plastic waste directly affects the fulfilment of SDGs 3, 6, 11, 12, 13, 14, and 15. The keyword "recycling" is linked to "mechanical properties" because it is known that polymers are generally degraded during recycling, especially mechanical recycling, due to high shear rates and high temperatures. Some subjects present in the word cloud will be discussed in later sections.

The keyword co-occurrences network is presented in Figure 7.3. The network presents 35 keywords with a minimum number of occurrences of 3.

In the network, the size of the circles is proportional to the occurrence of the keyword. For example, the keywords "sustainable development goals" and "circular economy" have the highest frequency, in agreement with the word cloud in Figure 7.2.

In the network, eight clusters can be visualized, with 121 links and a total link strength of 186. Each cluster represents a direction in the analysed research field, and each has a different colour in the network.

- The red cluster has six items, representing the positive role of plastics concerning the achievement of the SDGs (some keywords are "SDG 7 (Affordable and clean energy)", "SDG 9 (Industry innovation and infrastructure)", and "environment").

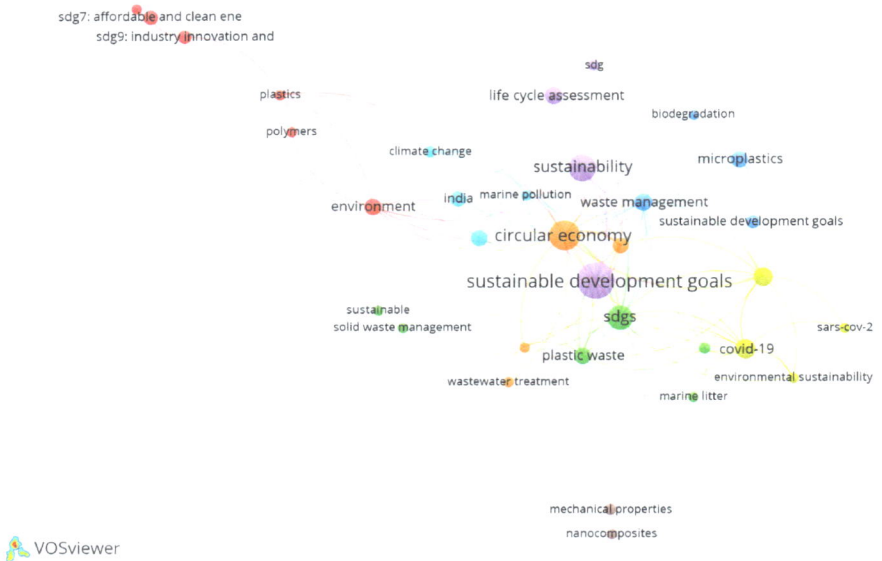

Figure 7.3   Keyword co-occurrences network.

- The green cluster depicts the importance of solid waste management in preventing plastic pollution and the consequent achievement of SDGs. This cluster has six items. Some keywords present in this cluster are "marine litter", "plastic waste", "SDGs", and "solid waste management".
- The blue cluster deals with the waste management of biodegradable plastics, with four items. The keywords present in this cluster are "biodegradation", "microplastics", "sustainable development goals", and "waste management".
- The light-green cluster is about plastic pollution due to the COVID-19 pandemic. It has four items: "COVID-19", "environmental sustainability", "plastic pollution", and "Sars-Cov-2".
- The purple cluster shows the life cycle assessment's importance in achieving SDGs. It has four items: "life cycle assessment", "SDG", "sustainability", and "sustainable development goals".
- The light-blue cluster deals with climate change. It has four items, two of them being "sustainable development" and "climate change".
- The orange cluster deals with the circular economy, with four items: "circular economy", "recycling", "resource recovery", and "waste water treatment".
- The brown cluster is about nanocomposites, with two items: "nanocomposites" and "mechanical properties".

As previously observed in the word cloud, "circular economy" has a high frequency of occurrence, being a relevant keyword in the role of plastics in relation to the SDGs. This keyword has 19 links, a total link strength of 46, with 25 occurrences, and it is linked to all the other clusters apart from the brown node. The keyword "sustainable development goals" has 19 links, a total link strength of 33, and 37 occurrences. There are other keywords with the same meaning as "sustainable development goals", such as "SDGs". However, the values previously cited refer to the keyword "sustainable development goals" (purple circle in the centre of the network), which has the highest number of occurrences.

The connections between the keywords "sustainable development goals" and "circular economy" with other keywords are shown in Figure 7.4.

In Figures 7.3 and 7.4, distances between the keywords depict relatedness in terms of co-occurrence links, that is, closer keywords point to closer relatedness.[18]

Regarding the keyword "sustainable development goals", it is important to observe the distance between the keywords:

- "sustainable development goals" (and/or "SDGs", green circle) and "plastic pollution";
- "sustainable development goals" (and/or "SDGs", green circle) and "pollution";
- "sustainable development goals" (and/or "SDGs", green circle) and "marine litter";
- "sustainable development goals" (and/or "SDGs", green circle) and "plastic waste";
- "sustainable development goals" (and/or "SDGs", green circle) and "climate change".

These distances are all less than the distance between the keywords "sustainable development goals" (and/or "SDGs", green circle) and "plastics". Bearing in mind that the red cluster presents some positive roles of plastic in the achievement of the SDGs, this result indicates that, in the literature, plastics have a more damaging impact than a positive one. In other words, the pollution caused by plastics (negative impact) outperforms the totality of their positive impact in achieving the SDGs.

In the same way, it is relevant to observe the distance between the keywords "sustainable development goals" and "COVID-19", and the distances between the keywords "COVID-19" and "pollution", "COVID-19" and "plastic pollution", and "COVID-19" and "marine litter". All these reduced distances identify COVID-19 as a massive barrier to achieving the SDGs, and a significant producer of polluting plastic waste due to the inappropriate disposal of PPE, resulting in a huge environmental challenge. For example, one study estimated that plastic pollution caused by the incorrect disposal of PPE in Bangladesh hindered the nation in achieving SDG 12.[19]

Concerning the keyword "circular economy", it is important to study the distances between the keywords "circular economy" and "recycling", "circular economy" and "marine pollution", and "circular economy" and

(a)

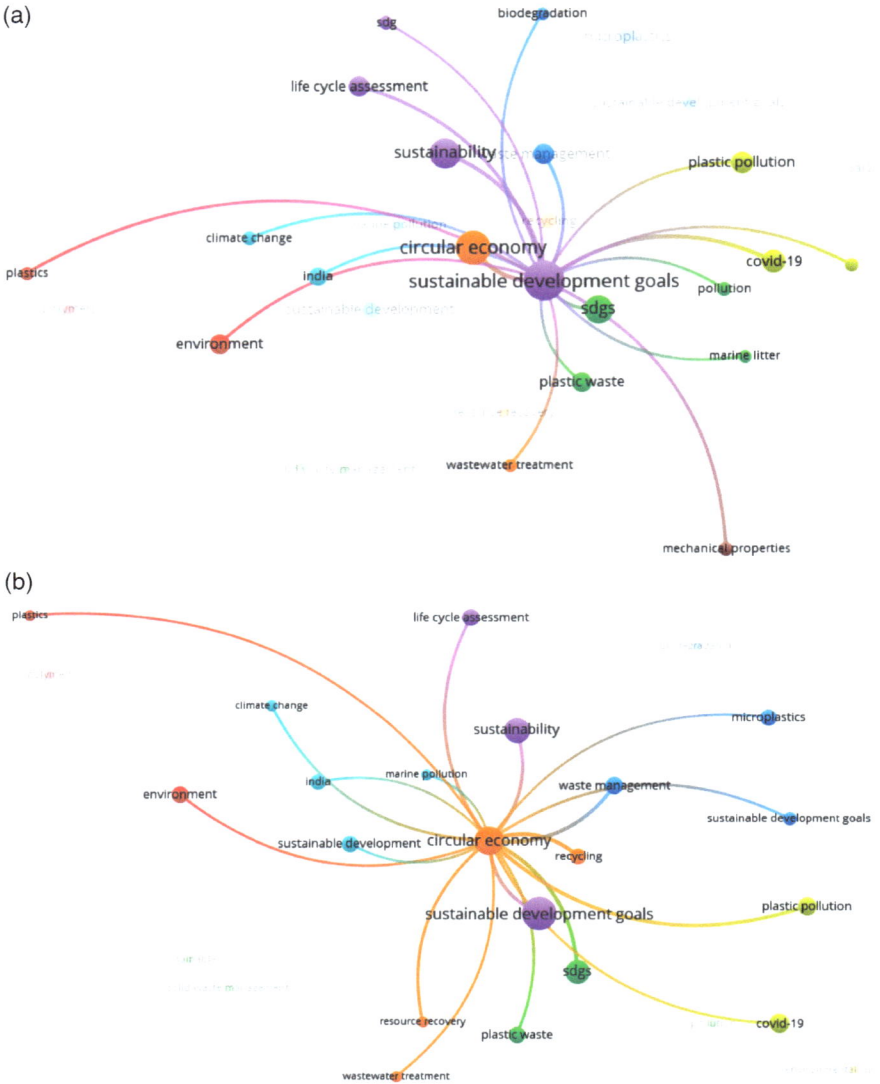

(b)

Figure 7.4   Connections of the keywords "sustainable development goals" (a) and "circular economy" (b) with other network keywords are presented in Figure 7.3.

"sustainable development goals". All these couplings have high relatedness and frequently co-occur in the publications. The literature considers the circular economy with strategies such as reduction and recycling as potential solutions to the socio-environmental problems that plastic may cause,[18] and as a key ally in achieving the SDGs, capable of mitigating/minimizing marine pollution.

The lines connecting the nodes in the network diagrams describe the strongest co-occurrences. Among all the lines connecting the keywords in the network present in Figure 7.4, the strongest is the line between the keywords "circular economy" and "recycling", showing their strong co-occurrence. Given the importance of the circular economy of plastics for achieving SDGs, these will be discussed later, based on the recent literature.

7.3.2 The Circular Economy of Plastic and the SDGs

As noted previously, plastics affect the development and achievement of the 2030 Agenda for Sustainable Development. Plastic presents pros and cons to the achievement of most SDGs. According to the current literature, the main pros and cons of plastic to SDGs 1–15 are presented in Table 7.1. Nevertheless, the role of plastic goes beyond the purposes mentioned in this chapter.

As observed in Table 7.1, plastic is essential for the achievement of several SDGs. For example, plastic reduces the fuel consumption of transportation in general, reducing the greenhouse gas emissions that contribute to climate change (SDGs 11, 13, and 15). Some other important applications of plastic are in the shelf-life extension of fresh fruits, vegetables, and animal-based proteins for safe food (SDGs 2 and 3), materials for medical applications (SDG 3), and membranes for water filtration (SDG 6), all of which present tangible positive contributions to the SDGs.[2]

On the other hand, the main detriment caused by the incorrect management of plastic residues is the pollution of water and its consequences, which is a barrier to achieving mainly SDGs 6 and 14. When considering pollution caused by microplastics, its impact is even worse, adversely affecting, directly or indirectly, at least 12 SDGs.[20] Consequently, the cons of plastic outweigh the pros concerning SDGs, as observed previously in the analysis of the word cloud and co-occurrence network in Figures 7.2–7.4. Nevertheless, Table 7.1 lists more positives than negatives.

As mentioned in Chapter 2, the origin of the term "circular economy" is unclear, and there is no consensus on its definition. In this sense, due to its varied definitions and consensual difficulty, it is also challenging to put it into practice. After analysing 114 definitions of circular economy, Kirchherr *et al.*[21] found the most prominent definition was the one developed by the Ellen MacArthur Foundation:[22]

> *An industrial system that is restorative or regenerative by intention and design. It replaces the 'end-of-life' concept with restoration, shifts towards the use of renewable energy, eliminates the use of toxic chemicals, which impair reuse, and aims for the elimination of waste through the superior design of materials, products, systems, and, within this, business models.*

By maintaining materials in circulation for as long as feasible through effective materials' usage that involves actions such as reduction, reuse, recycling, and recovering, as well as any necessary systemic change, the

Table 7.1 The role of plastic concerning the SDGs, according to the point of view of current literature, depicting its pros and cons.[12] Reproduced from ref. 12 with permission from Elsevier, Copyright 2021.

SDG	Benefits of plastic	Disbenefits of plastic
SDG 1	Generation of job and income Support for agriculture	Production and release of microplastics
SDG 2	Generation of job and income Support for agriculture Support for obtaining energy from renewable sources Water treatment, purification, disinfection, and decontamination	Production and release of microplastics High costs
SDG 3	Generation of job and income Evolution of medicine Support for agriculture Cheaper shoes for needy populations Avoid contamination through packaging Support for obtaining energy from renewable sources Water treatment, purification, disinfection, and decontamination Improved sanitation systems	Appropriate treatment of hospital waste Production and release of microplastics Skin diseases Migration of additives High costs
SDG 4	Environmental education (conscious use of the plastic, selective collection, recycling, *etc.*) Generation of job and income	
SDG 5	Generation of job and income Providing social integration Support for agriculture	Production and release of microplastics
SDG 6	Water treatment, purification, disinfection, and decontamination Improved sanitation systems	
SDG 7	Support for obtaining energy from renewable sources Research and development of new materials	High costs Difficulties inherent in research and development processes
SDG 8	Generation of job and income Providing social integration	

SDG	Benefits of plastic	Disbenefits of plastic
SDG 9	Research and development of new materials	Difficulties inherent in research and development processes
SDG 10	Generation of job and income Providing social integration Support for agriculture	Production and release of microplastics
SDG 11	Environmental education (conscious use of the plastic, selective collection, recycling, *etc.*) Research and development of new materials Support for obtaining energy from renewable sources Reduction of greenhouse gas emissions	Difficulties inherent in research and development processes High costs
SDG 12	Support for agriculture Use of packages Environmental education (conscious use of the plastic, selective collection, recycling, *etc.*)	Production and release of microplastics Socio-environmental problems when improperly disposed of Migration of additives
SDG 13	Reduction of greenhouse gas emissions Environmental education (conscious use of the plastic, selective collection, recycling, *etc.*) Support for obtaining energy from renewable sources Research and development of new materials	High costs Difficulties inherent in research and development processes
SDG 14	 Environmental education (conscious use of the plastic, selective collection, recycling, *etc.*)	Pollution of waters and its consequences
SDG 15	Research and development of new materials Environmental education (conscious use of the plastic, selective collection, recycling, *etc.*) Support for obtaining energy from renewable sources Reduction of greenhouse gas emissions	Difficulties inherent in research and development processes High costs

circular economy seeks to increase resource and energy efficiency.[2,21] Thus, among other activities, recycling is encouraged by the circular economy (Figure 7.4b).

Recycling can contribute significantly to the success of the SDGs by providing thousands of job opportunities. Employment is essential against extreme poverty (SDG 1), which, in turn, facilitates access to health care (SDG 3), gender equality (SDG 5) through women's empowerment in society, decent work and economic development (SDG 8), and reduction in inequality (SDG 10).[12] Avoiding the irregular and inappropriate disposal of plastic residues facilitates clean water and sanitation (SDG 6) and life below water (SDG 14); it supports industry, innovation, and infrastructure (SDG 9); assists sustainable cities and communities (SDG 11); combats climate change (SDG 13); encourages responsible consumption and production (SDG 12); and improves life on land (SDG 15). Recycling reduces dependence on fossil fuels and reduces the emission of greenhouse gases. In addition, recycling is an integral component of quality education (SDG 4).

Even considering all these benefits, the recycling rate of plastics worldwide is much lower than ideal. Around 6300 Mt of plastic waste had been generated up to 2015.[23] According to estimates, only 9% of the plastic garbage produced worldwide between 1950 and 2015 was recycled.[23,24] Data from 2018 indicate that the country that recycled the most was India (60%), and the country that recycled the least was the USA (9%).[25]

Some examples from the recent literature about the importance of recycling and of the use of plastics to achieve the SDGs are the following:

- Recycled plastic is used in the production of geopolymer concrete as green building materials[26] and in the production of reinforced concretes[27] with different types of recycled fibres such as polyethylene terephthalate (PET) from bottles[28] and bags,[29] low-density polyethylene (LDPE) from bags,[29] PVC from electronic waste,[30] PP,[31] and HDPE.[32] Comparing the mechanical performance of reinforced concretes with virgin PP and recycled PP, recycled PP fibre produced better reinforcement than virgin PP fibre.[31] According to the study, in the case of geopolymer concrete these recycled fibres contribute positively to 12 of the 17 SDGs.
- Studies have investigated the life cycle of plastics in non-durable goods, analysing four interventions for circularity and sustainability.[33] The highest potential for greenhouse gas emission reduction lies in plastic reduction, followed by replacement with recycled plastics, paper, and, finally, bioplastics.
- It is known that nutrient recovery is an important segment of the circular economy and a contributor to the SDGs. Wastewater is a significant source of nutrients. In one example,[34] an onsite pilot scale membrane contactor recovered ammonia and phosphorus from mesophilic digester reject water, using NPHarvest technology developed at Aalto University. Several parts of the system are made from plastic-based

polymers, such as in the pre-treatment unit where phosphorous and solids are removed through coagulation/flocculation.

- A project studied the utilization of molasses and sweet water from sugar cane and palm oil as carbon substrates for the polyester polyhydroxyalkanoate (PHA) and for biosurfactant production.[35] By-products from sugar cane and palm oil can be used to produce biosurfactants and PHA at a low cost. The use of agro-industrial residues to produce PHA and biosurfactant would significantly reduce the negative effects on the environment, improve the global economy, and reduce food waste, instilling the circular economy spirit in the agricultural industries in line with the SDGs. Another example is the effective valorization of lignin into valuable biosynthesized products, including PHA.[36]

Although not included in the analysis of keywords (Figures 7.2–7.4), the implementation of regulations and taxes relating to the use of items made of plastic, especially single-use products, is vital in achieving the SDGs and facilitating the transition toward a circular plastic economy.[37] Examples include the creation of the *impuesto verde* in Spain to reduce plastic residues[38] and the ban on plastic microparticles in personal care and hygiene items.[39,40] In some countries, it is illegal to use plastic microspheres in hygiene products.[41]

In addition to relying on the collective action of society concerning the correct management of plastic waste, another indispensable tool to achieve circularity and sustainability for plastics is environmental education.

7.4 Conclusions

The role of plastics in achieving the 2030 Agenda for Sustainable Development was identified as having both positive and negative impacts. The sheer scale of its use globally, coupled with poor disposal practices, results in plastics being considered a barrier to achieving some SDGs, mainly the ones relating to the quality and life in water (SDGs 6 and 14).

Noting the problems that the incorrect disposal of plastics may cause, the material contributes to the achievement of at least 15 SDGs, with the simplified bibliometric analysis and mapping of the authors' keywords highlighting the importance of a circular economy for plastic in achieving the 2030 Agenda for Sustainable Development. The COVID-19 pandemic has posed a considerable setback to achieving the SDGs due to the huge quantity of plastic residues generated and the consequences of their improper disposal.

References

1. WHO, WHO Director-General's opening remarks at the media briefing on COVID-19, 11 March 2020, https://www.who.int/dg/speeches/detail/who-director-general-s-opening-remarks-at-the-media-briefing-on-covid-19---11-march-2020 (accessed April 2020).

2. N. A. Tarazona, R. Machatschek, J. Balcucho, J. L. Castro-Mayorga, J. F. Saldarriaga and A. Lendlein, *MRS Energy Sustain.*, 2022, **9**(1), 28.

3. F. D. B. de Sousa, *Environ. Sci. Pollut. Res.*, 2021, **28**, 46067.

4. F. D. B. de Sousa, *Recycling*, 2020, **5**(4), 27.

5. R. Dias, *Sustentabilidade: Origem E Fundamentos Educação E Governança Global Modelo De Desenvolvimento*, Atlas Press, Brazil, 2015.

6. United Nations, 17 Goals to transform the world for persons with disabilities, New York, 2015, https://www.un.org/development/desa/disabilities/envision2030.html (accessed June 2020).

7. Plastics Europe, *Plastics: The Facts 2020*, https://www.plasticseurope.org/en/resources/publications/4312-plastics-facts-2020 (accessed April 2021).

8. X. Jiang, T. Wang, M. Jiang, M. Xu, Y. Yu, B. Guo, D. Chen, S. Hu, J. Jiang, Y. Zhang, and B. Zhu, *Resour. Conserv. Recycl.*, 2020, **161**, 104969.

9. O. S. Ogunola, O. A. Onada and A. E. Falaye, *Environ. Sci. Pollut. Res.*, 2018, **25**(10), 9293.

10. Plastics Europe, www.plasticseurope.org (accessed May 2021).

11. A. Chamas, H. Moon, J. Zheng, Y. Qiu, T. Tabassum, J. H. Jang, M. Abu-Omar, S. L. Scott and S. Suh, *ACS Sustain. Chem. Eng.*, 2020, **8**(9), 3494.

12. F. D .B. de Sousa, *Clean. Responsible Consum.*, 2021, **3**, 100020.

13. A. Nasir, K. Shaukat, I. A. Hameed, S. Luo, T. M. Alam, and F. Iqbal, *IEEE Access*, 2020, **8**, 133377.

14. M. Tripathi, S. Kumar, S. K. Sonker and P. Babbar, *COLLNET J. Sci. Inf. Manag.*, 2018, **12**(2), 215.

15. A. R. Shekhar, A. Kumar, R. Syamsai, X. Cai and V. G. *Pol, ACS Sustain. Chem. Eng.*, 2022, **10**(10), 3150.

16. H. B. Sharma, K. R. Vanapalli, B. Samal, V. R. S. Cheela, B. K. Dubey and J. Bhattacharya, *Sci. Total Environ.*, 2021, **800**, 149605.

17. R. Hossain, M. T. Islam, R. Shanker, D. Khan, K. E. S. Locock, A. Ghose, H. Schandl, R. Dhodapkar and V. Sahajwalla, *Sustain.*, 2022, **14**(8), 4425.

18. F. D. B. de Sousa, *Waste Manag. Res.*, 2021, **39**(5), 664.

19. P. Monolina, M. M. H. Chowdhury and M. N. Haque, *Heliyon*, 2022, **8**(7), e09847.

20. T. R. Walker, *Curr. Opin. Green Sustain. Chem.*, 2021, **30**, 100497.

21. J. Kirchherr, D. Reike and M. Hekkert, *Resour. Conserv. Recycl.*, 2017, **127**, 221.

22. Ellen MacArthur Foundation, *Toward the Circular Economy Vol 1,* Cowes, 2012, https://ellenmacarthurfoundation.org/towards-the-circular-economy-vol-1-an-economic-and-business-rationale-for-an (accessed December 2022).

23. R. Geyer, J. R. Jambeck and K. L. Law, *Sci. Adv.*, 2017, 3(7), e1700782.

24. B. Mrowiec, *Environ. Prot. Nat. Resour.*, 2018, **29**(4), 16.

25. Veolia UK, *Planet Magazine*, Aubervilliers, 2018, https://www.veolia.co.uk/sites/g/files/dvc1681/files/document/2018/10/Veolia UK _ Planet Magazine 16.pdf (accessed April 2020).

26. N. Shehata, O. A. Mohamed, E. T. Sayed, M. A. Abdelkareem and A. G. Olabi, *Sci. Total Environ.*, 2022, **836**, 155577.

27. R. Merli, M. Preziosi, A. Acampora, M. C. Lucchetti and E. Petrucci, *J. Clean. Prod.*, 2020, **248**, 119207.
28. E. L. Pereira, A. L. de Oliveira Jr and A. G. Fineza, *Constr. Build. Mater.*, 2017, **149**, 837.
29. M. Guendouz, F. Debieb, O. Boukendakdji, E. H. Kadri, M. Bentchikou and H. Soualhi, *J. Mater. Environ. Sci.*, 2016, 7(2), 382.
30. A. R. Kurup and K. S. Kumar, *J. Hazardous Toxic Radioact. Waste*, 2017, **21**(3), 06017001.
31. S. Yin, R. Tuladhar, J. Riella, D. Chung, T. Collister, M. Combe and N. Sivakugan, *Constr. Build. Mater.*, 2016, **114**, 134.
32. N. Pešić, S. Živanović, R. Garcia and P. Papastergiou, *Constr. Build. Mater.*, 2016, **115**, 362.
33. E. Jankowska, M. R. Gorman and C.J. Frischmann, *Sustainability*, 2022, **14**(11), 6539.
34. R. A. Al-Juboori, J. Uzkurt Kaljunen, I. Righetto and A. Mikola, *Sep. Purif. Technol.*, 2022, **303**, 122250.
35. S. H. Kee, K. Ganeson, N. F. M. Rashid, A. F. M. Yatim, S. Vigneswari, A. A. A. Amirul, S. Ramakrishna and K. Bhubalan, *Fuel*, 2022, **321**, 124039.
36. R. B. González-González, H. M. N. Iqbal, M. Bilal and R. Parra-Saldívar, *Curr. Opin. Green Sustain. Chem.*, 2022, **38**, 100699.
37. K. Syberg, M. B. Nielsen, L. P. Westergaard Clausen, G. van Calster, A. van Wezel, C. Rochman, A. A. Koelmans, R. Cronin, S. Pahl and S. F. Hansen, *Curr. Opin. Green Sustain. Chem.*, 2021, **29**, 100462.
38. El País, Spain resorts to a green tax to fight against plastic, 2 June 2020, https://elpais.com/sociedad/2020-06-02/el-gobierno-lanza-un-nuevo-impuesto-sobre-los-envases-plasticos-que-preve-recaudar-724-millones-de-euros.html?utm_source=Facebook&ssm=FB_CM&fbclid=IwAR2MWsqr6673k_Eb3e6c7AEwArp5grozy4sobZPIEHEvZBwCaMsUWRfycBY# Echobox (accessed June 2020).
39. A. Diggle and T. R. Walker, *Waste Manag.*, 2020, **110**, 20.
40. A. Leal, V. Cubas, R. T. Bianchet, I. Mariana, A. Souza Dos Reis and I. C. Gouveia, *Polymers*, 2022, **14**(21), 4576.
41. The Scottish Government, *The Environmental Protection (Microbeads) (Scotland) Regulations 2018*, SI 2018 No. 162, Edinburgh, 2018, https://www.legislation.gov.uk/ssi/2018/162/pdfs/ssi_20180162_en.pdf (accessed December 2022).

CHAPTER 8

Circularity and Sustainable Cities

PETER VANGSBO*

Arup Denmark, K2, Alextowers, 1609 Copenhagen, Denmark
*E-mail: petervangsbo@gmail.com

8.1   Introduction

Cities across the world are exploring the circular economy concept, as it is being recognised as a key driver for the much-wanted green transition, simultaneously enabling greater energy and material efficiency, enabling lower pollution and greenhouse gas emissions and promoting job creation. Cities acknowledge that it is time to accelerate the "reduce, reuse, recycle" paradigm by rethinking the approach to development. With their high densities, cities hold the potential to adopt circular, restorative economies in which we no longer consider anything to be waste.[1]

Cities and municipalities increasingly recognise the potential of the circular economy in serving as a catalyst for both efficiency and innovation, thereby providing benefits of both a strategic and an operational nature. Urban areas lend themselves particularly well to a circular economy system due to the close proximity of their citizens, producers, retailers and service providers. Research suggests that the circular economy could lead to more jobs and entrepreneurial activity within the areas of remanufacturing, repair, logistics and services.

For municipalities the methodology and metrics for measuring success are well established in conventional areas such as healthcare, education and

Issues in Environmental Science and Technology No. 51
The Circular Economy: Meeting Sustainable Development Goals
Edited by Sadhan Kumar Ghosh and Gev Eduljee
© The Royal Society of Chemistry 2024
Published by the Royal Society of Chemistry, www.rsc.org

transport. Here, the applications of specific goals and indicators are often deeply embedded and give them prominence in operational matters. When it comes to circularity and sustainability, however, things are not as clear-cut.[2,3] Chapter 8 addresses the following questions:

- How can cities create prosperity for growing populations whilst transitioning through systemic change to low-carbon economies?
- How can cities achieve continued prosperity whilst preserving and reducing their demands on natural resources, such as building and construction materials, food and fossil fuels, that are fuelling conventional economic growth and global warming?
- How can cities work and implement circularity and share a sustainable urban future?

The chapter describes areas under the jurisdiction of municipalities where there is significant potential to embed circular economy principles into systems and value chains in the urban environment. As a working hypothesis the chapter presupposes that researchers, entrepreneurs, businesses and municipalities around the world are empowered to co-develop circular cities of the future, create inspiration and exploit opportunities that can be identified in urban areas.

8.2   Defining Circularity

How do we transit from today's linear economy to a circular economy? Systems take inspiration from ecology and living systems, where both materials and nutrients are cycled to restore and regenerate the economic and ecological system. Without taking into account all circular economy steps prior to the activity of recycling, the recycling economy alone constitutes a transition solution to tackle today's huge amount of "waste" that cannot yet be optimised or recovered due to economic considerations. The linear economy is unable to provide for humanity in the near future; for this we need to go deep into the circular transition.[1,4] The model of production and consumption predominant since the Industrial Revolution relies on abundant natural resources and a linear economy. This model of development hastened progress and enabled a billion people to access material prosperity. The bedrock of the consumption society currently finds its limits with the environmental challenges, the progression of unemployment and the increase in the global population. Intakes of natural resources overwhelmingly exceed the biocapacity of the Earth: its ability to regenerate renewable resources, to provide non-renewable resources and to absorb waste.[2]

The Brundtland report from 1987 summarised the end goal of sustainability transitions – "development that meets the needs of the present without compromising the ability of future generations to meet their own needs" (see Chapter 2). Policymakers, scientists, businesses and civil society have been working for the past 30 years figuring out how to achieve this goal.[5] In 1989, Pearce and Turner[6] developed an early formulation of the concept of the circular economy that underlines the progression of the structure of the economy towards a reduction of waste outputs. Figure 8.1 illustrates the concept.[7] The circular economy uses theory and principles from industrial ecology. Theoretically, a circular economy can reach an optimal state of accomplishment and it rarely reflects on intergenerational equity, suggesting a diminished role for this goal and a shorter time horizon. However, there are many systemic circular economy approaches such as the Circular Humansphere.[8]

The concept of sustainability transitions, on the other hand, has developed over the last couple of decades. A prominent approach in the transition literature combines ideas from evolutionary economics, the sociology of innovation and institutional theory.[2] The circular economy model has its roots in concepts dating back to the 1970s, including the Club of Rome's "Limits to Growth" theory, Braungart and McDonough's "cradle to cradle" concept, Stahel's "performance economy" and Lyle's "regenerative design" model, to name a few.[9] The approach has gained attention recently thanks to the Ellen MacArthur Foundation, a charity dedicated to promoting the global transition to the circular economy. Drawing on these earlier works, the foundation developed the systems or "butterfly" diagram (Figure 8.2) based on the notion that material flows can be divided into two interacting loops: the technical and biological resource cycles.[10]

Within the biological cycle, renewable and plant-based resources are used, regenerated and safely returned to the biosphere – as in composting or anaerobic digestion. The bioeconomy is a growing sector with the potential to lower

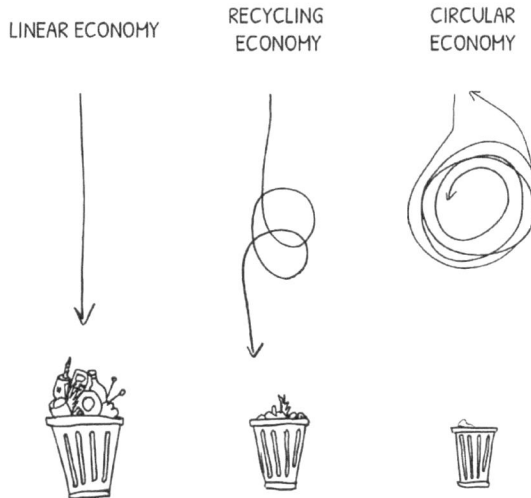

LINEAR ECONOMY   RECYCLING   CIRCULAR
                         ECONOMY   ECONOMY

Figure 8.1   From linear economy to circular economy. Reproduced from ref. 7 with permission from Circular Flanders.

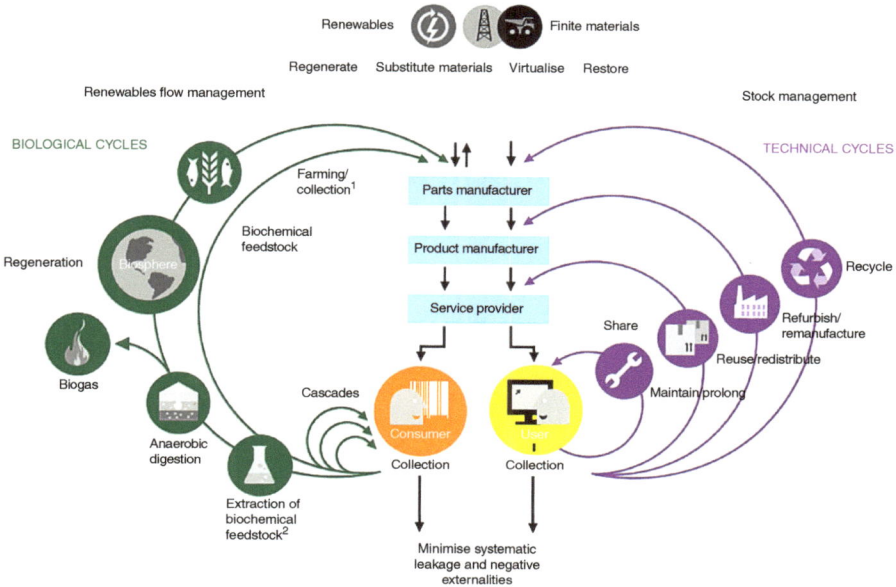

Renewables · Finite materials

Regenerate · Substitute materials · Virtualise · Restore

Renewables flow management · Stock management

BIOLOGICAL CYCLES · TECHNICAL CYCLES

Farming/collection[1]

Parts manufacturer

Biochemical feedstock

Regeneration · Biosphere · Recycle

Product manufacturer

Service provider

Biogas · Share · Refurbish/remanufacture

Reuse/redistribute

Anaerobic digestion · Cascades · Consumer · User · Maintain/prolong

Collection · Collection

Extraction of biochemical feedstock[2]

Minimise systematic leakage and negative externalities

Figure 8.2 The circular economy "butterfly" model. Reproduced from ref. 10 with permission from Ellen MacArthur Foundation, Copyright 2019.

the consumption of raw materials, reduce waste and generate higher-value products for sustainable biological re-use. Within the technical cycle, human-made products are designed so that at the end of their service life – when they can no longer be repaired and reused for their original purpose – their components are extracted and reused, or remanufactured into new products. This avoids sending waste to landfill and creates a closed-loop cycle.

In defining the concept of transition to a circular economy, it is useful to generate a shared understanding. A method of achieving such shared construction and bridging these differences may be found through bottom-up approaches that create a collective understanding of the socio-technical system that needs to be transformed.[11] The presence of a shared vision of the direction and potential means of the transition is critical. In this respect, using multi-stakeholder collaborative processes, in which practitioners' perspectives are gathered and analysed, is one of the key recommendations for supporting a transition process.

Cities can play a pivotal role in creating an enabling environment through regulations and incentives, but the private sector needs to collaborate and explore the cross-sectoral synergies required to achieve a circular model. There are immense opportunities for public–private collaboration in achieving goals that might not otherwise be possible for cities to accomplish alone.[1] Cities are embedding circular thinking in their utility processes, placing the onus on the private sector to come up with new business models that are both economically viable and ecologically sustainable. This could potentially result in a situation whereby circular products and services become the new market standard.[3,12]

8.3 The Built Environment in the Context of the Circular Economy

8.3.1 Concepts

The engineering and construction industry is the world's largest consumer of raw materials. It accounts for 50% of global steel production and consumes more than 3 billion tonnes of raw materials.[13] Global demographic and lifestyle changes are increasing the demand for these resources, many of which are becoming scarcer and harder to extract. Natural resources are currently being consumed at twice the rate they are produced. By 2050, this could be three times. Growth in the world's population and, in particular, its middle classes (which will expand from 2 to 5 billion by 2030, adding to existing demand for homes and services) is putting unprecedented pressure on natural resources.[13] As things stand, a global economy predicated on growth is helping to increase the world's middle class and its purchasing power. This trend is starkly at odds with the finite nature of our natural resources. Ongoing volatility in global commodity markets shows we need to reconsider how society consumes goods. Competition for resources and disruptions to supply are already contributing to volatile prices of materials, creating uncertainty in the short term and increasing costs overall.[11] Stricter global environmental regulations aimed at protecting fragile ecosystems are also making it harder and more costly to extract and use certain resources.

The holy grail is a world without resource scarcity, where human behaviour does not have a negative impact on our climate, where waste does not exist and where sustainability and growth are symbiotic. The challenge is to show that this can be achieved. Global population and resource consumption *per capita* are rising, as well as the demand for food, power and housing, requiring a change in the sustainability narrative as well as the structures of modern society. Digitalisation, globalisation, urbanisation, the fourth Industrial Revolution and climate change are transforming our world. The global climate is also changing rapidly – these trends are closely related. Currently, more consumption of virgin resources, traffic and buildings equates to more climate change.

A world without resource scarcity provides examples for replication, and inspiration that collective action can be harnessed to change the narrative. Throughout the world there is an abundance of already existing material. Concrete, bricks, wood, plastic, glass and metal can all be recirculated and become new materials with new functions and increased value and prosperity. Encouragingly, the opportunity and potential are not limited to the built environment; there is huge potential across all industries. Viewing waste as a resource will change the paradigm while making good business sense. This change of mindset requires a fundamental transformation in the mindset of professionals, industries, businesses, politicians and citizens alike, thinking differently to create businesses where economic growth and sustainability are mutual perquisites and where consumption does not have to come at the expense of the environment.

The built environment sector is a major consumer of natural resources, and is under increasing pressure to minimise its impact. Municipalities recognise the need to fundamentally evolve the processes, components and systems they utilise to eliminate waste and increase efficiency. A circular approach could help the sector to reduce its environmental footprint, and to avoid rising costs, delays and other consequences of volatile commodity markets. This will create a huge opportunity across the entire supply chain. Some manufacturers, for example, are already designing products that can be reused or repurposed. The EUR 1.8 trillion opportunity revealed by the Ellen MacArthur Foundation reaffirms the economic rationale of moving towards a circular economy. Realising and capturing the benefits of this systemic transition requires a cross-industry, cross-performance and multidisciplinary approach.[12,14] Arrangements to incentivise the return of products are starting to emerge, but there is as yet no clear articulation of exactly how individual companies and processes will need to change across the industry. The impetus required to catalyse this shift has not yet materialised. The structure of the industry also hinders transformative change.[15] The circular economy concept offers a chance to make the step change needed. It aims to decouple economic growth from resource consumption. Instead, products and assets are designed and built to be more durable, and to be repaired, refurbished, reused and disassembled. This maintains components and their materials at the highest useful purpose as long as feasible, which minimises resource waste.[16]

By moving away from the linear model to an ecosystem where natural capital is preserved and enhanced, renewable resources are optimised, waste is prevented and negative externalities are designed out. Instead, materials, products and components are held in repetitive loops, maintaining them at their highest possible intrinsic value.[14]

8.3.2 Challenges

As is the case with the adoption of any new operating method, the adoption of circular building should also be subject to critical discussion. Buildings are in many ways a high-risk subject of procurement: people spend the majority of their waking hours in buildings, so it is clear that buildings need to be safe, durable and healthy. By adopting new detailed building codes globally, the digitalisation of building will be a megatrend in the future.[17] For example, the renovation of certain buildings such as day-care centres and schools is often challenging, even with new materials, due to concerns related to building materials.

Turning to circular building materials, the construction sector is still in the middle of a development phase in which carrying out pilots is essential for assessing the positives and negatives of the concept. This sector will be one of the fast-growing stakeholders in the circular economy transition in the future (also see Chapter 9). Much more experience will need to be accumulated about the use of recycled materials before circular building can become a standard procedure in the sector, as there are many preconceptions associated with recycled materials with regard to their purity, health impacts, durability and life-cycle carbon footprint.

The use of recycled materials does not automatically guarantee the best possible outcome in building projects. The reuse of a material is problematic when the material is used for a lower purpose than what it could be used for. For example, this is the case with wood used for energy after its first life as a building material. Furthermore, recycled materials that cannot themselves be reused at the end of their life cycle (such as blended fabrics) are problematic despite being recycled. Here the digital regime will play a role in the future. The use of recycled materials can also require additional procedures, such as manual cleaning or transportation, which may generate unexpected emissions and costs.

As such, it is important to incorporate emissions and cost calculations into pilot projects in order to make it easier to justify the benefits of circular building materials in the future. This could be done for example by using Life-cycle Assessment (LCA) methods that take into consideration recycled materials and are able to calculate the carbon footprint of buildings. One such method is One Click LCA.[18] Such calculations can also reveal if a material or operating method is not the most economically advantageous or viable due to its life-cycle environmental impacts: for example, the carbon footprint of some recycled materials may be greater than that of equivalent virgin materials.

In time, there will be many more circular economy-marked recycled materials on the market whose viability in construction projects has been proven. Such materials include crushed concrete and glass wool, which have been utilised for such a long time that their use is not necessarily even considered to be part of a circular economy.

8.4   Transforming Linear City Systems

8.4.1   Challenges

An increasing number of municipalities have an ambitious vision and strategy to become minimal or zero-waste cities in the near future. These visions are utilised to help shape and direct to more specific plans and policies, at both city and district level. A well-written and defined vision makes it clear to all departments of the administration what they need to keep in mind when putting their development strategies into practice on a daily basis.[19-21]

Municipalities usually quantify their success through the application of specific goals and indicators, generally designed and selected to match their overall strategies. For strategies within conventional municipal fields – education, health and transport – goals and indicators are well engrained into the municipal psyche, granting them *de facto* prominence in operational matters. For strategies concerning the relatively new fields of sustainability and the circular economy, goals and indicators are less defined. The critical questions for cities are as follows:

- What needs to be done on a municipality level to turn theory into practice, and what hurdles need to be overcome?
- How can city municipalities engage business and civil society to contribute to the transition?
- Which areas within the city should be focused on to begin the transition towards a circular economy?
- What are the key benefits of incorporating circular economy principles and how do we engage the youth in this endeavour?

In order to become a circular economy, it is vital to identify and initiate change in areas where there is strong political and economic rationale. The circular economy is a strategy that can be applied to help municipalities achieve their overall vision and top-level goals regarding the city's economy, environment and quality of life for citizens. Furthermore, due to its sustainable nature, the circular economy principles can provide a city's strategic overview with a greater element of long-termism and an opportunity to collaboratively form and implement solutions with citizens, for the future of the city. In its essence, the circular economy is about how things can be made smarter, cheaper and more resource efficient. It can create savings, new income streams, jobs and social cohesion.[15]

8.4.2 Transforming Cities from Linear to Circular

It is estimated that 67% of the world's greenhouse gas emissions are related to the consumption of materials.[22] As such, it is clear that operations based on circular economy principles, which involve minimising the consumption of virgin natural resources, also play a major role in efforts to combat climate change. As noted previously, the shift to a circular economy requires organisations and individuals not only to rethink their use of materials, products and assets, but also to redesign and adopt new business models. Such models should be based on dematerialisation, longevity, refurbishment, remanufacturing, capacity sharing and increased reuse and recycling.[23] Offering products as a service rather than sale is a central circular business model.[3] For this circular transition, a city needs to establish circular city functions, services, infrastructure and tools that facilitate circular business models. The city needs to change the way urban systems are planned, designed and financed, and how they are made, used and repurposed.[1,23]

8.4.3   Public Procurement

Cities have many ways of promoting circularity, one of the most effective tools being public procurement. Circular public procurement is an approach of 'greening' procurement, which recognises the role that public authorities can play in the transition towards a circular economy. Circular procurement can be defined as the process by which public authorities purchase works, goods or services that seek to contribute to closed energy and material loops within supply chains whilst minimising – and in the best case avoiding – negative environmental impacts and waste creations across their life cycle.

Circularity principles can potentially play a key role in procurement practices to help public sector buyers choose a more holistic approach to sustainability. A holistic approach would consider the life cycle of a product, from the early stages of the procurement process to the end of product life, while achieving potential savings. Circular procurement can therefore be the instrument to leverage many of the pressing resource challenges facing growing cities today, including waste reduction and resource depletion, while simultaneously securing sustainable growth.

What needs to be done at a municipal level to turn theory into practice and how can current policies, alongside business and civil society initiatives, contribute to a circular transition? If cities increase their demand for circular solutions, it becomes more attractive for designers and producers to offer circular products and services. However, procurers often lack the knowledge of how to incorporate relevant circular requirements for suppliers and how to design tender documents to promote circularity. In addition, procurers often lack knowledge of the economic and environmental benefits associated with procuring circular solutions.

Public procurement applies across a variety of goods, services and building projects. In Europe, the volume of public procurements is large, accounting for approximately 14% of gross domestic product.[24] This translates to annual purchases of EUR 1800 billion, made by over 250 000 European public sector actors. The Nordic countries alone spend EUR 171 billion in public procurement every year.[25] For example, Helsinki spends approximately EUR 2.2 billion on public procurements a year, accounting for as much as 40% of the city's expenditure.[26]

Circular procurement means the procurement of goods, services or contracts in a way that minimises the consumption of virgin resources and does not result in the generation of waste. Instead, the raw materials and the value bound in them are kept in "circulation" for as long as possible. In practice, circular procurement can be defined as procurements that support the five principles of the circular economy and circular economy business models (see Table 8.1).[15]

1. **Preventing the generation of waste by means of planning** by improving the material efficiency of products, production processes and services;

Table 8.1 Circular economy business models supported through procurement. Data from ref. 15.

Product as a service	Circular supply chain	Life-cycle extension	Sharing platforms	Returns and recycling
• Total cost of ownership is borne by producers and retailers • Leasing and paying for use • Performance over quantity; sustainability over disposal	• Renewable, recyclable and biodegradable materials • Successive life cycles	• Active maintenance of products • Repair • Upgrading • Remanufacturing • Remarketing	• Increasing the utilisation of goods and resources • Renting • Sharing • Exchange	• Waste is valuable raw material • Recycling and use for other purposes • Waste to energy

designing products to be re-used, refurbished, remanufactured and recycled, for example focusing on usage and service orientation instead of ownership.

2. **Extending life cycles** by allowing products to be modified, repaired and updated, for example multiple successive life cycles for different purposes enabled by high-quality and sustainable materials.
3. **Relying on renewable** energy and minimising energy consumption in all stages of the production chain.
4. **System-level thinking,** examining the parts and different material streams of the system as parts of a larger whole in order to perceive the various opportunities offered by a circular economy (industrial symbioses, other cooperation and business opportunities, avoiding partial optimisation). The realisation of this principle requires adaptability and agility on the part of the operating environment in particular, such as the procurement unit.
5. **Recycling biological material streams** as efficiently as possible and ultimately returning them to the nutrient cycle.

In addition to these five principles there are some other factors that may help define circular procurement:

- product design that enables dismantling
- recycling of products and materials
- minimisation of value destruction
- promotion of new business models
- elimination of hazardous chemicals and harmful substances.

Procurement that promotes the circular economy can therefore consist of the procurement of products by leasing instead of ownership, the prioritisation of renewable and easily recyclable raw materials in products or the utilisation of sharing services relying on digitality, which can help improve the utilisation rate of products and thus reduce the need to, for example, manufacture new products.

The system-level transition to the circular economy, including procurement based on circular economy principles, often requires procuring units to adopt new operating methods and/or procurement practices. The key is to engage in proactive dialogue with the market and other interest groups, as the service or product needed may not have yet been developed. Thus, circular procurements can be considered to share many similarities with clean technology (cleantech) procurements and innovation procurements – the subject of innovation procurement is usually a new or clearly improved product or service that has not yet entered widespread use. An innovation procurement can also be a new way of carrying out procurement or a project, such as a construction contract.

Public procurement carried out by municipalities can help promote the development and scaling of sustainable products and services. By driving demand for circular and sharing economy services and products, public procurement can also accelerate circular business activities overall. Through their own responsible procurement practices, municipalities can also serve as examples to others and spur more sustainable solutions. For example, since it participated in the construction project of Stockholm municipality town hall, the construction company BAM has adopted circularity in its operations. The company was one of the first to become a part of Ellen McArthur Foundation's network CE100. In addition, municipalities throughout the Netherlands have been inspired by this example and started to use circular criteria in their construction projects.

The entire supply chain needs to engage to identify overlapping obstacles, remove barriers, identify the opportunities and discuss how to work together. Contractors might see the benefits of circular economy practices but mainly they see the risks; these need to be removed and converted into opportunities – where their materials are sourced, how much is used and what can be reused. By driving demand for circular and sharing economy services and products, public procurement can also accelerate circular business funding activities overall.

8.4.4   The EIB Circular City Funding Guide

The European Investment Bank (EIB) developed The Circular City Funding Guide, which provides information for municipalities, businesses and other actors that want to create sustainable cities and implement circular initiatives and projects.[27] The guide provides information on the circular economy in the urban context, funding instruments and sources, and how to set up programmes for circular funding and financing. The aim of the guide is to share knowledge, best practices and information on circular solutions, and on ways to finance the preparation and implementation of such solutions. As part of the European Investment Advisory Hub, more than 40 cases from around the world have been collected for inspiration and replication in relation to strategy development, urban refurbishment, procurement, utilities and civic waste.[1] Three example case studies are provided in Table 8.2.

8.4.5   Adopting UN sustainability goals

Policymakers have an important role to play un the transition to circular economy. The public sector is uniquely placed to take the long-term perspective required for setting ambitious goals and driving positive change. It is challenging to measure the positive impact of the circular economy on the environment, resources and the economy within current economic frameworks, but this is crucial for securing resources and recognition of circularity to move forward. Adopting United Nations sustainability goals can aid this process. The UN's 17 Sustainable Development Goals (SDGs) are an ambitious, globally shared framework for a better, cleaner, more just and sustainable world by 2030 (see Chapter 2 for details). By implementing the principles of circular economy in the built environment, cities, companies and residents/housing associations are able to use the SDG framework as a call to action. In particular, the following SDGs are relevant here:

- **SDG 3. Good health and well-being.** The circular economy can be a driver for a cleaner and healthier built environment, where design for disassembly and material filters place focus on high-quality products and a reduced reliance on chemicals, ensuring healthy lives and promoting societal well-being.
- **SDG 8. Decent work and economic growth.** Promote continued, inclusive and sustainable economic growth, full and productive employment and decent work for all. Circular business development relies on local loops and promotes job creation and better integration of businesses in local communities.
- **SDG 11. Sustainable cities and communities.** Make cities and human settlements inclusive, safe, resilient and sustainable. A circular economy in the built environment drives a new wave of adaptable and resilient urban development mixing green and built environments to the benefit of all residents in our cities.
- **SDG 12. Responsible consumption and production.** Ensure sustainable consumption and production patterns. The circular economy is vitally engaged with creating frameworks for integrating sustainable behaviours directly into the business models and practices of companies, municipalities and residents.

A clear linkage can be seen between circularity and the attainment of sustainability in a city environment. Through the functionality of the SDG framework with its associated targets, indicators, metrics and assessment

Table 8.2 Example circular city cases. Data from ref. 1.

	City-wide circular strategy	Urban refurbishment	Civic waste
Summary	Amsterdam was the first city to carry out a comprehensive scan of the city's material flow and economic benefits of becoming more circular. Over 70 projects have commenced, including those related to the built environment with a roadmap, knowledge networks, living labs and procurement criteria	The City of Houston Reuse Warehouse accepts donations of reusable building material, diverts it from landfills and makes it freely available to non-profit organisations, schools, universities and government agencies	In Quezon City, plastic bags amount to 12% of the total waste composition. A primary challenge faced was improperly discarded bags ending up in the waterways, landfills and the environment. Regulations have therefore been enforced
Time period	2015–2018	Established in 2009	2012 – ongoing
Municipal levers	Research, procurement, provision of information, knowledge networks, planning policy	Online marketplace, start-up grant, planning	Fiscal policy, tax
Scale	The initial city-wide circle scan showed that the implementation of material reuse strategies had the potential to create a value of EUR 85 million per year within the construction sector through increased efficiencies. Under the construction value chain, Amsterdam includes all activities related to demolition, renovation, transformation and new construction of buildings, civic and hydraulic engineering and the public space, within the city's boundaries. The chain of subcontractors also falls within this scope, even if they are located outside Amsterdam	The programme has diverted 4500 tonnes of material from landfills and involved over 700 non-profit organisations. A start-up grant of approximately USD 150 000 helped with the purchase of equipment and restoration of existing city property, and covered part of the salary of one staff member for the first year. The city continues to support this initiative, which is free to users	This city-wide regulation has raised at least EUR 4.8 million in fees for plastic bag use for Green Funds held by the retailers that must be put towards community environmental initiatives in the city

| CO_2 reduction | The initial scan estimates that, in the construction chain alone, material savings of 500 000 tonnes are possible, significant when compared with the current annual import of 1.5 million tonnes of materials. This would save 0.5 million tonnes of CO_2 per year, 2.5% of the current annual CO_2 emissions of the city | The net impact on greenhouse gas emissions is unclear. However, with documented historical and monthly tonnages of diverted and reused materials in 12 categories (ceramics, concrete, wood, *etc.*), published industry sources of accurately measured savings and sequestration of greenhouse gases and embodied energy will be identified |

methodologies, the efficacy of circular economy measures can be measured and assessed. Construction and the built environment is further examined in Chapter 9.

8.4.6   Designing for Disassembly

Associated with SDG 3 (see Section 8.4.5), the greatest opportunity for delivering buildings and products that fit within the circular economy is at the design stage. Design for disassembly is a cornerstone of the circular economy. It allows resources to fit into looped material cycles where they can be reused, reassembled and recycled at similar or higher value. In cities or more generally in the built environment, this requires a strategic approach to building components during the design phase, consideration of the differing life cycles of construction elements and careful thought about what happens to elements at the end of their life (also see Chapter 9).

There are a number of ways to enable easy disassembly of products irrespective of scale, the key point being that all connections between components must be reversible, without causing damage to the parts. This means that screws, splints, nuts and bolts are favoured over nails, and that binders such as glue should be avoided to allow for easier deconstruction. Connections must be easy to access and preferably visible. Designing for disassembly in this way increases the possibilities for effective reuse of building components and materials, as well as the possibilities for integrating reused elements from former buildings or other industries in construction projects. Collaboration across the full construction value chain is the key to success – the complexity and different scales and life cycles in the built environment require systemic solutions involving many products and partners.

Design for disassembly and life-cycle design across urban scales hold large potential gains for people, businesses and environment. The world's cities provide an unprecedented concentration of resources, capital, people, ideas and talent. This concentration enables economies of scale and critical mass for material loops across the urban function. It can be the basis of a new form of green and social urban development though resource sharing, upcycling and localising links between produces and users.

Design for disassembly is not solely about making already existing products more effective, nor is it about developing sustainable add-ons. Circular design thinking is about changing design methods and rethinking products and business models. The goal is to completely disable the connection between the individual citizen, company and society from the consumption of virgin resources and material, without compromising on aesthetics, quality, the economy or the environment. Many medium-sized and large cities have the necessary scale to enable new markets and new collaborative business models. The circular economy has been steadily developing in many cities and regions around the world that hold a large and varied supply of both products and producers for the built environment, regional development, job creation and urban transformation.

8.4.7   Supporting Young Entrepreneurship Though the Circular Economy Transition

8.4.7.1   The Concept

As noted earlier, cities echo the "reduce, reuse, recycle" paradigm and call for acceleration and rethinking the circular economy approach at all levels of society. Cities around the world hold the potential to adopt circular and restorative economies where nothing is considered waste. But most cities have yet to engage the younger cohort of the community, especially youth from socially challenged neighbourhoods, in order to develop the connection between the circular economy and social enterprise. This is the connection that circular economy design schools plan to make.

The circular economy design school is a proven concept that has been rolled out in the city of Aarhus for 60 young people from five public schools, and for 45 schoolchildren in the city of Copenhagen, in Denmark in 2021. The programme was more than an education exercise on reuse and redesign of clothing; it was also a mentoring and coaching session to educate young entrepreneurs that "one person's waste can be another person's treasure". The circular economy design school is a low-cost initiative by which almost any waste stream (road and construction waste, organic material, plastics, textiles, metals, furniture, *etc.*) can be rethought into new business models. Through circular economy design schools, and together with education and coaching, cities around the world are an ideal platform to link waste management to start-up creation, education and job creation – the motto being "just do it".

In this context the circular economy question presented in Section 8.4.1 that cities address is: What are the key benefits of incorporating circular economy principles and how do we engage young people? In order to succeed in the circular economy transition in cities, it is vital to identify and initiate change in areas with a strongly established political, social and economic rationale. This demonstrates how the circular economy is a strategy that can be applied to help municipalities achieve their overall vision and top-level goals regarding the city's economy, environment and quality of life for citizens. Furthermore, due to its sustainable nature, the circular economy principles can provide a strategic long-term overview for the city. In its essence, the circular economy is about things being made in a smarter, cheaper, more accessible and resource-efficient way, creating new incomes, jobs and social cohesion.

As an example, the city of Aarhus has adopted an approach of engaging young people in a facilitated entrepreneurship programme where the student develop their own business model in the Aarhus Municipality Circular Economy Design School. The aim of the circular economy design school is to create a platform that is straightforward and accessible for young entrepreneurs between the ages of 13 and 19 through youth clubs across the city. The specific tasks for setting up and operating the circular economy design school could include the five work packages described in the following subsections.

8.4.7.2   Task 1 – Education and Entrepreneurship Coaching Manual

In order to ensure a harmonised entrepreneurship coaching approach, a manual needs to be developed in which the municipality partners will share experiences from other education and entrepreneurship programmes. It is envisioned that all the engaged youth at the circular design schools will receive a certified documentation/knowledge passport showing that they have participated in the programme. Furthermore, if these individual "knowledge passports" are collected on an online platform, it will be possible for external parties to observe the circular economy community and their experience. This will secure optimal utilisation of the resources available and create more synergies between other circular networks and individuals, and perhaps enhance collaboration and synergies to other circular economy initiatives. This would include the following steps:

1. outline the learning approach, key performance indicators and success criteria
2. outline how to measure the performance and success of the programme
3. capture lessons learned – inspiration from other coaching programmes
4. online prototyping of the coaching manual
5. outline of the criteria of the knowledge passport to ensure that the participating youth has a valid certificate on course completion.

8.4.7.3   Task 2 – Circular Economy Material Inventory

The available material resources and the associated costs are key to a successful execution of the project and will be collected for each youth club included in the programme. The material inventory will ensure a continuous material stream to the young entrepreneurs, which will be the key aspect in redesigning new products. Before the programme starts, a material inventory will be established to get an overview of resources, location, ownership and cost. The material inventory will include an explanation of the source from the different waste streams in order to ensure transparent access to resources and to investigate potential contamination sources in the products. This includes the following tasks:

1. Investigate how the waste reduction initiatives support the local municipality climate strategy.
2. Work with recycling facilities in the city and discarded products from the children's homes. Persuade parents to donate items they discard from homes and offices.

8.4.7.4   Task 3 – Establishing the Physical Learning Space and Material

For each of the youth clubs, it is a prerequisite that they identify the physical learning space for the entrepreneurship programme to accommodate the youth and for the programme to potentially grow beyond the first 6 months. This includes the following tasks:

1. identifying the venue, location and local pedagogues
2. overview of material, tools and software needed to work with the different waste streams.

8.4.7.5   Task 4 – Execution of the Circular Economy Design Schools

The project will execute the circular economy design schools in collaboration with the Entrepreneurial Discovery Process (EDP) programme held over 6 months in at least three different youth clubs in Aarhus. The schools will focus on mentorship, creating a thinking culture of sustainability, for example commercialising and recirculating urban products by redesigning second-hand clothes, textiles, electronics, furniture or building materials. Sustainable practices will be encouraged through the following platforms:

1. Storytelling using real-life examples. Craig Native shares his story of becoming a designer in a South African township and developing a brand that launched him as a world-renowned clothes designer.
2. Mentorship and design sessions.
3. Designing products with the mentality of long-term potential profitability and sustainability.
4. Campaigns for branding and marketing.
5. Evaluation, assessment and next steps.

8.4.7.6   Task 5 – Circular Design Award

All solutions produced from the three youth clubs will participate in the newly established circular design award which will be a one-time award "Oscars" for engaged young entrepreneurs promoting design and art from second-hand products, inspiring other young entrepreneurs to replicate the best ideas. All partners involved in the project will participate in the circular design award. This includes the following tasks:

1. design of the award programme, online platform, call for solutions, evaluation criteria, prize
2. curate an inspirational catalogue of new designs
3. secure the jury panel, award venue, branding, communication and outreach.

8.5   Business Models for the Future

8.5.1   Concepts

The current, linear system of limited resources and opportunities that exist will straitjacket operation, preventing cities and society from making a positive change. A circular economy paves the way for a new and radical vision whereby current obstacles are future opportunities. The tool to value innovation in the future is the circular economy, in which the only way to create value as a business is to decouple value from the use of materials and of energy. To move beyond a benchmark approach of doing "less bad", three types of business are important to evaluate and catalyse circular innovation:

1. **The regenerative business**, where the business model requires that the use of materials and products enables the business and the environment to be mutually beneficial.
2. **The responsible business,** which endeavors to do less bad, but still operates within a linear business model, making it difficult to decouple its growth and value from greenhouse gas emissions and the use of virgin raw materials.
3. **The stranded business**, which has not yet had the need or the opportunity to operate differently – it is stranded in a traditional structure where business and the environment are still at odds with each other.

Circular business models are a key to the transformation from a linear to a circular economy, specifically in the building and construction sector, and must work alongside design strategies, governance and regulations for the transition to be a success. Circular business models generate new ways to develop and grow a business while improving planning, creating saving and leading to responsible material choices. The business models are based on a comprehensive life-cycle approach and seek to forge new productive partnerships in the construction value chain. The principal business models identified when moving forward are the following:

1. **Circular supply:** replacing virgin raw material with materials that are renewable or biodegradable.
2. **Resource recovery:** recovering discarded production or by-products to recycle or upcycle the material.
3. **Life extension:** extending the life cycle of a product, or parts of a product, while preserving the original function.
4. **Sharing platforms:** increasing the use of a product though new models for sharing, accessibility and ownership.
5. **Products as a service:** optimising productivity of a resource or product while maintaining ownership.

Figure 8.3 illustrates how the five business models integrate within the building life cycle.

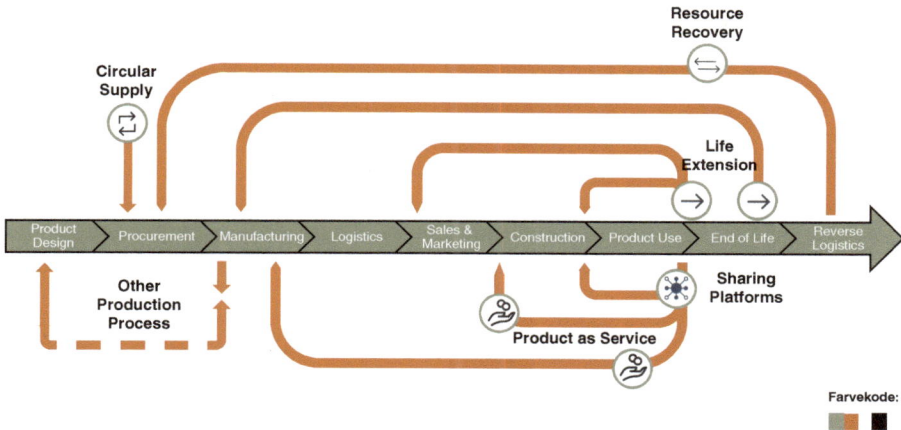

Figure 8.3 Five business models integrated into the building life cycle.

The models can be implemented in cities by creating consortia and innovation hubs that include designers, suppliers, service providers, contractors and demolition companies, working closely together to forge robust business partnerships.[28]

8.5.2 Circular Supply

In this business model, the focus is on supplying fully renewable recyclable or biodegradable resources as input to support circular production and consumption systems (see Figure 8.4). In this way, materials keep their quality and value, securing a steady supply of raw materials for new construction, while companies replace linear approaches and reduce the use of scare resources and the production of waste.

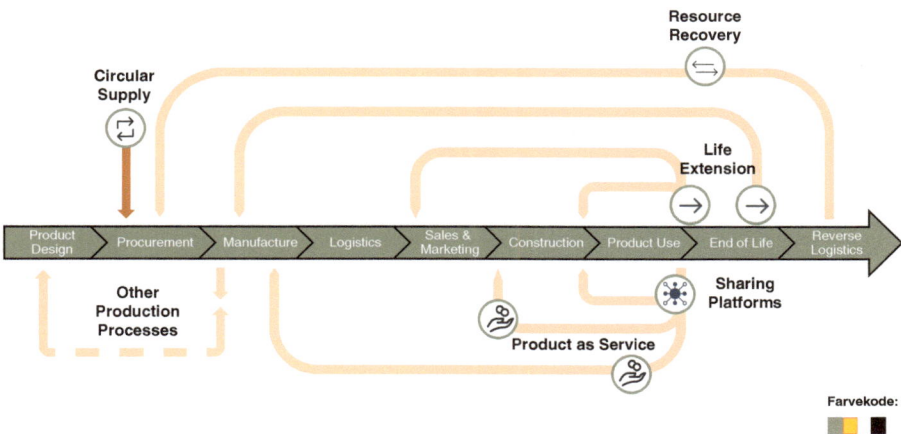

Figure 8.4 The circular supply business model.

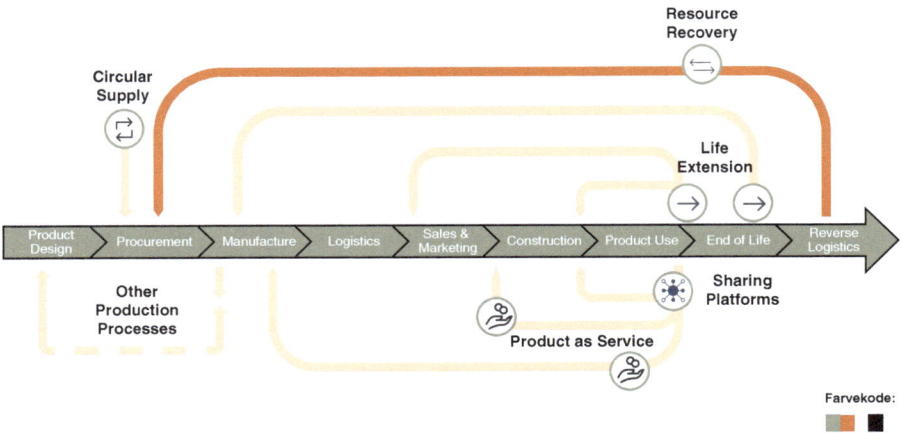

Figure 8.5   The resource recovery business model.

8.5.3   Resource Recovery

This business model relies on reverse cycles for the recovery of embedded value at the end of product life cycles (see Figure 8.5). High-quality resources lend themselves to recycling and upcycling processes that maintain, or even increase, resource value as they are used as input for new construction material.

8.5.4   Life Extension

Life-extension business models allow for capturing additional revenue based on extending the life cycle and use of products and assets (see Figure 8.6). Value that would otherwise be lost can be maintained or improved through

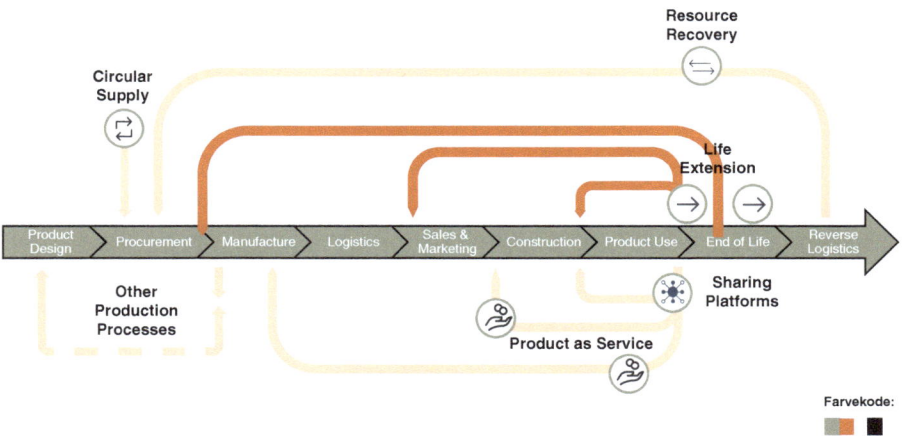

Figure 8.6   The life extension business model.

direct reuse or resale; repair and/or upgrade for resale; or separation of products into parts for remanufacturing and/or refurbishing of the products in an upgraded version for resale.

Design for disassembly and adaptive reuse of the built environment can help extend the life of building and infrastructure, but demolition companies also play a key role in life extension. They are in the position to ensure a product's life extension though new markets in reused and upcycled products.

8.5.5   Sharing Platforms

This business model promotes sharing platforms to facilitate cooperation among users, either individuals or organisations (see Figure 8.7). Virtual sharing platforms enable distribution of the surplus supply of material and utilisation of underused equipment and services. The model facilitates this by either enabling or offering shared use, access or ownership. Emerging technologies for additive manufacturing and automated fabrication makers' design and fabrication blueprints also offer a shareable resource.

8.5.6   Product as a Service

This business model is an alternative to the traditional model of "buy and own". The focus is on performance rather than products, and ownership usually stays with the service providers (see Figure 8.8). Through various service arrangements, such as pay for use, leasing, renting or performance arrangements, the product is then used by one or more customers. The customer pays for the actual, metred use of the product by buying a particular performance, with supplier and customer agreeing on the right to use the product for a defined period of time.

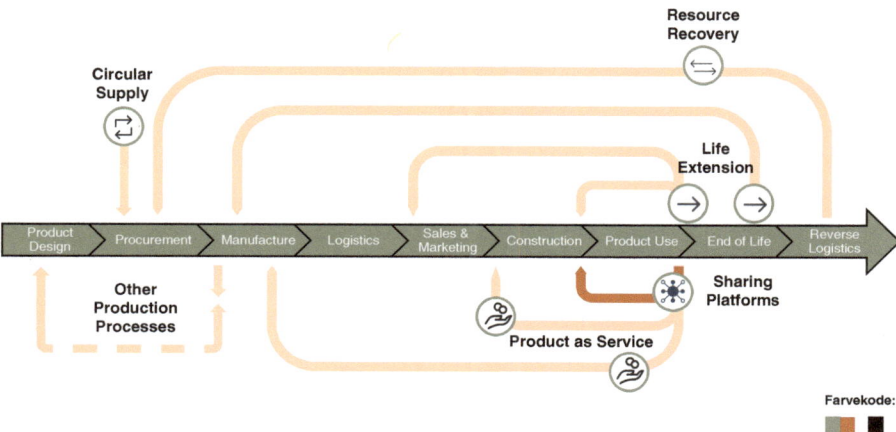

Figure 8.7   The sharing platform business model.

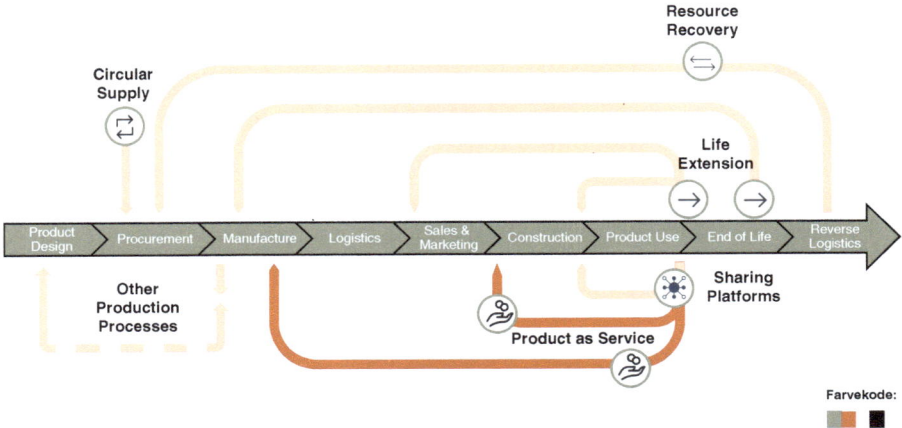

Figure 8.8   The products as a service business model.

8.6   Cities of the Future

Half of the population of the Earth – 3.5 billion people – currently live in cities. The number is expected to reach 5 billion by 2030. Although cities cover approximately 3% of the world's land area, they consume 60–80% of the world's energy and produce 75% of global carbon dioxide emissions.[29] Understandably, cities have been recognised as one of the key factors in mitigating climate change.[30] The challenges of climate change, together with the decrease in natural biodiversity, force society to explore any and all potential ways of addressing them in order to build a more sustainable future. In this regard it is unquestionable that circular economy solutions show significant potential in mitigating these adverse environmental impacts.

As discussed previously, the transition from linear to circular systems is a complex journey, demanding the involvement of several different stakeholders and, perhaps most importantly, the engagement and involvement of cities.[23,31] The growing urban areas around the world and continuous population growth give cities the highest potential for such change. The proximity to citizens, producers and entrepreneurs enables municipalities to work as strategic change agents for a new sustainable and circular future.

Companies, entrepreneurs and cities successfully adapting to new circular business models will have a competitive advantage in the circular economy transition. However, their growth potential will differ, depending on their placement and role in the emerging circular value chain for the built and construction sectors. Both public and private supply chain stakeholders in circular cities are key to changing practices in the construction value chain and apply circular business models. First adopters will largely benefit and save value in operation and extended use and closed material cycles to enable new forms of real-estate investment.

The focus and *modus operandi* of consultants such as architects and contractors will change, and companies that are able to offer strategic advice on how best to meet circular requirements will have an advantage. End-of-life contractors will play a key role in resource recovery and life extension of building materials. As such they have an expanding business potential, provided they redefine their position as facilitators of resource loops within the construction sector.[32]

If material producers and wholesalers in the built environment understand how to redefine their role from simply being retailers to also becoming service providers, a new prosperous business path will emerge. For the suppliers of building materials and products, there are gains and losses. The new circular business models will seriously challenge suppliers of simple products and material based on virgin resources that can readily be replaced by reused products and materials.

To conclude, the trend in cities around the world indicates that in the next 5–7 years developers, urban planners, investors and civil socity will have differentiated needs and demand sustainable solutions. The construction supply chain will need to gear up to meet this demand.

The circular economy will shape our sustainable urban future.

References

1. P. Vangsbo, *Municipality-led Circular Economy Case Studies*, EIT Climate/KIC Circular Cities Project, C40 Cities, London, 2018, https://nordic.climate-kic.org/wp-content/uploads/sites/15/2018/05/Municipality-led-circular-economy-case-studies.pdf (accessed May 2023).
2. Ellen MacArthur Foundation, *Towards the Circular Economy – Economic and Business Rationale for an Accelerated Transition Vol. 1*, Cowes, 2013, https://www.ellenmacarthurfoundation.org/assets/downloads/publications/Ellen-MacArthur-Foundation-Towards-the-Circular-Economy-vol.1.pdf (accessed May 2023).
3. P. Vangsbo, *Municipalities as Drivers for Circular Economy in Refurbishment and Construction Projects*, EIT Climate/KIC Circular Cities Project, C40 Cities, London, 2018, https://nordic.climate-kic.org/wp-content/uploads/sites/15/2018/05/Municipalities-as-drivers-for-circular-economy.pdf (accessed May 2023).
4. Creative Denmark, *Designing the Irresistible Circular Society*, Creative Denmark, Copenhagen, 2021, https://cms.creativedenmark.com/media/Creative-Denmark_white-paper_Designing-the-irresistible-circular-society.pdf (accessed May 2023).
5. World Commission on Environment and Development, *Our Common Future*, Oxford University Press, Oxford, 1987.
6. D. W. Pearce and R. K. Turner, *Economics of Natural Resources and the Environment*, Prentice Hall, London, 1989.

7. Circular Flanders, Infographics, https://vlaanderen-circulair.be/en/infographics (accessed May 2023).

8. World Economy Forum, The circular economy imperative, Video blog from WEF, 2023, https://www.weforum.org/about/circular-economy-videos (accessed Jan 2023).

9. E. Achterberg, J. Hinfelaar and N. Bocken, *The Value Hill Business Model Tool: Identifying Gaps and Opportunities in a Circular Network*, 2018, https://docplayer.net/86718304-The-value-hill-business-model-tool-identifying-gaps-and-opportunities-in-a-circular-network.html (accessed May 2023).

10. Ellen MacArthur Foundation, *The Butterfly Diagram: Visualising the Cirular Economy*, https://ellenmacarthurfoundation.org/circular-economy-diagram (accessed May 2023).

11. A. Lendager and D. Vind, *A Changemaker's Guide to the Future*, Lendager Group, 2018, https://lendager.com/project/a-changemakers-guide/ (accessed May 2023).

12. K. Webster, *The Circular Economy – A Wealth of Flows*, Ellen MacArthur Foundation Publishing, Cowes, 2015.

13. World Economic Forum, Can the circular economy transform the world's number one consumer of raw materials? Geneva, 2016, https://www.weforum.org/agenda/2016/05/can-the-circular-economy-transform-the-world-s-number-one-consumer-of-raw-materials/.

14. C. Lemmens, *The Circular Economy in the Built Environment*, Arup, London, 2016, https://www.arup.com//media/arup/files/publications/c/arup_circulareconomy_builtenvironment.pdf.

15. P. Vangsbo, *The Challenges and Potential of Circular Procurements in Public Construction Projects*, EIT Climate/KIC Circular Cities Project, C40 Cities, London, 2018, https://www.climate-kic.org/wp-content/uploads/2019/06/Procurements-in-Public-Construction-v2.pdf (accessed May 2023).

16. Ellen MacArthur Foundation. *Cities in the Circular Economy: An Initial Exploration*, Cowes, 2017, https://www.ellenmacarthurfoundation.org/assets/downloads/publications/Cities-in-the-CE_An-Initial-Exploration.pdf.

17. M. Van Sante, *Digitalisation must be top priority for construction companies*, ING Bank N. V., 2022, https://think.ing.com/articles/digitalisation-must-be-top-priority-for-construction-companies (accessed May 2023).

18. One Click LCA, https://www.oneclicklca.com/ (accessed May 2023).

19. TNO Delft, *Circular Amsterdam – A Vision and Agenda for the City and Metropolitan Area*. Delft, 2016, https://repository.tno.nl/islandora/object/uuid%3Af7d0eaf1-8625-4439-ae8e-2168bfc20e95 (accessed May 2023).

20. Circle Economy, *Circular Prague*, Amsterdam, 2019, https://www.circle-economy.com/insights/circular-prague (accessed May 2023).

21. Copenhagen Solutions Lab, Circular Copenhagen, Online strategy development forum, Copenhagen, 2023, https://circularcph.cphsolutionslab.dk/cc/home (accessed Jan 2023).

22. UN Development Programme, *Circular Economy Strategies for Lao PDR – A Metabolic Approach to Redefine Resource Efficient and Low-carbon Development*, New York, 2017.

23. European Investment Bank, *The EIB Circular Economy Guide – Supporting the Circular Transition*, Luxembourg, 2018, https://www.eib.org/attachments /thematic/circular_economy_guide_en.pdf (accessed May 2023).

24. European Commission, Public procurement, European Commission, Brussels, 2019, https://ec.europa.eu/growth/single-market/public-procurement_en (accessed May 2023).

25. Nordic Ecolabelling, Green public procurement, Stockholm, 2019, https://www.nordic-ecolabel.org/why-choose-ecolabelling/green-public-procurement (accessed May 2023).

26. City of Helsinki, *The Carbon-neutral Helsinki 2035 Action Plan*, Helsinki, 2018, http://carbonneutralcities.org/wp-content/uploads/2019/06/Carbon _neutral_Helsinki_Action_Plan_1503019_EN.pdf (accessed May 2023).

27. European Investment Bank (EIB), *The Circular City Funding Guide*, EIB, Brussels, 2020, https://www.circularcityfundingguide.eu/ (accessed May 2023).

28. P. Vangsbo, *Transforming Municipality Districts into Learning Centres of Circular Economy*, EIT Climate/KIC Circular Cities Project, C40 Cities, London, 2018, https://nordic.climate-kic.org/wp-content/uploads/ sites/15/2018/05/Transforming-Municipality-Districts-into-Learning-Centres-of-Circular-Economy-.pdf (accessed May 2023).

29. United Nations, Sustainable development goals, https://www.un.org/sustainabledevelopment/cities/ (accessed Dec 2022).

30. M. Christis, A. Athanassiadis and A. Vercalsteren, *J. Clean. Prod.*, 2019, **218**, 511.

31. European Investment Bank, *The 15 Circular Steps for Cities*, Luxembourg, 2018, https://www.eib.org/attachments/thematic/circular_economy_15_ steps_for_cities_en.pdf (accessed May 2023).

32. World Economic Forum, *Shaping the Future of Construction: A Breakthrough in Mindset and Technology*, Geneva, 2016, https://www3.weforum. org/docs/WEF_Shaping_the_Future_of_Construction_report_020516. pdf (accessed May 2023).

CHAPTER 9

Construction and the Built Environment

PURVA MHATRE-SHAH AND AMOS NCUBE*

EarthShift Global LLC, Kittery, ME 03904, USA
*E-mail: amos@earthshiftglobal.com

9.1 Introduction

Infrastructure, including construction and the built environment, is necessary to sustain and support human life on the planet. Demand for infrastructure continues to grow in response to growth in human population and economic activities.[1–3] Recent literature identifies the construction sector as one of the biggest industrial sectors contributing to waste and greenhouse gas (GHG) emissions causing climate change.[4,5] According to Hertwich,[6] from 1995 to 2015 GHG emissions from material production alone increased by 120%, with 11 billion tonnes of CO_2-equivalent emitted in 2015 and two-fifths of this overall carbon footprint attributed to the construction sector. The accelerated increase in construction waste and its heterogeneous composition has highlighted the need for efficient value recovery systems.[7] Application of circular economy (CE) approaches in the built environment has the potential to reduce GHG emissions and other environmental impacts.[8] According to Lima et al.,[9] CE approaches can potentially achieve a carbon-neutral construction sector.

In 2015, the 2030 Agenda for Sustainable Development was adopted by all United Nations Member States and provided a "shared blueprint for peace and prosperity for people and the planet, now and into the future." The agenda introduced 17 Sustainable Development Goals (SDGs), making an

Issues in Environmental Science and Technology No. 51
The Circular Economy: Meeting Sustainable Development Goals
Edited by Sadhan Kumar Ghosh and Gev Eduljee

"urgent call for action by all countries – developed and developing – in a global partnership. The SDGs recognize that ending poverty and other deprivations must go hand-in-hand with strategies that improve health and education, reduce inequality, and spur economic growth – all while tackling climate change and working to preserve our oceans and forests."[10]

The connection between a CE and SDGs has not been extensively probed by the recent literature, yet both have broader agendas aimed at ensuring balanced gains within the three pillars of sustainability, *i.e.* economic, social, and environmental dimensions.[11] The SDGs are a set of global indicators developed to provide powerful aspirations to improve our world within a common blueprint aimed at creating peace and prosperity for the people and planet. The CE global agenda is aimed at tackling some of the existing global challenges such as climate change, biodiversity loss, waste, and pollution largely caused by the dominant linear economy. The urgent call for action requires all countries to act in a solid global partnership towards a CE. This will help transform the take–make–waste industrial mindset by creating regenerative closed-loop systems capable of utilizing resources and materials for as long as possible.[12,13] The adoption of a CE can contribute to the attainment of SDGs and a thriving future within the construction sector.[14]

Chapter 9 examines how CE approaches in the construction and built environment sector can support the attainment of SDGs and improve the sustainability of the sector. Reviewed information is presented, based on the following subsections: (1) definition and application of a CE in the built environment, (2) contributions of a CE to SDGs to inform and build upon a sustainable construction sector, (3) lessons learnt and the way forward based on actionable insights for future perspectives, and (4) conclusions and recommendations.

9.2 Circular Economy in Construction and the Built Environment

A successful business model builds on the understanding of "the rationale of how an organisation creates, delivers, captures value".[15] The core ideology of the CE is based on an industrial economy that is restorative and regenerative by design, and thus captures and utilizes maximum value from resources across their useful life.[16] The Ellen MacArthur Foundation summarizes CE as having five primary characteristics:[10]

- designing out waste (by optimization of technical and biological resources *via* repair and reuse)
- building resilience using diversification (by developing different potential solutions to ensuring a sustainable supply and value chains)
- dependence on renewable energy
- systems thinking (interlinking parts to a whole)
- thinking in cascades (extracting added value from resources).

The built environment sector encompasses a wide spectrum of structures – from residential and commercial units (houses, buildings, skyscrapers, factories, and industries) to transportation (roads, bridges, airports, and harbours) and utility-based services (water and sewage networks, electricity, supply lines, *etc.*). The incorporation of a CE has the potential to minimize resource consumption and to reduce waste and emissions across the life cycle of construction of different structures. In addition, economic benefits, resource savings, and social wellbeing are some of the many advantages of adopting a CE.[17]

9.2.1 Principles and Concepts of a CE in Construction and the Built Environment

The incorporation of a CE in the construction and built environment sector builds on multidimensional elements across the life cycle of construction projects. Product longevity[10] or increasing lifespan of a structure is ensured by efficient design, regular repair, maintenance, using durable materials, and regular replacement of damaged components. Efficient structural design is integral to waste minimization and encompasses design for deconstruction or disassembly, flexible spaces, and other versatile features to enable diversified utility of similar structural components. Durability and economic viability are maintained across the lifespan of a structure by adhering to construction standards and specifying use of durable materials, so that the structures last longer.[18]

The circular input model is based on the substitution of finite resources with renewable or recovered ones, thereby generating value or revenue from the resale of secondary resources. This model builds on CE-based strategies such as reuse, remanufacture, and design for disassembly, among others. The concept of adaptive reuse[19] fits the circular input model as adaptive reuse is based on the reuse, remanufacture, refurbishment, and recycling of components and resources from obsolete buildings and cycling them back into the construction ecosystem. For enabling the circular input model, standardization of structural components is actively practised.[20] Further, developing regulations and technical standards for strength and serviceability criteria for reuse of structural components, along with non-destructive testing mechanisms can aid in component reuse.

Different construction materials have the potential to be reused in similar or different types of structural constructions.[21] For instance, secondary or used concrete aggregates can be recycled to produce recycled aggregates for use in new constructions. Timber-based systems can be used to build sustainable structures.[22] Lean constructions using Building Information Modelling (BIM) systems have proven to be efficient in waste elimination.[23] Prefabrication and modularization[24] of structural components are known to build processes for efficient use of resources, while facilitating structural flexibility, ease of deconstruction, and reduced wastage of construction materials.[25] Design for disassembly aids in waste elimination, durability, and technical upgrade of structural components.[26] Selective disassembly[19] has proven beneficial in cases of large-scale structures such as bridges and skyscrapers. Designing structures for deconstruction[27] also aids in easing component recovery and reuse, thus facilitating waste elimination.

In the product as a service model, the producers are liable for product maintenance and ownership. This model is largely incorporated in public infrastructures such as transportation routes – roads, highways, railways – and utility-based infrastructure – electricity lines, water supply, and similar. The ownership of such infrastructures primarily lies with the regional authority or governments, with the users paying a utilization fee (toll for using highways, water bills for water consumption, electricity bills, *etc.*) for using these infrastructures. Another example of product as a service model in the built environment sector is that of commercial complexes where offices or commercial spaces are rented for short terms to serve the requirements of the users, while the onus of maintaining the space lies with the owner of that building or structure. A similar model is observed in residential buildings for tenants.

The collaborative consumption technique is based on stakeholder coordination and connection to facilitate sharing of resources and to maximize the value of available resources. Urban resource cadastres[28] (a register with details of ownership, boundaries, and property values) have the potential to track resources used, their utility, and reuse or recycling potential across the regional real-estate properties and different built environment constructions. This can be paired with development of web-based platforms for sharing information on material stocks[29] and facilitating stakeholder coordination and supply-chain integration,[30] to understand resource demand and availability.

Finally, the waste-to-value model reinforces the need to recover waste using technologies to optimize the value of resources and to maximize the economic benefits. For instance, wastes from agricultural, industrial, and municipal units are used to produce geopolymer concrete.[31] Bitumen pavements can be repaired using polymers and metal shavings.[32] Internal recycling strategies such as recycling Construction and Demolition Waste (CDW) into recycled aggregates and recycling steel, iron, aluminium, and other construction materials are gaining traction. Furthermore, external recycling strategies, such as inputs from other sectors (agricultural sector, textiles, mining, chemicals, and more) for the manufacturing of construction materials, also fosters environmental performance in the sector.

It can thus be inferred that CE-based strategies are essentially designed to increase the life of construction materials, parts, and components in buildings and infrastructures and to aid in closing the resource loops at the end of life (EoL) of these structures. A CE-based transition can benefit the construction industry environmentally as well as economically by minimizing resource consumption and emissions and by increasing the productivity of resources through lower construction costs by reusing or recycling materials through a closed-loop system.

Figure 9.1 presents a graphical representation of CE adoption in the construction or the built environment sector across the different life-cycle stages of design, construction, and EoL of a structure.

Design for durability
Design for modularity
Design for standardization
Design for disassembly
Design for remanufacturing
Design for use of secondary materials

Design

Stakeholder coordination
Resource traceability
Take-back schemes

Life-cycle stages

Stakeholder coordination
Resource traceability
Process efficiency

End of life   **Construction**

Resource data management
Resource reverse logistics
Resource recycling
Resource recovery and segregation
Determination of reusable and recyclable materials

Stakeholder coordination
Resource traceability

Material banks
Waste as a resource
Adaptability/flexibility

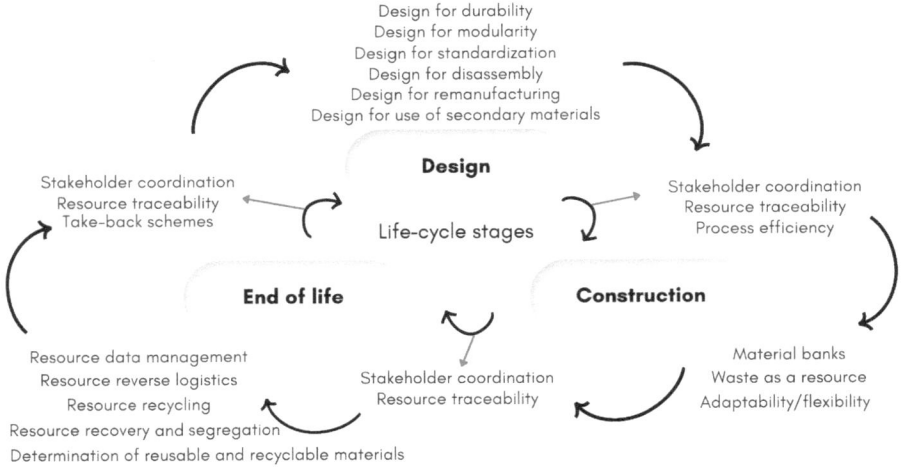

Figure 9.1   Circular strategies in the construction and built environment sector.
Data from refs. 36, 58 and 80.

9.2.2   Circular Economy in Practice

Implementation of a CE in the construction sector has a number of advantages, such as minimized pollution, reduced use of materials and resources,[33] economic savings, and promotion of social equality.[34,35] Thus, many governments across the globe, as well as construction firms, have consciously pledged to include the CE as a part of their development plans.

For instance in Seattle, USA, municipalities have adopted 'Salvage Assessment' as a measure to reduce waste generation and enforce material reuse.[36] China has passed legislation such as Value Added Tax Policy on Comprehensive Resource Utilization and Related Products (2008), Interim Management Method Circular Economy Development Funds (2012) and Circular Economy Promotion Law of the People's Republic of China to foster the CE in the built environment as well as in other sectors.[37] Europe's Regional Law for Circular Economy 2016–2020 has built the foundation for CE-based policies and frameworks for EU Member States across different sectors of the economy. Furthermore, the introduction of certifications such as Leadership in Energy and Environmental Design (LEED) have fostered CE pathways in the construction sector. For instance, LEED-certified buildings with their modular and prefabricated units are more circular in their composition.[38] Economic instruments such as taxation on landfills have also fostered resource recovery efficiently in a number of regions and countries.[39]

CE-based practices have also been incorporated in the construction of diverse structures across the globe. These include use of plastic waste and jet grouting waste in the production of asphalt mixes for construction of flexible pavements.[40] Concrete aggregates have been used in the construction of urban pavements.[41] Construction aggregates are partially replaced by CDW aggregates,[42] mining wastes[43] and mineral wool,[44] depending on the structural requirements based on the strength and serviceability of the concerned structure. Research has been undertaken to perform structural feasibility assessments for replacing binding materials such as cement by Sewage Sludge Ash (SSA),[45] fly-ash from thermal incinerators, and Municipal Solid Waste (MSW) ash from waste incinerators. The construction sector has also actively incorporated the practices of prefabrication and modularization owing to the cost- and resource-saving benefits associated with these processes. In addition, construction firms in developed regions have also started accounting for and reporting the GHG emissions associated with the construction of different structures.[46]

Incorporation of the CE in the built environment sector has gained pace across the globe, in terms of research, as well as practical adoption and policy implications. The following sections assess and discuss the impact of transitioning to a CE in the construction and built environment sector on achieving the SDGs.

9.3 Sustainable Development Goals

Adoption of the CE in the construction and the built environment sector can be viewed as a condition for achieving sustainability in the sector: studies and research have identified the CE as a tool and implementation technique to achieve significant gains in achieving the SDGs.[47] Incorporation of circular design and intention in construction practices can aid in restoring and regenerating resources.[48] Another outcome of adopting the CE is enhanced waste minimization and reduced mining for extraction of virgin resources as materials and resources are recirculated in a closed-loop system, thus minimizing emissions and pollution.[13] Furthermore, restructuring of the construction value chain for incorporation of CE-based practices has the potential to generate increased employment opportunities, thus fostering social and economic advantages.[35] Thus, adoption of the CE is perceived among intergovernmental organizations, governments, and policymakers as an effective approach to achieve the UN sustainable development goals at global, national, and local levels.[49]

In the case of the construction and built environment sector, studies have shown that there are direct and indirect linkages between the adoption of a CE and the SDGs. Figure 9.2 illustrates how CE-based strategies can aid in achieving the SDGs.

The following subsections discuss the direct and indirect linkages of CE adoption in the built environment as summarized in Figure 9.2, while highlighting the impact of achieving the SDGs.

9.3.1 Direct Contribution of a CE to Achieving the SDGs

The available literature and experiences from current practices have demonstrated that CE adoption has direct implications on at least 10 of the 17 SDGs as identified by the UN Development Programme.[50] These include SDGs 3, 6, 7, 8, 9, 11, 12, 13, 15, and 17.[47]

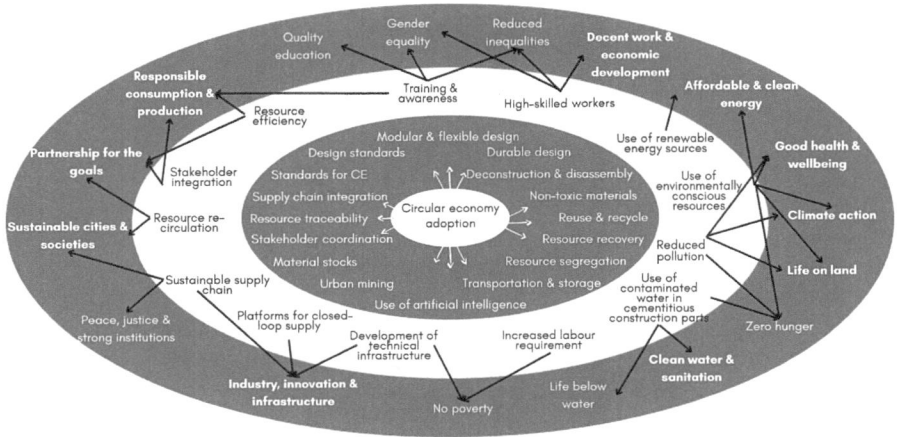

Figure 9.2   Impact of CE adoption in the construction and built environment sector on achieving the SDGs.

SDG 3: Good health and wellbeing: An ideal case scenario of transitioning to a CE in the built environment sector implies construction of infrastructures and buildings designed for resource and energy efficiency, along with the choice of environmentally conscious materials, to facilitate reuse of structural components and to eliminate toxicity and exposure to hazardous substances.[51] These aid in ensuring safe living conditions for the occupants or users of the residential and commercial buildings or other infrastructural structures. Furthermore, the CE-based structural configurations will foster use of renewable energy sources, thus minimizing pollution and other emissions associated with energy generation from fossil-based resources.

SDG 6: Clean water and sanitation: The core idea of a CE is efficient use of available resources and ensuring elimination of waste emissions and toxins in the external environment. This premise of waste elimination ensures non-contamination of water sources in and around the geographical boundaries of the construction sites and other allied processes. Across the different life-cycle stages of the construction of structures, operational process efficiencies and other CE-based practices facilitate maintenance of clean water quality and sanitation. For instance, the process of mining produces water contaminated with mining waste. This waste water can be used to manufacture ceramic bricks. In this way, requirement for fresh water is reduced, and the contaminating elements of mining waste are encapsulated in a solid ceramic mix.[43]

SDG 7: Affordable and clean energy: A transition to a CE has the potential to offer opportunities and dependable outputs for improving energy and environmental performance across the built environment sector.[5] The continuous pursuit for using renewable energy sources facilitates technological innovation and builds resilience and reliability for using cleaner energy sources. In certain cases, resource recovery processes can aid in energy generation and recovery. For instance, recycling of steel can be coupled with energy recovery *via* gasification.

SDG 8: Decent work and economic development: Research and practical examples of the CE in businesses across the different sectors of the economy have confirmed the economic and social benefits of adopting a CE.[52] In the construction and built environment sector, adoption of the CE has the potential to create employment opportunities, particularly in the manufacturing and recycling sectors.[49] Complete utilization of resources at the EoL of structures offers unique opportunities for supply chain coordination and collaboration, thus requiring more managerial resources.[53] Resources will also be required to deliver models and estimates on the material flows, and available material stocks to predict the urban mining potential using Artificial Intelligence (AI) or blockchain technology. Furthermore, jobs are estimated to be created for resource and component recovery, disassembly of structural components, and resource segregation at the EoL of structures.[54] Since these jobs require a higher skill set and knowledge, the employment opportunities generated would provide greater economic benefits and thus foster better social and living conditions for the employees.[48]

SDG 9: Industry, innovation, and infrastructure: Incorporation of a CE in the construction and built environment sector has the potential to foster AI and virtual reality-based tools for identification of material recovery potential and enforcing material circularity for profit maximization.[25] The material stocks and resource flows in the existing constructions can be tracked using blockchain technology,[55] thus aiding in proactive planning for reusability of resources. Research has shown that using blockchain for resource modelling can provide comprehensive material and energy traceability solutions, thus enabling value chain actors to predict the reuse and recycling potential of different materials and components. Development of platforms with technical and environmental standards can aid in facilitation of market-based infrastructural and technical capabilities for closed-loop resource supply.[56] For instance, the Adaptive Reuse of Cultural Heritage buildings (ARCH) tool – for CE-based investment[21] – has been developed to restore heritage buildings and aids in determination of investment opportunities to foster the CE. Development of similar platform or business models for enabling a CE will aid in technological innovation and transition the industry and infrastructure towards a sustainable built environment.

SDG 11: Sustainable cities and communities: Construction resources stocked in existing built environment systems constitute a major portion of the global resources and embodied carbon. Development of sustainable cities and communities is one of the many benefits of transitioning to a CE.[57] Adoption of the regenerative approach across the construction processes helps increase the value of construction materials by facilitating material reuse and minimizing environmental impacts by waste elimination and reduction in mining for new materials.[58] Supply chain integration fosters construction resource sharing and material recirculation across the construction and the built environment value chains.[59] Furthermore, energy and resource efficiency, along with maintenance of environmental quality standards, can be achieved using CE-based practices. These CE-based practices set the foundation for a resource-efficient and sustainable built environment ecosystem, wherein resource sharing and recirculation aids in judicious use of available resources.

SDG 12: Responsible consumption and production: A conscious and collaborative effort among different stakeholders across the different life-cycle stages of a construction project – design, material supply, construction, and use phase – can minimize waste and foster efficient use of resources.[49] According to the World Economic Forum,[60] adoption of CE-based strategies in the construction industry can result in savings worth USD 100 billion per year. Research has shown that it is possible to refurbish a building through the 83% of walls and ceilings that can be disassembled and recovered during the operation phase[26] and reused in a different structure, with a waste generation potential of less than 10%. Off-site construction techniques, such as prefabrication of structural components and precast units, are important CE-based strategies that successfully reduce waste generation, while enabling reuse of different standard-size components.[25] Prefabricated reinforced concrete structures have the potential to reduce the material footprint by about 50%.[29] At the EoL of structures, recovering at least 50% of the global construction concrete aggregates to produce recycled aggregates has the potential to save approximately 179 million tonnes of sand and gravel by 2060.[29]

SDG 13: Climate action: Adoption of the CE in the construction sector has the potential to reduce GHG emissions in two broad ways: (1) facilitation of resource recovery, reuse, and recycling of construction materials, which leads to reduced emissions caused by mining or extraction of virgin resources for construction, and (2) recirculation of construction materials and structural components, which reduces the emissions caused by disposal of used construction materials in landfills. It is reported that reuse and recycling of construction waste has the potential to save up to 54% and 6.22% of the embodied energy, respectively.[61] Retaining the foundation, load-bearing structure and ceiling elements for subsequent building can aid in reducing 14% of the total emissions associated with the construction of a building.[62] Partial replacement of cement or aggregates using recycled glass, CDW aggregates, fly ash, *etc.*, has the potential to reduce GHG emissions by 19%, energy consumption by 17% and also enable significant cost savings.[63] Across the life cycle of a structure, strategies such as light and efficient design, material substitution, extended service life using durable materials, increased service efficiency, regular maintenance, reuse, and recycling can potentially reduce the global GHG emissions by 50 Gt CO_2 eq. by 2050.[64]

The FastCarb project – an initiative to build a scalable and sustainable model for carbon capture and storage – has been using recycled concrete aggregates to decrease the carbon dioxide impact of its structures.[65] It has been estimated that, for every tonne of concrete, about 10–50 kg of CO_2 equivalent can be stored in a structure using recycled aggregates. This amount of carbon storage is dependent on the natural carbonation capacity of the aggregates, water content, regional temperature, and size of the recycled aggregates.

SDG 15: Life on land: Efficient use of resources in the construction of the built environment can aid in achieving SDG 15 through biodiversity conservation.[66] Diverting construction wastes from landfills to construction systems *via* reuse, refurbishment, and recycling can aid in preserving the land-based ecosystem by reduced emissions and minimizing interaction of toxic materials with the local biosphere.

SDG 17: Partnership for the goals: The substance of the CE lies in creating avenues for developing mutual trust and partnerships among diversified stakeholders such as consumers, designers, manufacturers, as well as policymakers, business owners, and governments.[67] Development of regional, nationwide, and even global standards for material recirculation *via* reuse or refurbishment in the construction sector can pave the way to opportunities for partnership and collaboration among different stakeholders globally. A global pathway to achieving a circular built environment would aid in standardization of structural parameters and facilitate ease of reuse of secondary components and construction materials. Finally, since the adoption of the CE in the built environment sector is in its infancy, setting examples and guidelines using practical examples and sharing the same with other practitioners is a novel and unique approach to build strong partnership with other nations.

9.3.2 Indirect Contribution

The remaining SGDs (1, 2, 4, 5, 10, 14, and 16) are impacted by CE adoption in a consequential manner – CE-based actionable insights have indirect impacts on achieving these SDGs. CE adoption aids in generating more employment opportunities, resulting in economic and social development, thus reducing the impact of existing inequalities. Moreover, the development of a wide spectrum of businesses and employment opportunities around the circular supply chain creates opportunities for cultivating gender-neutral roles, and training and education to meet the required skill sets.

SDG 1: No poverty: This SDG focusses on ending poverty in all its forms. An important aspect of CE adoption in the construction and built environment sector is recirculation of construction materials and structural components back into the construction ecosystem. This closed-loop supply of resources requires human input in terms of physical labour – for deconstruction of structures, resource recovery and segregation, and refurbishment and recycling of materials. CE-based operations also need human coordination for supply chain integration and collaboration, and thus require managerial and tech-based support to identify and predict material stocks and cater to the supply–demand balance in the regional construction ecosystem. This increased requirement for human assistance to facilitate CE strategies has the potential to create more jobs across the life cycle of construction projects and the circular supply chain of resources in the built environment sector. Increased employment opportunities, along with new CE-based businesses, will foster economic development and reduce regional, as well as global, poverty.

SDG 2: Zero hunger: The aspect of food security with respect to the adoption of a CE in the built environment can be perceived in two dimensions. First, CE adoption will foster increased employment opportunities, the majority requiring higher-level skill sets, thus facilitating economic development among different strata of the societies. Improved financial security will enable people to afford food and build resilience against fluctuations in food prices. The second dimension to food security *via* CE adoption is that of ecological conservation. Reduced mining activities by procuring materials *via* urban mining will reduce land and water pollutions. Lower levels of biosphere toxicity and preservation of freshwater sources will aid in maintaining a sustainable agricultural sector for global food requirements.

SDG 4: Quality education: Resource recovery, segregation, and reuse and refurbishment of structural components require skilled and trained staff. Furthermore, CE adoption is synchronous with awareness, training, and education about efficient designs, product longevity, maintenance, repair, and EoL resource recovery. Thus, CE adoption facilitates and builds lifelong learning opportunities for different stakeholders – from workers and value chain actors to governments and policymakers.

SDG 5: Gender equality: CE adoption will create new and increased employment opportunities across the built environment supply chain encompassing the Internet of Things (IoT), AI, and blockchain-based services. Furthermore, there will be a continuous requirement for technical and managerial support for resource recovery, reuse, and material recirculation in the built environment ecosystem. These new opportunities will increase the ratio of female engagement in the sector, thus fostering gender equality, with equitable distribution of work opportunities, empowering individuals to select work based on their preferences and skill sets.

SDG 10: Reduced inequalities: Incorporation of CE-based construction practices has the potential to significantly improve the social conditions of the workers engaged in this sector. Resources and different structural components are recovered at the EoL of different structures. This process of resource recovery – involving disassembly of structural components, segregation of resources, recycling, *etc.* – is crucial to recirculation of resources back into the construction economy and employs large-scale skilled labour. The demand of skilled labour warrants higher wages, and, consequentially, improved social conditions. Additional employment opportunities will also be created in the allied sectors of transportation, storage, recycling facilities, *etc.*, further fostering the social and economic development across communities and regions.

SDG 14: Life below water: CE-based strategies of efficient resource use and elimination of waste and emissions help in conserving the aquatic ecosystem. Reduced mining for virgin materials will create fewer toxic emissions from the mining process and no hazardous substances will infiltrate and pollute the groundwater, rivers, and oceans. Furthermore, minimization of disposal and landfilling of CDW will reduce leaching of chemicals and inorganic substances from the waste and prevent contamination of water bodies. Efficient structural designs will also aid in optimized resource use, and efficient use of water. For instance, using lean construction practices and prefabrication of reinforced concrete structures can aid in reducing the water consumption by a factor of 10 relative to water consumption in wet processing of concrete.[29,68]

SDG 16: Peace, justice, and strong institutions: An essential dimension of maintaining peace and justice across the globe is the availability of resources for nationwide social and economic development of communities. Efficient resource use and closed-loop supply of resources – outcomes of CE – have the potential to safeguard resources for judicious use in the future. This will aid in the equitable distribution of construction resources, at least regionally, and reduce dependency on imported resources from distant locations. In addition, global CE adoption will require accountable institutions at different levels of the society and thus aid the building of strong collaboration across regions, localities, and nations.

9.4   The Way Forward

A sustainable construction is appropriately described "as the means by which it is possible to create and responsibly manage a built environment able to ensure the healthiness of inhabitants and, at the same time, of natural environment, by saving its resources and controlling pollution."[69] To realize sustainable development in the construction and the built environment sector, every activity – from the design of the structure to its EoL – needs to be coordinated and executed based on the sustainable development goals.[70] The CE provides an approach towards achieving the SDGs. Integration of circular design across the life cycle of the construction projects is a prerequisite for realizing the closed-loop material systems-based model, an indispensable component of the CE.[17]

Adoption of the CE in the built environment sector has gained traction in the past few years and its penetration into the sector continues to grow globally. However, to achieve the SDGs, the transition to a CE needs to be carefully and comprehensively planned throughout the different life-cycle stages of the construction and the built environment ecosystem.

In the design phase, structures and their components are required to be designed for multiple phases of use, thus fostering reuse of structural components and resources. The design phase is an important phase for adopting lean designing and the manufacture of structures for disassembly and deconstruction, incorporating advancing and innovation dimensions of the IoT, AI and 3D printing to foster the CE.[71]

In the construction phase, measures are required to ensure efficient and optimum material usage and waste elimination. Use of secondary materials for reuse of construction materials requires prioritization over recycling.[42] Development of International Organization for Standardization (ISO) standards for reuse would aid in increasing the percentage of reuse of construction materials.[72]

At the EoL phase of any structure, maximum resource recovery could be facilitated by determining the economic and residual value of resources at different stages of its life cycle.[73] Resource recovery and segregation can be facilitated by hands-on training for construction professionals. Development of reverse networks and availability of markets for secondary materials would aid in increased resource recovery from structures at their EoL.[16] Development of storage facilities also provides safe storage of secondary materials and ensures the supply of the required quantity of uniform construction materials,[74] incentivizing consumers and construction firm managers or owners to use more secondary materials.

Apart from the above approaches, CE-based strategies require continuous and consistent application to embed the recirculation of materials and resources. Stakeholder coordination is the first and most important approach to enabling CE in the built environment.[75] This includes capacity building and knowledge sharing among different stakeholders.[76] Developing transparency in material and resource use will help structural engineers and material contractors to determine the strength and serviceability estimates of recovered materials. This can be supplemented with information brokerage among different value chain actors.[77] Using material flow analysis or material stock assessment for traceability information on resources across the life-cycle stages of different structures will aid in predicting their urban mining potential at different time periods. AI or blockchain technology can be employed for this purpose. Green supply chain management or tracking of building elements can be performed using BIM and can be used through the supply chain.[23]

Furthermore, development and implementation of resource mapping procedures,[78] which includes determination of carbon dioxide offsets achieved by material recirculation, will support the prediction of the environmental impacts of CE adoption. Life-cycle Assessment (LCA) can also be used to determine the environmental impacts associated with the use of different resources as well as reuse or recycling of structural components.[79,80] Other enabling conditions for CE adoption include active steps from governments and construction project developers to foster consumer awareness; development of infrastructure to facilitate reuse and recycling of construction materials; eco-clustering of businesses; framing of laws and legislations for reuse, remanufacture, and recycling of construction materials; and building financial capabilities to reduce risks in the markets for secondary materials. Finally, strong policies and economic instruments such as extended producer responsibility, Pay-As-You-Throw (PAYT), landfill bans, and construction waste disposal charging schemes, will foster the mainstreaming of CE practices.

9.5   Conclusions

A comprehensive transition to the CE in the construction sector – encompassing economic, environmental, technical, and social aspects of the system – can be a potent enabler to achieve the UN SDGs. CE-based strategies applied to the construction sector lead to environmental benefits and economic savings, thus achieving the SDGs of climate action, and responsible production and consumption. Important aspects of CE adoption are closed-loop supply chains, fostering resource reuse, supply chain integration, and the development of technical and physical infrastructure for resource recirculation. These CE strategies help achieve several SDGs, in particular: sustainable cities and communities; partnership for achieving goals; industry and innovation; and responsible consumption and production. The reduced pollution and emissions from the sector will ensure the protection of life on land and life in water. This in turn will aid in sustaining the agricultural sector and thus ensure a sufficient food supply, addressing the SDG of zero hunger, with other consequential benefits.

The inclusion of circular strategies and pathways in the construction sector has the potential to reduce the global environmental impacts of the sector as well as contributing to the achievement of SDG targets. What is now needed as a way forward and a call to action is the involvement of all the stakeholders to develop a roadmap with set standards and regulations that can readily be enforced in order to achieve the SDGs and create a sustainable global construction industry. While the inter-relationship between a CE and achievement of the SDGs has highlighted numerous positive synergies, an all-inclusive achievement of SDGs is dependent on the willingness and motivation of different stakeholders engaged in the construction and the built environment sectors. This challenge is further complicated by motivations driven by profit and less for the environment and society, further widening the gap between economic, environmental, and social gains.

Although the path towards achieving the SDGs requires all-inclusive global action between governments and industries, and different stakeholders and societies, CE adoption provides a clear path towards targetable and achievable insights. Efficient policies and standards, coupled with technical and regulatory assistance and supported by the necessary capital and other resources can pave way towards the actualization of a CE-based sustainable built environment.

References

1. W. Chen, R. Jin, Y. Xu, D. Wanatowski, B. Li, L. Yan, Z. Pan and Y. Yang, *Constr. Build. Mater.*, 2019, **218**, 483.
2. UN Environment and IEA, *Towards a Zero-emission, Efficient, and Resilient Buildings and Construction Sector*, Global Status Report 2017, Nairobi, 2017.
3. M. H. A. Nasir, A. Genovese, A. A. Acquaye, S. C. L. Koh and F. Yamoah, *Int. J. Prod. Econ.*, 2017, **183**, 443.

4. M. S. Bhat, Q. S. Afeefa, K. P. Ashok and A. G. Bashir, *J. Ecol. Nat. Environ.*, 2014, **6**(1), 1.
5. P. Ghisellini, A. Ncube, G. D'ambrosio, R. Passaro and S. Ulgiati, *Energies (Basel)*, 2021, **14**(24), 8561.
6. E. G. Hertwich, *Nat. Geosci.*, 2021, **14**(3), 151.
7. R. Andersen and K. Negendahl, *J. Build. Eng.*, 2022, **65**, 105696.
8. M. Drewniok, C. Dunant, J. Allwood, T. Ibell and W. Hawkins, *Ecological Economics*, 2022, **205**, 107725.
9. A. T. Lima, S. G. Simoes, D. Aloini, P. Zerbino, T. I. Oikonomou, S. Karytsas, C. Karytsas, O. S. Calvo, B. Porcar and I. Herrera, *Resour. Conserv. Recycl.*, 2023, **190**, 106808.
10. Ellen MacArthur Foundation, *Transitioning to a Circular Economy Business*, Cowes, 2013.
11. I. E. Nikolaou, N. Jones and A. Stefanakis, *Circular Economy and Sustainability* 2021, **1**(2), 783.
12. S. K. Ghosh, in *Circular Economy: Global Perspective*, ed. S. K. Ghosh, Springer, Geneva, 2019, pp. 1–23.
13. R. Merli, M. Preziosi and A. Acampora, *J. Clean. Prod.*, 2018, **178**, 703.
14. Republic of South Africa, *Sustainable Development Goal in South Africa: Voluntary National Review 17 July 2019*, https://sustainabledevelopment. un.org/content/documents/24474SA_VNR_Presentation__HLPF_17_ July_2019._copy.pdf (accessed May 2023).
15. A. Osterwalder and Y. Pigneur, Business Model Generation: A Handbook for Visionaries, Game Changers, and Challengers, John Wiley & Sons, Oxford, 2010.
16. M. Lewandowski, *Sustainability (Switzerland)*. 2016, **8**(1), 43.
17. S. Geisendorf and F. Pietrulla, *Thunderbird International Business Review*, 2018, **60**(5), 771.
18. M. Esposito, T. Tse and K. Soufani, *Calif. Manage. Rev.*, 2018, **60**(3), 5.
19. E. Eray, B. Sanchez and C. Haas, *Buildings*, 2019, **9**(5), 1.
20. K. Anastasiades, J. Goffin, M. Rinke, M. Buyle, A. Audenaert and J. Blom, *J. Clean. Prod.*, 2021, **298**, 126864.
21. G. Foster and R. Saleh, *Resour. Conserv. Recycl.*, 2021, **175**, 105880.
22. J. H. Arehart, F. Pomponi, B. D'Amico and W. Srubar, *Resour. Conserv. Recycl.*, 2022, **186**, 106583.
23. C. Wu, J. R. Rzasa, J. Ko and C. C. Davis, in *Laser Communication and Propagation through the Atmosphere and Oceans VI*, ed. J. P. Bos, A. M. J. van Eijk and S. M. Hammel, Proceedings of SPIE, 0277-786X, V. 10408, 2017.
24. B. Mignacca, G. Locatelli and A. Velenturf, *Energy Policy*, 2020, **139**, 111371.
25. T. M. O'Grady, N. Brajkovich, R. Minunno, H. Y. Chong and G. M. Morrison, *Energies (Basel)*, 2021, **14**(13), 1.
26. T. O'Grady, R. Minunno, H. Y. Chong and G. M. Morrison, *Resour. Conserv. Recycl.*, 2021, **175**, 105847.
27. L. A. Akanbi, L. O. Oyedele, K. Omoteso, M. Bilal, O. O. Akinade, A. O. Ajayi, J. M. D. Delgado and H. A. Owolabi, *J. Clean. Prod.*, 2019, **223**, 386.

28. M. Lanau and G. Liu, *Environ. Sci. Technol.*, 2020, **54**(7), 4675.
29. C. Mostert, C. Weber and S. Bringezu, *Recycling*, 2022, 7(2), 13.
30. O. O. Akinade and L. O. Oyedele, *J. Clean. Prod.*, 2019, **229**, 863.
31. N. Shehata, O. A. Mohamed, E. T. Sayed ET, M. A. Abdelkareem and A. G. Olabi, *Sci. Total Environ.*, 2022, **836**, 155577.
32. P. Mhatre, V. Gedam and S. Unnikrishnan, *Resources Policy*, 2021, **74**, 102446.
33. É. Mata, S. Harris, A. Novikova, A. F. P. Lucena and P. Bertoldi, *Resour. Conserv. Recycl.*, 2020, **158**, 104817.
34. K. Skanberg, M. Berglund and A. Wijkman, *The Circular Economy and Benefits for Society: Jobs and Climate Clear Winners in an Economy Based on Renewable Energy and Resource Efficiency – A study Pertaining to Finland, France, the Netherlands, Spain and Sweden*, The Club of Rome, 2014.
35. M. Geissdoerfer, P. Savaget, N. M. P. Bocken and E. J. Hultink, *J. Clean. Prod.*, 2017, **143**, 757.
36. B. C. Guerra, S. Shahi, A. Molleai, N. Skaf, O. Weber, F. Leite and C. Haas, *J. Clean. Prod.*, 2021, **324**, 129125.
37. W. Li and W. Lin, in *Towards a Circular Economy: Corporate Management and Policy Pathways*, ed. V. Anbumozhi and J. Kim, ERIA, Jakarta, 2016, pp. 95–112.
38. G. Tokazhanov, O. Galiyev, A. Lukyanenko, A. Nauyryzbay, R. Ismagulov, S. Durdyev, A. Turkyilmaz and F. Karacaet, *J. Clean. Prod.*, 2022, **362**, 132293.
39. J. Freire-González, V. Martinez-Sanchez and I. Puig-Ventosa, *Waste Management*, 2022, **139**, 50.
40. F. Russo, C. Oreto and R. Veropalumbo, *Resour. Conserv. Recycl.*, 2022, **187**, 106633.
41. M. Contreras-Llanes, M. Romero, M. J. Gázquez and J. P. Bolívar, *Materials*, 2021, **14**(21), 6605.
42. M. Wahlström, J. Bergmans, T. Teittinen, J. Bachér, A. Smeets and A. Paduart, *Construction and Demolition Waste: Challenges and Opportunities in a Circular Economy*, European Environment Agency, Copenhagen, 2020.
43. J. Suárez-Macías, J. M. Terrones-Saeta, A. Bernardo-Sánchez, A. Ortiz-Marqués, A. M. Castañón and F. A. Corpas-Iglesias, *Materials*, 2022, **15**(3), 1076.
44. A. B. López-García, M. Uceda-Rodríguez, S. León-Gutiérrez, C. J. Cobo-Ceacero and J. M. Moreno-Maroto, *Constr. Build. Mater.*, 2022, **345**, 1.
45. L. M. Ottosen, D. Thornberg, Y. Cohen and S. Stiernström, *Resour. Conserv. Recycl.*, 2022, **176**, 105843.
46. M. Rangelov, H. Dylla, A. Mukherjee and N. Sivaneswaran, *J. Clean. Prod.*, 2021, **283**, 124619.
47. Netherlands Enterprise Agency (NEA) and Holland Circular Hotspot (HCH), *Circular Economy & SDGs: How Circular Economy Practices Help to Achieve the Sustainable Development Goals*, 2020, https://holland circularhotspot.nl/publications/ (accessed May 2023).
48. A. Murray, K. Skene and K. Haynes, *J. Bus. Ethics*, 2017, **140**(3), 369.

49. O. E. Ogunmakinde, T. Egbelakin and W. Sher, *Resour. Conserv. Recycl.*, 2022, **178**, 106023.

50. United Nationa Development Programme. *Transforming Our World: The 2030 Agenda for Sustainable Development*, UN, New York, 2019.

51. V. Forti, C. P. Baldé, R. Kuehr, G. Bel, *The Global E-waste Monitor 2020: Quantities, Flows, and the Circular Economy Potential*, United Nations University, Tokyo, 2020. Available at: https://ewastemonitor.info/gem-2020/ (accessed May 2023).

52. P. Mhatre, R. Panchal, A. Singh and S. A. Bibyan, *Sustain. Prod. Consum.*, 2021, **26**, 187.

53. Y. A. Villagrán-Zaccardi, A. T. M. Marsh, M. E. Sosa, C. J. Zega, de N. Belie and S. A. Bernal, Resour. *Conserv. Recycl.*, 2022, **177**, 105955.

54. M. U. Hossain, S. T. Ng, P. Antwi-Afari and B. Amor, *Renewable and Sustainable Energy Reviews*, 2020, **130**, 109948.

55. A. Shojaei, R. Ketabi, M. Razkenari, H. Hakim and J. Wang, *J. Clean. Prod.*, 2021, **294**, 126352.

56. A. Luciano, L. Cutaia, F. Cioffi and C. Sinibaldi, *Environ. Sci. Poll. Res.*, 2021, **28**(19), 24558.

57. J. Li, W. Sun, H. Song, R. Li and J. Hao, *Sustain. Cities Soc.*, 2021, **71**, 102956.

58. L. A. López Ruiz, X. Roca Ramón and S. G. Domingo, *J. Clean. Prod.*, 2020, **248**, 119238.

59. Q. Chen, H. Feng and B. G. Garcia de Soto, *J. Clean. Prod.*, 2022, **335**, 130240.

60. World Economic Forum, *Platform for Accelerating the Circular Economy*, 2017, https://www3.weforum.org/docs/WEF_PACE_Platform_for_Accelerating_the_Circular_Economy.pdf (accessed May 2023).

61. W. Y. Ng and C. K. Chau, *Energy Procedia*, 2015, **75**, 2884.

62. H. Kröhnert, R. Itten and M. Stucki, *Build. Environ.*, 2022, **222**, 109409.

63. M. Nodehi and V. M. Taghvaee, *Glass Structures and Engineering*, 2022, **7**(1), 3.

64. S. Pauliuk, N. Heeren, P. Berrill, T. Fishman, A. Nistad, Q. Tu, P. Wolfram and E. G. Hertwich, *Nat. Commun.*, 2021, **12**(1), 5097.

65. J. M. Torrenti, O. Amiri, L. Barnes-Davin, F. Bougrain, S. Braymand, B. Cazacliu, *et al.*, *Case Studies in Construction Materials*, 2022, **17**, e01349.

66. A. Opoku, *Resour. Conserv. Recycl.*, 2019, **141**, 1.

67. O. Persson, *What is Circular Economy? The Discourse of Circular Economy in the Swedish Public Sector*, Uppsala University, Department of Earth Sciences, 2015.

68. C. Mostert, H. Sameer, D. Glanz and S. Bringezu, *Resour. Conserv. Recycl.*, 2021, **174**, 105767.

69. E. Conte, *Sustainability (Switzerland)*, 2018, **10**(6), 2092.

70. T. Guerin, *Solar Energy*, 2017, **146**, 94.

71. F. Setaki and A. van Timmeren, *Build. Environ.*, 2022, **223**, 109394.

72. S. Marinelli, M. A. Butturi, B. Rimini, R. Gamberini and M. A. Sellitto, *Sustainability (Switzerland)*, 2021, **13**(18), 10257.

73. L. Jiang, S. Bhochhibhoya, N. Slot and R. de Graaf, *Resour. Conserv. Recycl.*, 2022, **186**, 106541.
74. P. Antwi-Afari, S. T. Ng and M. U. Hossain, *J. Clean. Prod.*, 2021, **298**, 126870.
75. R. Charef, *Clean. Eng. Technol.*, 2022, **8**, 100454.
76. F. Bucci Ancapi, K. van den Berghe and E. van Bueren, *J. Clean. Prod.*, 2022, **373**, 133918.
77. M. K. C. S. Wijewickrama, R. Rameezdeen and N. Chileshe, *J. Clean. Prod.*, 2021, **313**, 127938.
78. T. B. Christensen, *Resources, Conservation and Recycling Advances*, 2022, **15**, 200104.
79. J. Devènes, J. Brütting, C. Küpfer, M. Bastien-Masse and C. Fivet, *Structures*, 2022, **43**, 1854.
80. A. Sáez-de-Guinoa, D. Zambrana-Vasquez, V. Fernández and C. Bartolomé, *Energies (Basel)*, 2022, **15**(13), 4747.

CHAPTER 10

Circular Biowaste Management and its Contribution to the Sustainable Development Goals

ZOË LENKIEWICZ*

The Global Waste Lab, 128 City Road, London EC1V 2NX, UK
*E-mail: zoe@globalwastelab.com

10.1 Introduction and History of Biowaste Management

In terms of a circular economy being vital to sustaining life on earth, maintaining and restoring the biological cycle is perhaps the most important.

Humans have been managing biowaste since the Stone Age: evidence from Scotland suggests farmers were planting crops in compost at least 12 000 years ago.[1] One of the oldest existing references to the use of "manures" in agriculture was found on a set of clay tablets from the Akkadian Empire in the Mesopotamian Valley (in modern-day Iraq), around 2300 B.C.[2] The tribes of Israel were reported to have composted manure with street sweepings and organic refuse outside the city walls.[3] The ancient Greeks and Romans made use of agricultural "waste" to feed livestock,[4] and by the third century B.C., Chinese farmers were using anaerobic digestion techniques to fertilise rice paddies.[5]

Although these farmers were unlikely to have used the term "circular biowaste management", they had clearly mastered the essence. This type of regenerative agriculture continued throughout history with farmers using

Issues in Environmental Science and Technology No. 51
The Circular Economy: Meeting Sustainable Development Goals
Edited by Sadhan Kumar Ghosh and Gev Eduljee
© The Royal Society of Chemistry 2024
Published by the Royal Society of Chemistry, www.rsc.org

excess organic matter as mulch for crop planting, thereby replenishing their soils.

Between the 1930s and the late 1960s, the Green Revolution introduced new methods and technologies that increased agricultural production worldwide, creating and driving the market for synthetic fertilisers and pesticides. These products, originating from the oil and gas sector, have progressively replaced natural biowaste fertilisers. While they deliver certain productivity benefits, they rely upon linear resource use and result in the degradation of soils and the pollution of aquatic ecosystems.

This chapter discusses how circular biowaste management, in both the traditional and modern senses, can restore the natural balance, preserve soils, protect water sources and prevent major sources of anthropogenic greenhouse gas emissions, and how such approaches are vital to delivering the Sustainable Development Goals.

10.2   Major Threats to Soil Health

Soil is a vital, non-renewable resource for ecosystems, playing an essential role in services such as water purification and food production. It is also a major carbon sink, with the ability to remove greenhouse gases from the atmosphere.

The notion of preserving soil functionality has been embedded in the land-degradation-neutrality concept as part of the United Nations' Sustainable Development Goals (SDGs).[6] Soil provides 95% of our food, plays a vital role in the water cycle, is home to about a quarter of all the planet's biodiversity and is the largest carbon store on the planet.[7] Soil is also a finite resource, meaning its loss and degradation is not recoverable within a human lifespan.[8] Worryingly, 90% of all soils are likely to be degraded by 2050,[9] and soil loss and declining fertility have been identified as the main threats to sustainable development.[10] In England and Wales, soil degradation was calculated in 2010 to cost GBP 1.2 billion every year.[11]

10.2.1   The Importance of Soil Microbiology

One of the key causes of soil loss and degradation is the widespread application of synthetic nitrogen, phosphorous, potassium (NPK) fertilisers.

As a plant grows, it releases exudates (organic acids, amino acids, proteins, sugar, phenolics and other secondary metabolites produced through photosynthesis) into the soil through its roots. The cocktail of chemicals released is influenced by plant species, soil and climactic conditions, which together shape and are shaped by the microbial community within the rhizosphere (the area surrounding the roots).[12] The exudates attract "symbiotic plant growth promoting rhizobacteria" that may enhance plant growth either directly, for example by increasing nutrient availability, or indirectly by stimulating plant defences against pests or pathogens[13] (see Figure 10.1). It is for these reasons that "no dig" gardening techniques have grown in popularity over recent years.

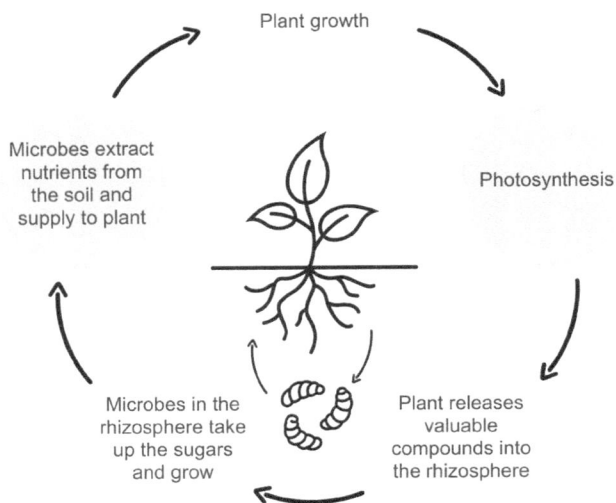

Figure 10.1   The symbiotic relationship between plants and soil microbes.

Synthetic fertilisers are designed to provide the plant with the nutrients it needs to grow in the short-term. As a result, the plant does not invest in the production of exudates to attract symbiotic microbes, leading to a significant decrease in bacterial diversity in the soil.[14] Over time, the fertilisers wash away, leaving no nutrients and no microbes to exchange nutrients for sugar and other vital compounds.

The plant starts exhibiting signs of nutrient deficiency, resulting in chlorosis (yellowing of leaf tissue) among other symptoms. The plant then puts all its energy into the seed to have the best chance of reproduction. The absence of soil microbes renders the plant completely dependent on fertilisers. It can take only a few seasons for the soil microbiology to become almost irreversibly degraded.[15]

Conversely, the application of compost and manures replenishes the soil with healthy populations of bacteria, fungi and other microorganisms that can exchange nutrients for sugar and thereby support plant growth. Research has demonstrated that organic fertilisers made from biowaste provide similar nutrient elements to synthetic fertiliser, while reducing issues such as soil degradation, nitrogen leaching, soil compaction, reduction in soil organic matter and loss of soil carbon.[16]

10.2.2   Global Phosphorous Shortage

Not only are synthetic fertilisers harming soils worldwide, but they are also increasingly in short supply. One of the key ingredients of synthetic fertilisers is phosphorous, an element upon which life on earth depends. Phosphorous promotes growth and reproduction and is necessary for all major building blocks of life, including ATP (the energy carrier of life), RNA, DNA and cellular membranes.[17,18]

Excessive use of chemical fertilisers is depleting phosphorous reserves to the point where global food production could be disrupted. Phosphate occurs naturally in rock found most abundantly in Morocco and Western Sahara, followed by China and Algeria. Reserves in the USA have fallen to 1% of previous levels, while Britain has always had to rely on imports.[19] Scientists are now warning of "phosphogeddon" as critical fertiliser shortages loom, leaving many nations struggling to feed their own people.

At the same time, excess phosphate fertiliser washed from soils, alongside sewage inputs into rivers, lakes and seas, is causing widespread algal blooms. When these aquatic plants die, their decay in anaerobic conditions at the bottom of the water body leads to methane emissions – a major contributor to the climate crisis.

In 2019, over 500 scientists signed the Helsinki Declaration calling for transformation across food, agriculture, waste and other sectors to deliver urgent improvements to global phosphorous sustainability.[20] The authors of *Our Phosphorous Future* recommended 10 key actions, including the following:

- Increase the use of recycled phosphorous in fertiliser and other chemical industries, as an alternative or supplement to phosphate rock.
- Increase appropriate application of manures, other phosphorous-rich residues, and recycled fertilisers to soils, to complement appropriate mineral fertiliser use.
- Implement national to global strategies to increase recovery and recycling of phosphorous from solid and liquid residue streams.
- Ensure sufficient access to affordable phosphorous fertilisers (mineral, organic and recycled) for all farmers.

10.2.3 Plastic Pollution

While much attention has been paid of late to the leakage of plastic waste into the oceans, much less research has been undertaken into its fate on land. The accumulation of plastic waste is causing one of the most widespread and long-lasting changes to the earth's surface, and its degradation into smaller pieces (particularly microplastics <5 mm in diameter) is a cause of international concern.

Soils represent a significant sink for microplastics,[21,22] and evidence suggests that pollution from microplastics can alter the coupling between carbon and nutrient cycling through significant increases in nutrients in dissolved organic matter and CO_2 fluxes.[23] The extent to which microplastics could impair a range of ecosystem processes mediated by soil organisms, such as organic matter decomposition and nutrient cycling, is largely unknown.[24] In 2020, the first-ever field study to explore how the presence of microplastics can affect soil fauna was published in *Proceedings of the Royal Society*. The paper notes that terrestrial microplastic pollution has led to the decrease of species that live below the surface, such as mites, larvae and other tiny creatures that maintain the fertility of the land.[25] The authors called for a reduction in the use of plastics and to avoid burying plastic wastes in soils, to prevent as yet unknown consequences to soil communities and biogeochemical cycles.

10.3   Introduction to Biowaste

10.3.1   Definitions and Compositions

The use of synthetic fertilisers and agricultural plastics are contributing to the degradation of soils worldwide, threatening global food supply and contributing to climate change. After the impacts of agriculture, and the transport of food from farms to cities, comes a whole array of new challenges around managing municipal biowaste. Rapid urbanisation and poor solid waste management has resulted in an accumulation of biowaste in cities, far from the agricultural soils that would benefit from its nutrients and microbes. Instead of being ploughed back into fields, biowaste is filling landfills and dumpsites in every country on every continent, creating unhealthy conditions and contributing further to climate change.

Biowaste can be defined as any waste capable of undergoing anaerobic or aerobic decomposition, including plant and animal waste arising from:

- agriculture and aquaculture,
- the food and drink industry,
- municipal sources such as retail, restaurants, homes, parks and gardens.

Biowaste comprises both food (plant and animal) waste and woody biomass. While the composition and parameters of biowaste vary, biowaste of plant origin contains cellulose, hemicelluloses and lignin, while biowaste of animal origin contains unabsorbed fats, proteins and carbohydrates. Unsurprisingly, biowaste has an elevated Biological Oxygen Demand (BOD) and Chemical Oxygen Demand (COD), and its presence in large quantities can disrupt terrestrial and aquatic ecosystems.

10.3.2   Sources of Biowaste

10.3.2.1   *Food Waste*

Globally, it is estimated that a third of all food grown is lost or wasted, equating to 1.3 billion tonnes per year,[26] impacting food security, food safety, economies and environmental sustainability. Around 14% of food is lost between harvest and the retail market, valued at USD 400 billion,[27] while a further 17% of food is wasted at the retail and consumer levels.[28] Food loss and waste represent not just the forfeiture of nutritional value, but also a waste of the land, water, labour and energy used in food production.

Across the world, biowaste accounts for the largest fraction of municipal waste, comprising 40% of the municipal waste stream in Europe, and 60–70% in lower-income countries.[26] While some of this is captured and returned to beneficial use, the vast majority remains as waste. SDG 12 (Sustainable consumption and production) Target 12.3 aims to halve *per capita* food waste and reduce food loss by 2030. A significant reduction in food loss and waste will also have a positive impact on many other SDGs, not least SDGs 2 (End hunger), 13 (Climate action) and 15 (Life on land). Furthermore, the collection and management of biowaste from urban centres will contribute to SDGs 6 (Clean water and sanitation) and 11 (Sustainable cities and communities). Most countries, however, have very little data on food loss and food waste, hindering meaningful progress towards SDG Target 12.3.

Case Study 1: Reducing Food Waste in the UK

As part of its commitment towards the Sustainable Development Goals (specifically Target 12.3), the UK government has committed to halving the UK's *per capita* food waste by 2030. According to the most recent report by the charity Waste and Resources Action Programme (WRAP),[29] the UK produced around 9.5 million tonnes of food waste in 2018:

- 69% from households (6.6 million tonnes),
- 16% from manufacturers (1.5 million tonnes),
- 12% from hospitality and food service (1.1 million tonnes),
- 3% from retail (0.3 million tonnes).

Some 70% was intended to be consumed by people, while 30% was classified as "inedible parts". WRAP estimated that this waste had a value of over GBP 19 billion a year and would be associated with more than 25 million tonnes of greenhouse gas emissions. Of this waste, 6.4 million tonnes could have been eaten, equivalent to over 15 billion meals.

WRAP concluded that, in order for the UK to meet SDG Target 12.3, further reductions in food waste of 1.8 million tonnes are needed, comprising 1.3 million tonnes from homes and over half a million tonnes from across the supply chain.

10.3.2.2   Woody Biomass

Woody biowaste can include scrap wood, sawmill residues, forestry and horticultural residues (such as branches and leaves), agricultural residues such as cassava stems, nut and coconut shells, straw and corn stalks, and packaging.

From a circular economy perspective, the use of natural biomass products in construction reduces the carbon footprint of buildings, while wood, cellulose and its derivatives (such as lignin) are a viable substitute for non-renewable materials.[30] The diversion of woody biowaste from disposal may be achieved through lumber salvage, paper recycling, composting and the production of biochar.

10.4   Problems from Mismanaged Biowaste

For the value of biowaste to be recovered, contributing to soil health, global food security and reduced greenhouse gas emissions, it needs to be separately collected and processed. Without such treatment, an accumulation of biowaste causes a wide range of negative impacts that prevent progress towards the Sustainable Development Goals.

From the spread of disease to poor water quality, land degradation and climate change, the impacts of mismanaged biowaste are severe, requiring urgent action throughout all of society.

10.4.1   Open Dumping and Burning

While waste collection may be a service that many take for granted, its absence across the Global South is contributing to the spread of disease. Universal waste collections are therefore a necessity to achieve SDG 3 (Good health).

Data compiled for the World Bank's 2018 publication *What A Waste 2.0*[26] suggests that, while high-income countries such as the UK have waste collection rates of 96%, the average rate across low-income countries is a mere 39%. Without a regular collection service, residents and business owners have little option other than to dump or burn their waste. Particularly in urban areas, the negative environmental and health impacts of these poor waste management practices are significant. SDG 11 (Sustainable communities and cities) Target 11.6 calls for special attention to be paid to waste management to reduce the adverse *per capita* environmental impact of cities.

Food waste discarded in the open attracts house flies (*Musca domestica* L.), a common mechanical vector of a diverse range of pathogens including bacteria, fungi, viruses and parasites.[31] In a slum district of Islamabad in Pakistan, researchers at the London School of Hygiene and Tropical Medicine found that the number of flies was five times higher in houses where waste is not collected regularly than in those where it is properly managed,[32] inflicting a wide range of health impacts upon local residents.

The alternative form of waste disposal is the open burning of waste, often carried out in a backyard, which generates black carbon (PM2.5 and PM10, short-lived climate forcers) and a wide range of gaseous air pollutants known to harm human health and contribute further to climate change. The widespread practice of open dumping and burning affects the respiratory, digestive, reproductive and nervous systems and can harm congenital and infant health.[33,34,35,36] However, due to its informal nature, little data is available on the scale of open dumping and burning of waste worldwide, hindering political, technical and financial investment.

10.4.2 Landfill Disposal

10.4.2.1 Landfill Gas

When waste is disposed of to landfill, it soon becomes compacted beneath fresh waste. In the absence of oxygen, methanogenic bacteria decompose the biowaste, releasing methane-rich landfill gas into the atmosphere. Methane (CH_4) has 21 times the global warming potential of carbon dioxide (CO_2) and is a short-lived climate forcer,[37] so efforts to divert biowaste from landfill disposal and into circular management systems will have a relatively rapid positive impact on climate cooling.

Landfills and dumpsites (unlined and/or unmanaged disposal sites) are the third largest source of anthropogenic methane globally, contributing around 15% of global methane emissions.[38] While on modern, well-run landfills, landfill gas may be collected and combusted to generate power (see Section 10.5.6), the majority of historic and active landfill sites (including uncontrolled dumpsites) around the world have no gas control at all. Pockets of landfill gas can spontaneously combust, posing a threat to people working or living nearby and emitting a cocktail of particulate matter and greenhouse gases into the atmosphere. Over just one month in 2022, the Ghazipur landfill in east Delhi experienced three fire outbreaks, sending a dense plume of smoke into the sky and causing local residents to complain of eye irritation and breathing difficulties.[39] The combination of explosive methane gas and significant quantities of combustible plastic waste mean many dumpsites around the world are smouldering continually.[40]

Understanding the extent of methane emissions from waste disposal sites has presented a range of challenges for the international community, not least the cost and complexity of visiting and detecting emissions on site. To overcome these barriers to data collection, the Netherlands Institute for Space Research (SRON) is using a combination of satellite systems to detect methane plumes and identify super-emitters. The TROPOspheric Monitoring Instrument (TROPOMI[41]) on board the Copernicus Sentinel-5 Precursor satellite provides a near-real-time data stream with the ability to detect methane plumes. Broad geolocations are then inputted to GHG Sat,[42] which enables emissions monitoring from landfills and dumpsites at a resolution of 25m.

This level of accurate, independent and verifiable methane emissions data is adding to the evidence that governments and municipalities have to understand the emissions from waste disposal sites, and to integrate measures to prevent emissions into national policies, strategies and planning (SDG Target 13.2).

The Paris Agreement (adopted by 196 countries at the UN Climate Change Conference COP21 in Paris, France, in 2015) to limit global warming to 1.5 °C required countries to develop and implement strategies for reducing greenhouse gas emissions. The separate collection and diversion of biowaste from landfill has since been recognised as a climate mitigation strategy in a significant number of national climate policies.

10.4.2.2 Leachate

The water content in biowaste contributes significantly to the production of landfill leachate, and unlined landfills and uncontrolled dumpsites are a common point source of groundwater pollution.[43,44,45] Rainwater percolates through the waste and mixes with liquids from biowastes and other wastes (such as detergents, paints and battery acids), polluting surface water and groundwater systems.

Landfill leachate is characterised by high levels of ammonia, COD, heavy metals and salinity (chlorides and sulphates) as well as pathogenic micro-organisms, persistent organic pollutants and emerging contaminants such as pharmaceuticals that may bioaccumulate in the food chain and cause harm to human and environmental health.[46,47] In addition, leachate often contains an excess of nutrients causing eutrophication and acidification.[48,49,50]

Leachate can remain in rock, gravel and sand systems beneath a landfill for decades, continuing to pollute groundwater long after a site has been closed and even excavated.[51] The extent of this harm is unknown as very little monitoring takes place, although the growing body of evidence points towards a fast-growing crisis.[52,53,54] Controlled, geo-engineered landfill sites have impermeable membranes that collect leachate and direct it to treatment systems. While it is possible to retrofit such systems to existing sites, disturbing the unknown contents of landfill can result in additional environmental and health risks.

The environmental risk of landfill leachate is addressed in SDG 6 (Clean water and sanitation) Target 6.3, which calls for improved water quality by reducing pollution, eliminating dumping and minimising the release of hazardous chemicals and materials.

10.5   Sustainable Biowaste Management and a Circular Economy

10.5.1   Circularity and the SDGs

Sustainable biowaste management follows the cascading waste hierarchy to maximise the recovery of environmental, social and economic value (see Figure 10.2).

Preventing food loss and waste, redistributing edible food and using unavoidable food waste to feed animals can reduce significantly the quantity of biowaste that needs to be managed. For unavoidable biowaste arisings, a separate collection system that prevents contamination and recovers its value is

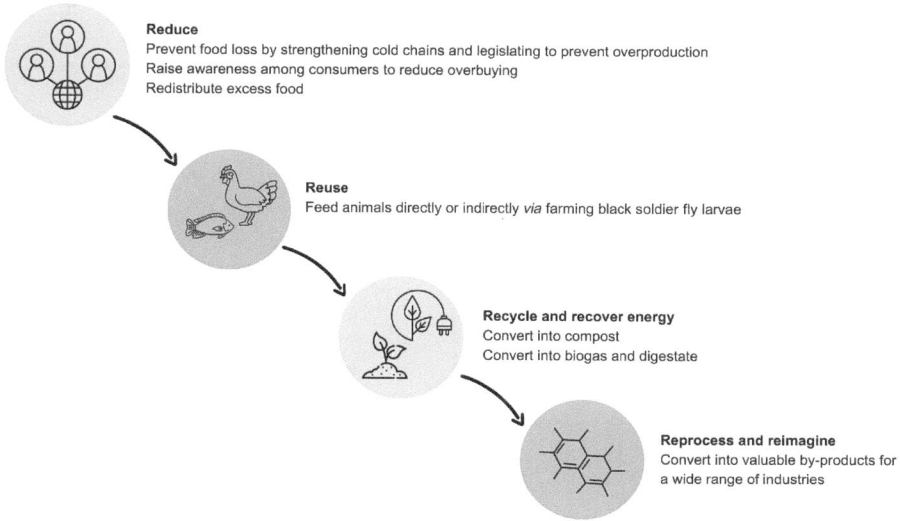

Reduce
Prevent food loss by strengthening cold chains and legislating to prevent overproduction
Raise awareness among consumers to reduce overbuying
Redistribute excess food

Reuse
Feed animals directly or indirectly *via* farming black soldier fly larvae

Recycle and recover energy
Convert into compost
Convert into biogas and digestate

Reprocess and reimagine
Convert into valuable by-products for a wide range of industries

Figure 10.2   Cascading hierarchy of sustainable biowaste management.

fundamental to its sustainable management. Importantly, by returning biowaste to the soil and helping to regenerate agricultural soils, natural habitats and ecosystems, the biological cycle plays a crucial role in the circular economy.

As discussed in the introduction to this chapter, humans have understood the benefits of applying biowaste to agricultural land since prehistoric times. The challenge today, however, is in the separate collection and transfer of biowaste from urban to rural areas, without contamination from other (non-bio) wastes. While a seemingly simple endeavour, the source-separated collection of food waste poses a specific set of challenges. It is bulky, wet, odorous and of little immediate economic value. However, the long-term rewards of capturing and recovering the value from biowaste are numerous and contribute to many of the Sustainable Development Goals, as shown in Table 10.1.

10.5.2   Reducing Food Loss and Waste

Sustainable Development Goal 2 (End hunger) aims to achieve food security and improved nutrition, and to promote sustainable agriculture. While there are structural reasons for hunger beyond the supply chain, food loss and waste remains a priority to address in the fight against global hunger. During agricultural production, food may be lost to pests and disease, floods or drought, harvesting too early or late, or due to oversupply or produce being misshapen. Across the Global South, smallholders are gradually gaining access to vital data through their mobile phones to help them respond to supply and demand and reduce production losses.[55]

Table 10.1   Sustainable biowaste management solutions to meet the Sustainable Development Goals and associated targets.

Sustainable Development Goal and relevant targets	Sustainable biowaste management solution
2. End hunger, achieve food security and improved nutrition and promote sustainable agriculture 2.1, 2.2 End hunger and malnutrition, and ensure universal access to sufficient food all year 2.3, 2.4 Double agricultural productivity and implement resilient agricultural practices that improve land and soil quality	Prevent food loss through strengthening cold chains. Redistribute unwanted food Convert food waste into animal feed to boost agricultural resilience Convert biowaste into compost or digestate and return to agricultural land, displacing the need for chemical fertilisers and pesticides
3. Ensure healthy lives and promote well-being for all 3.3 End water-borne and other communicable diseases 3.4 Reduce by one third premature mortality from non-communicable diseases	Provide universal waste collection to prevent open dumping and burning and associated health impacts
6. Ensure availability and sustainable management of water and sanitation for all 6.6 Improve water quality by reducing pollution, eliminating dumping and minimising release of hazardous chemicals and materials	Divert biowaste from disposal sites Introduce leachate management systems at existing disposal sites
7. Ensure access to affordable, reliable, sustainable and modern energy for all 7.1, 7.2, 7.3 Ensure universal access to affordable, reliable and modern energy services, including renewable and sustainable energy services for all in developing countries	Utilise biowaste to generate renewable energy *via* anaerobic digestion
11. Make cities and human settlements inclusive, safe, resilient and sustainable 11.6 Reduce the adverse *per capita* environmental impact of cities, including by paying special attention to air quality and municipal and other waste management	Collect food waste separately to reduce open dumping and burning
12. Ensure sustainable consumption and production patterns 12.3 Halve *per capita* food waste at the retail and consumer levels and reduce food losses along production and supply chains, including post-harvest losses 12.5 Substantially reduce waste generation through prevention, reduction, recycling and reuse	Prevent food loss through strengthening cold chains Redistribute edible food Collect and sustainably manage unavoidable food waste through a range of measures, including innovations in the bioeconomy

Sustainable Development Goal and relevant targets	Sustainable biowaste management solution
13. Take urgent action to combat climate change and its impacts 13.2 Integrate climate change measures into national policies, strategies and planning	Divert biowaste from landfill disposal Detect methane plumes and identify super-emitters to prioritise for landfill gas capture Recognise the value of sustainable biowaste management in climate policies and nationally determined contributions Utilise biochar for carbon sequestration
15. Protect, restore and promote sustainable use of terrestrial ecosystems, and halt and reverse land degradation 15.2 Halt deforestation 15.3 Restore degraded land and soil	Convert food waste into animal feed, reducing the need for deforestation to grow soy-based products Return treated biowaste to agricultural soils in the form of compost or digestate Utilise biochar to improve soil qualities

Another particular challenge in the Global South is the lack of access to sustainable cooling and cold chains, threatening to put SDG 2 out of reach and stunt the ability of agricultural economies to grow sustainably.[56] A lack of sustainable cold chains is estimated to cause 526 million tonnes of food production loss annually,[57] and contributes to a 15% reduction in the income of smallholder farmers.[58] Access to affordable and clean energy to supply a cold chain is also closely related to SDG 7 (Affordable and clean energy), since refrigeration is a key enabler for scalable food production. A reliable cold chain would enable farmers to grow more nutritious, perishable crops, with the confidence that they will raise a good price at market. This would increase employment in the agricultural sector (a major employer of women in

---

Case Study 2: ColdHubs Provide Cooling-as-a-Service

ColdHubs in Nigeria was the winner of the Cooling as a Service (CaaS) Prize due to its CaaS model that saved 20 400 tonnes of food from spoilage in 2019 alone. ColdHubs provides a "plug and play" modular, solar-powered walk-in cold room, for 24/7 off-grid storage and preservation of perishable foods. The insulated ColdHubs are powered by roof-mounted solar panels with high-capacity batteries feeding an inverter, which in turn feeds the refrigeration unit. ColdHubs installed in farms and markets are extending the freshness of fruits, vegetables and other perishable food from 2 days to about 21 days. Farmers take advantage of the flexible pay-as-you-store subscription model, paying a daily flat fee for each crate of food they store, reducing post-harvest loss by up to 80%.

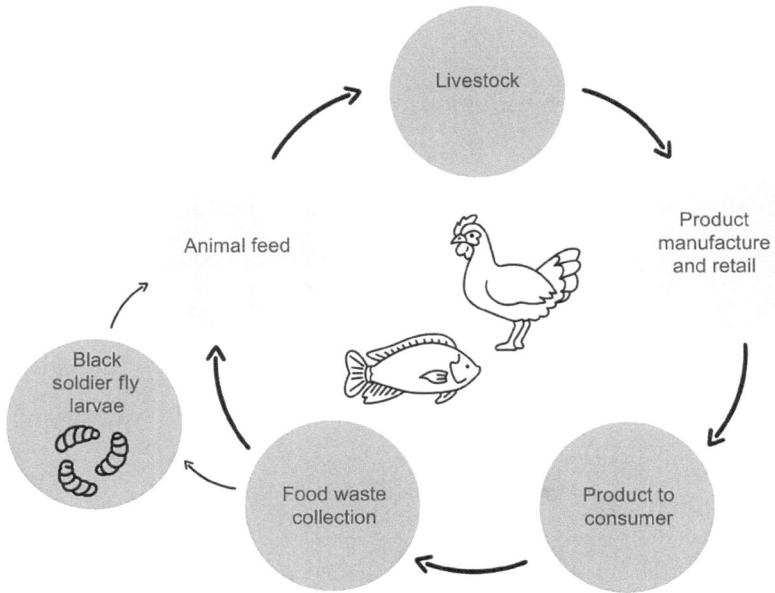

Figure 10.3 Producing animal feed within a circular bioeconomy.

developing markets, thereby supporting SDG 5, Gender equality), and feed up to 950 million more people.[56]

Meanwhile, in the Global North, legislation is targeting the behaviour of buyers to reduce production losses, while imperfect produce is increasingly available for sale.[59] In addition, some governments are enacting policies to encourage and even require the redistribution of unsold food from the retail and food service and hospitality sectors, an activity that has hitherto been delivered by the non-profit sector.

10.5.3   Animal Feed

The circular bioeconomy for animal feed production is shown in Figure 10.3.

Currently, a third of total farmland is used to grow animal food, and producing, processing and transporting this feed contributes to about 45% of the sector's emissions.[60] Using plant and dairy waste as an alternative to soy-based animal feed could see a big drop in agricultural emissions (SDG 13 Climate action) and prevent deforestation (SDG 15 Life on land).

As part of its *No Food Left Behind* programme, WWF and Quantis, Penn State University, assessed data from farms, universities and retailers to explore pathways for transforming food waste into feed for egg-laying hens.[61] The research compared three options against standard feed production:

1. Food waste from retail outlets fed to black soldier fly larvae, processed into meal.

2. Food waste from retail outlets processed into a feed ingredient.
3. Bakery by-products from food manufacturing plants.

A life-cycle assessment considered four environmental impacts: global warming potential, land use, water consumption, and marine eutrophication, and found that each of the three alternatives helped avoid additional land conversion for growing feed crops. The research did not factor in the reduction in greenhouse gas emissions from diverting waste from landfill, which would add further to the environmental benefits of using food waste for animal feed.

Case Study 3: Farming Black Soldier Fly Larvae in Kenya

Black soldier fly larvae (BSFL), *Hermetia illucens*, have been increasingly employed as an environmentally friendly and inexpensive means of transforming biowaste into nutrient-rich animal feed for fish, poultry and pet feeds, as well as fertiliser for soil amendment.[62,63] BSFL thrive in various ranges of organic matter composition and with simple rearing systems.[64]

Hydro Victoria Fish Hatchery Farm in Port Victoria, Kenya, rears fish to sell to aquaculture farmers, hotels, supermarkets and restaurants. Due to the increased costs of feed and fertiliser during the COVID-19 pandemic, the farm diversified vertically into farming black soldier fly. The larvae are made up of 40–60% crude protein, providing an alternative source of feed for fish, poultry and pig farmers. Each tonne of food waste collected from markets is converted into 250 kg (wet weight) of larvae.

The farm buys larvae from contracted women and youths for USD 1 *per* kg to formulate animal feed and also to feed its own Tilapia stock in cages anchored in Lake Victoria. The initiative has improved waste collection, reduced biowaste disposal to the local dumpsite, improved access to quality feeds and enabled farmers to diversify their incomes.

10.5.4 Compost

Composting is the act of collecting and storing organic material so it can decay and be added to the soil to improve its quality (see Figure 10.4). Through separate collection and composting systems, significant biowaste waste can be diverted from disposal and greenhouse gas emissions prevented.

The addition of compost to soils brings a range of benefits, including:

• enhancing water retention in soils,
• improving soil structure and microbiology,

- providing carbon sequestration,
- promoting higher yields of agricultural crops,
- reducing, and in some cases eliminating, the need for chemical fertilisers.

In addition to agricultural and horticultural applications, compost can be used to remediate contaminated soils, and help aid reforestation, wetlands restoration and habitat revitalisation by improving contaminated, compacted and marginal soils. Perhaps most importantly, composting is low cost (with little to no set-up costs), low tech and scalable. Globally, compost is the dominant form of biowaste management, and in Europe over 90% of the separately collected food and garden waste is processed into compost.[6]

The composting process can be undertaken outside (in windrows, or piles, that are turned periodically for aeration) or inside a building or chamber, sometimes with forced aeration (in-vessel composting). In Europe, it is a legal requirement to process any biowaste containing animal by-products in an in-vessel system, to prevent the spread of pathogens from food waste to wild animals.

When making compost, it is vital that the feedstock material has not been mixed with other non-biological wastes to prevent contamination with plastic residues, heavy metals or other pollutants that could bioaccumulate in the food chain[65,66,67] and cause irreversible damage to soils.[68,69] A good-quality compost product also depends on the balance of system inputs, requiring the correct ratio of carbon to nitrogen, as well as aeration and moisture.[70]

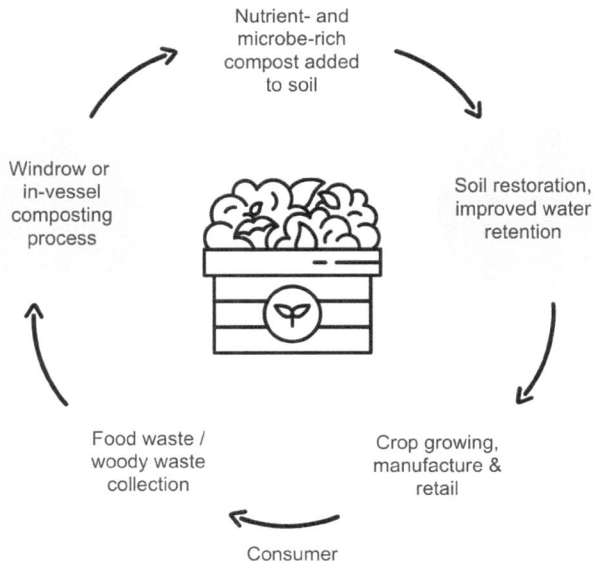

Figure 10.4   The compost cycle.

Many organisations and public agencies have established compost quality programmes or standards to ensure or encourage the production of high-quality compost, especially targeting safety and the environment.[70]

Case Study 4: National Organic Waste Strategy in Chile

The national organic recycling programme in Chile was launched in 2007 with the support of climate finance investment from the Canadian government. The programme has seen the development of 15 central-ised composting facilities, as well as two biogas plants and five landfill gas recovery systems. In 2021, a National Organic Waste Strategy was approved with the goal to recover 30% of municipal organic waste by 2030 and 66% by 2040. More than 40 municipalities have participated in the programme, establishing a range of pathways for the diversion of biowaste from landfill. In addition to infrastructure, the Chilean govern-ment has made clear progress in engaging the population in both food waste reduction and source-separation of biowaste, enabling the pro-duction of a quality compost product fit for agricultural and horticul-tural applications.

Residents have also shown an interest in home composting and ver-micomposting, with demand for home composters outstripping supply. The broad uptake of home composting has reduced the overall costs of waste management, saving space at landfills, reducing greenhouse gas emissions and reducing transport costs for municipalities.[71,72]

10.5.5 Biochar

Biochar, also known as black carbon, is a product derived from plant-derived organic matter (biomass) rich in carbon. Found in soils in very stable solid forms, biochar can persist for thousands of years.[73] Biochar is a fine-grained, highly porous material similar to charcoal, produced from the decomposi-tion of biomass in a low- or zero-oxygen environment, a process known as pyrolysis.[74]

Biochar occurs in soils around the world as a consequence of naturally occurring fires and, in the Amazon basin, as a result of its deliberate addition by past human populations. These so-called *terra preta* soils are renowned for enhancing the fertility of soils and are a highly valued soil amendment.[75] While the properties of biochar depend on its source feedstock and processing con-ditions, it can variously improve overall soil quality, microbial and enzymatic activity, soil organic carbon content, and nutrient retention and availability.[76] When added to soils, the highly porous structure of biochar can act like a slow-release "sponge" for water and beneficial soil nutrients, positively impacting microbial populations and biogeochemistry.[77] These beneficial characteristics have led to the application of biochar for the remediation of contaminated soils and wastewater.[78] Carbon remains sequestered in biochar for centuries, and so sustainable biochar production could be a powerful tool in the fight against anthropogenic climate change.[79] The biochar cycle is illustrated in Figure 10.5.

The UK Biochar Research Centre based at the University of Edinburgh defines a "sustainable biochar system" as one that:

- produces and deploys biochar safely and without emitting non-CO_2 greenhouse gases (such as methane and nitrous oxide);
- reduces net radiative forcing;
- does not increase inequality in access to and use of resources.

Despite being an exothermic process, manufacturing biochar is energy intensive, requiring conditions between 300–1000 °C depending on the process, as well as a tightly controlled environment to eliminate oxygen from the combustion process. The pyrolysis of biomass releases gases including carbon monoxide and dioxide, as well as methane. In a sustainable biochar system, these gases are captured and combusted, with the energy released being used to power the pyrolysis process itself, or alternatively to generate heat and electricity.

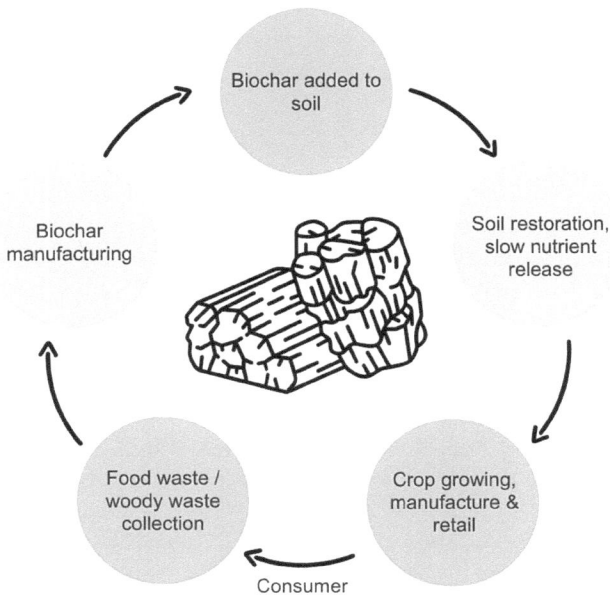

Figure 10.5   The biochar cycle.

Pyrolysis of biomass also generates bio-oil, which can be used to replace fossil fuels in boilers and engines. It is technically feasible to upgrade bio-oil for use as a transport fuel, though the process is currently prohibitively expensive.[80]

In 2021 Suez Energy announced plans to turn sawmill by-products into biochar and other products on an industrial scale,[81] while some companies are supporting biochar in their carbon removal portfolios (Microsoft has contracted carbon removal credits through six biochar projects with a total of >10 000 MtCO$_2$).[82] At around USD 100 per tonne, biochar projects offer one of the lowest-cost engineered carbon-removal systems available.[83]

A comprehensive literature survey in 2023 found that the global biochar market is projected to reach USD 368.85 million by 2028,[84] contributing to net zero emissions and a circular economy in the framework of the UN Sustainable Development Goals.

Case Study 5: Biochar Applications in Construction

Biochar improves the mechanical properties of concrete used in construction materials.[85] Its large surface area and amorphous silica content has been found to increase the compressive strength and tensile strength of conventional concrete.[86] Since biochar does not react with cement, it can be employed as a cement filler to reduce the porosity of cement and increase the durability of concrete.[87] The addition of biochar also improves hydration for cement mortars, thereby enhancing fire resistance and thermal stability.[88] The water retention qualities of biochar have also been found to have beneficial qualities within the bottom layer of green roof substrates,[89] delaying wilting,[90] retaining nutrients,[91] adjusting the substrate temperature, changing the microbial community structure and increasing plant growth,[92] as well as enhancing storm water management and minimising urban flooding.[93]

10.5.6 Biogas

Biogas typically refers to the gas produced from the decomposition of wet biowaste in the absence of oxygen.[94] The process, known as anaerobic digestion (AD), has been in use for centuries. Microorganisms break down the organic materials and produce a methane-rich gas and nutrient-rich digestate. When the process happens in a landfill or dumpsite the resultant gas is referred to as landfill gas (see Section 10.4.2.1). Figure 10.6 illustrates the cycle.

The chemical process of AD consists of four stages: hydrolysis, acidogenesis, acetogenesis and methanogenesis. At each stage, microorganisms progressively break down the long hydrocarbon chains in biowaste, resulting in

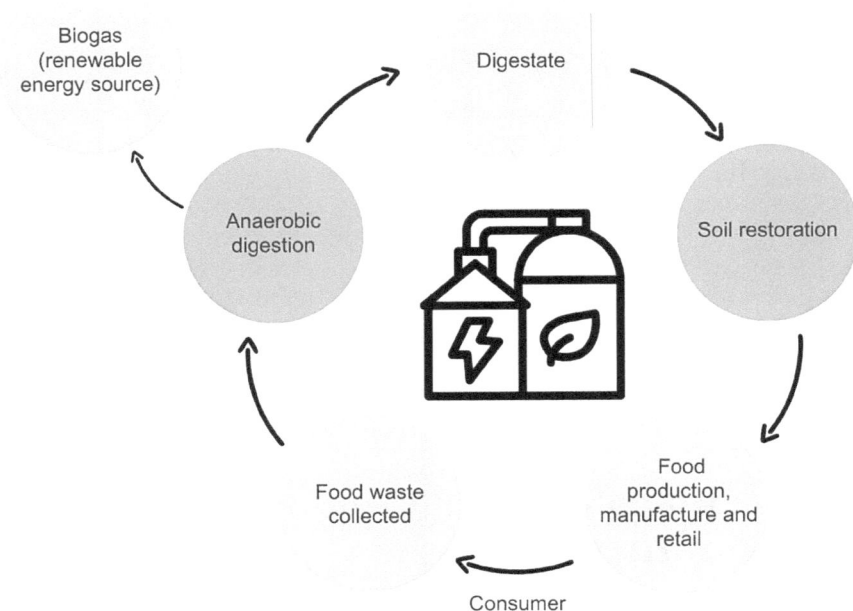

Figure 10.6   The anaerobic digestion cycle.

the production of a digestate and raw biogas, which typically comprises 50–75% combustible methane; 25–50% carbon dioxide; water vapour; and trace amounts of hydrogen sulphide, nitrogen, oxygen and ammonia.[95]

The composition of biogas is largely determined by the feedstock, which may include food waste, sewage, abattoir wastes, industrial residues, manures and crop residues. Anaerobic digestion systems are available in 25-litre kitchen-sized units, producing enough gas for a small family to cook with, through to large industrial-scale facilities. The largest AD plant in the UK, at Poplars in Staffordshire, is able to process 120 000 tonnes of commercial and industrial food and organic waste per year, generating 6 MW of renewable power. While designs vary significantly, all AD systems comprise a feedstock inlet, a sealed chamber for digestion, a sealed chamber for biogas collection and outlets for the biogas and digestate.

Through avoided methane emissions to the atmosphere from the uncontrolled decomposition of biowastes, AD provides climate mitigation (SDG 13) while at the same time providing clean, renewable energy (SDG 7) and a viable treatment technology for waste management (SDG 11 Sustainable communities and cities). Currently, 2000 AD facilities across the European Union, Switzerland, Norway and the UK are treating 29 million tonnes of biowaste every year.[6]

The World Biogas Association estimates that there is sufficient feedstock globally to produce enough renewable energy to meet up to 9% of current global energy needs and prevent up to 13% of global greenhouse gas emissions.[96] In addition, the resulting digestate could substitute up to 7% of global synthetic fertiliser – enough to fertilise sustainably the combined arable land of Indonesia and Brazil (SDG 15).

Biogas can be utilised in a number of applications, as shown in Figure 10.7. It can be used directly for cooking, heating or in an alternating current (AC) generator. Alternatively, it can be upgraded by removing non-combustible elements to produce biomethane. Biomethane can be injected into a national gas grid or compressed to produce Compressed Natural Gas (CNG) or Liquified Natural Gas (LNG).[97] A key feature of biogas is that it is stored energy and can be transported to where there is demand. A further benefit is that, since feedstock is readily available wherever humans live, AD systems can help to overcome issues of energy poverty.

Globally, anaerobic digestion has a vital role to play in creating a more circular economy, treating unavoidable biowaste while preventing greenhouse gas emissions, delivering energy security, creating job opportunities and contributing to multiple Sustainable Development Goals.

10.5.7 Biorefineries and the Circular Bioeconomy

Since the early 21st century, ethanol biorefineries converted food crops into fermentable sugars for fuel. By the turn of the century, however, the expansion of the bioethanol industry with its food crop feedstock had become highly controversial (known as the food *vs.* fuel debate).[101] Utilising unavoidable biowaste as a feedstock for biorefineries and other biobased products therefore brings many advantages both for sustainable development and a circular bioeconomy.[102]

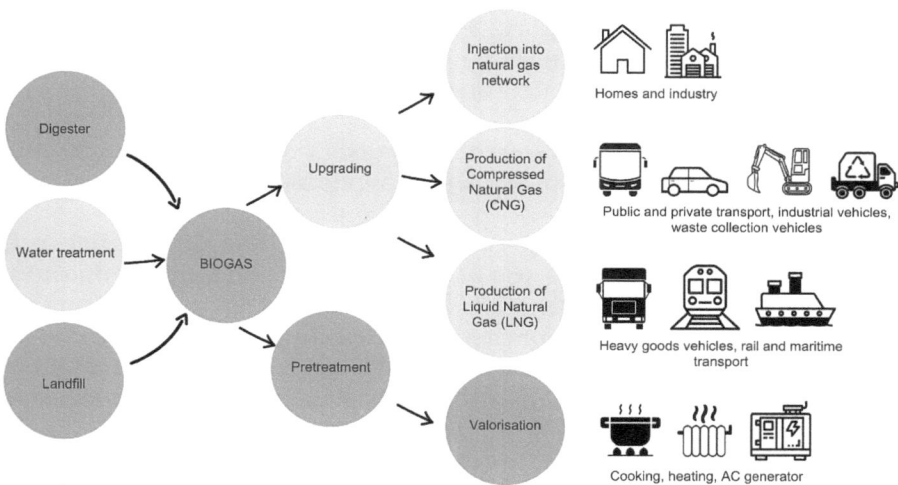

Figure 10.7   Potential sources and applications for biogas.

Case Study 6: Promoting Biogas in India

For the last 30 years, the Government of India has subsidised small domestic-scale biogas systems for rural communities. As of 2019, more than 5 million families had benefitted from the scheme, which provides biowaste and sewage disposal (bringing health benefits associated with SDG 3), alongside clean fuel for cooking and heating.[98] A complementary scheme providing support for off-grid biogas power generation has seen 389 projects completed generating a total of 9000 kW of electricity. Large-scale bioenergy projects utilising municipal food waste and crop residues (providing an alternative to stubble burning, the cause of significant air pollution in many parts of India) are also underway,[99] alongside a target to build 5000 compressed biogas plants across the country by 2024.[100] These plants would provide 15 million tonnes of LNG to enable clean and affordable transport, reducing smog in Indian cities (SDG 11) and bringing positive health outcomes for urban residents (SDG 3).

In biorefineries, biomass (and, as the focus of this chapter, biowaste) is transformed into chemicals, energy, enzymes, fuels and fibres. Biorefineries often work closely with local farms that supply the biowaste feedstock, linking the sustainable chemical industry with agriculture, and giving long-term security for the supplier farms (SDG 8, Decent work and economic growth; SDG 9, Industry, innovation and infrastructure; and SDG 12, Sustainable consumption and production).[103] Scientific and technological advances within the bioeconomy are enabling a new range of applications for biowaste within and across all economic sectors,[104] including biofuels, biochemicals and biomaterials, displacing the need for products with virgin and/or fossil fuel origins.[105] Other examples include the following:

- eggshells and urine can be used as precursors to produce hydroxyapatite (HAp) biomaterial used in biomedical applications, such as orthopaedic and dental implants;[106]
- by-products from poultry, cattle, sheep, goat and pig farming and slaughterhouses can be used to fabricate biopolymers, composites, heart valves, collagen, scaffolds, pigments and lipids, among other industrially important biomaterials;[107]
- squalene from broken rice and rice waste can be used in skincare products, medical applications (wound healing), food (jelly fibre), faux leather and textile fibre;[108]
- lignin from woody biowaste can be upgraded to produce vanillin, benzene, toluene, xylene, quinones and eugenol for use in the chemical, pharmaceutical, cosmetic and textile industries.[109]

Due to its value and economic potential, sustainable and circular biowaste management is likely to attract a lot of public attention and funding in the coming decade. The "owners" of the feedstock are typically municipalities that also have the obligation to manage it.[110] While these scientific developments hold great potential for the circular bioeconomy to contribute to the Sustainable Development Goals, clear regulation and financial incentives remain essential to enable innovations to scale from *niche* to large-scale market applications.[111]

Case Study 7: BioPreferred Programme in the USA

The US Department of Agriculture runs the "BioPreferred Program" to "increase the purchase and use of biobased products, which spurs economic development, creates jobs, reduces the nation's reliance on petroleum, reduces adverse climate and health impacts, and provides new markets for farm commodities".[112] The programme defines biobased products as products, other than food, feed or fuel, that are "Composed, in whole or in significant part, of biological products, including renewable domestic agricultural materials, renewable chemicals, and forestry materials; or an intermediate ingredient or feedstock."

The bioeconomy was estimated to have added USD 470 billion to the US economy in 2019, displaced 9.4 million barrels of oil a year, and supported 4.6 million jobs. Promoted through a Federal Purchasing Preference Program for federal agencies and their contractors, and a voluntary labelling initiative for biobased products, the programme now has more than 130 designated biobased product categories. For example, biobased products for the construction sector include wood and concrete sealers, erosion control materials, metal cleaners, and roof and floor coatings. While not all products are manufactured from biowaste (some originate from renewable biomass), the government support is giving confidence to US manufacturers in the future of the circular bioeconomy.

10.6   Summary and Conclusions

Despite humans' long history of sustainable biowaste management, activities over the last century have disrupted natural cycles and led to the catastrophic loss of soil health. Meanwhile, the disposal of biowaste in landfills in towns and cities across the world is contributing significantly to anthropogenic climate change and other negative public and planetary health outcomes. Restoring the biological cycle by collecting and returning biowaste to nature has thus become a priority for sustainable development.

The sustainable management of biowaste supports soil restoration, carbon sequestration, sustainable agriculture, and offers a renewable source of feedstock for animal feed, renewable energy and bioproducts. Innovations in technology and bioengineering (SDG 9), coupled with a greater awareness of resource scarcity and the need for ecosystem restoration, are increasingly driving international cooperation (SDG 17) and investment in the emerging bioeconomy market.

Many of the targets within the UN SDGs rely heavily on our willingness to address the biowaste challenge. From tackling food security (SDG 2), energy poverty (SDG 7), climate change (SDG 13) and pollution (SDGs 3, 6 and 15), to building liveable cities (SDG 11) and displacing the need for fossil fuels for fertiliser, energy and chemicals (SDG 12), the sustainable bioeconomy offers a multiplicity of promise.

References

1. E. B. A. Guttman, *World Archaeology*, 2005, **37**(2), 224.
2. J. I. Rodale, *The Complete Book of Composting*, Rodale Books, Emmaus, PA, 1960.
3. J. M. Lopez-Real, High Temperature Composting as a Resource Recovery System for Agro-industrial Wastes, in *Biomass Utilization*, ed. W. A. Côté, NATO Advanced Science Institutes Series, vol. 67. Springer, Boston, 1983, pp. 529–536.
4. L. Foxall, *Environmental Archaeology*, 1998, **1**, 35.
5. Q. Wen, *Organic Matter and Rice*, International Rice Research Institute, Los Banos Laguna, Philippines, 1984.
6. European Compost Network, *Compost and Digestate for a Circular Bioeconomy*, ECN Data Report 2022.
7. FAO (Food and Agriculture Organization of the United Nations), *Campaign Report, World Soil Day*, FAO, Rome, 2020.
8. FAO, *Soil is a Non-renewable Resource*. International Year of Soils, 2015, https://www.fao.org/documents/card/en/c/ec28fc04-3d38-4e35-8d9b-e4427e20a4f7/ (accessed May 2023).
9. FAO, *Saving our Soils by all Earthly Ways Possible. Highlighting Some Key Achievements of the FAO-led Global Soil Partnership*, FAO, Rome, 2022.
10. F. Leonardis, The ecosystem services provided by soil and the importance of its protection: the essential role of organic waste. *Rivista Quadrimestrale di Diritto Dell'ambiente*, Opinioni E Segnalazioni, 2022, Numero 1.
11. Environment Agency. *The State of the Environment: Soil*. London, June, 2019.
12. J. R. McNear Jr., *Nature Education Knowledge*, 2013, **4**(3), 1.
13. U. Conrath, C. M. Pieterse, and B. Mauch-Mani, *Trends Plant Sci.*, 2022, 7, 210.
14. R. Sun, X. X. Zhang, X. Guo, D. Wang, H. Chu, *Soil Biol Biochem*, 2015, **88**, 9.

15. E. Wiedmer. How do synthetic fertilizers kill soil microbes?, 2023, https://www.linkedin.com/feed/update/urn:li:activity:7040386422276562945/ (accessed May 2023).

16. W. Lin, M. Lin, H. Zhou, H. Wu, Z. Li and W. Lin, *PLoS One*, 2019, **14**(5), e0217018.

17. D. Schulze-Makuch, "Phosphorous: You Can't Have Life Without It, at Least on Earth", *Smithsonian Magazine*, 17 November 2017, https://www.smithsonianmag.com/air-space-magazine/phosporus-you-cant-have-life-without-it-least-earth-180967243/ (accessed May 2023).

18. C. Gibard, S. Bhowmik, M. Karki, E-K. Kim and R. Krishnamurthy, *Nature Chemistry*, 2018, **10**, 212.

19. R. McKie, "Scientists Warn of 'Phosphogeddon' as Critical Fertiliser Shortages Loom", *The Observer*, 12 March 2023.

20. W. J. Brownlie, M. A. Sutton, K. V. Heal, D. S. Reay, B. M. Spears, eds., *Our Phosphorus Future*, UK Centre for Ecology and Hydrology, Edinburgh, 2022.

21. M. S. Helmberger, L. K. Tiemann, M. J. Grieshop, *Functional Ecology*, 2020, **34**, 550.

22. A. A. de Souza Machado, W. Kloas, C. Zarfl, S. Hempel and M. C. Rillig, *Global Change Biol*, 2018, **24**, 1405.

23. X. Ren, J. Tang, X. Liu and Q. Liu, *Environ. Pollut*, 2020, **256**, 113347.

24. M. C. Rillig, M. Ryo, A. Lehmann, C. A. Aguilar-Trigueros, S. Buchert, A. Wulf, A. Iwasaki, J. Roy and G. W. Yang, *Science*, 2019, **366**, 886.

25. D. Lin, G. Yang, P. Dou, S. Qian, L. Zhao, Y. Yang and N. Fanin, *Proceedings of the Royal Society B*, 2020, **287**, 20201268.

26. S. Kaza, *What a Waste 2.0*, 2018, The World Bank, New York, 2018.

27. FAO, *The State of Food and Agriculture. Moving Forward on Food Loss and Food Waste Reduction.* Rome, 2019.

28. UNEP, *Food Waste Index Report 2021*. Nairobi, 2021.

29. WRAP, *UK Progress against Courtauld 2025 Targets and UN Sustainable Development Goal 12.3*, Banbury, 2020.

30. United Nations Economic Commission for Europe, *Forests: Circularity*, 2022.

31. F. Khamesipour, K. B. Lankarani, B. Honarvar and T. E. Kwenti, BMC Public Health, 2018, **18**, 1409.

32. A. Gulland, "Regular Waste Collection Dramatically Reduces Number of Disease-carrying Flies, Study Finds", *The Telegraph*, 19 August 2019.

33. N. M. T. Faiza, N. A. Hassan, M. R. Farhan, M. A. Edre and R. M. Rus, *Journal of Wastes and Biomass Management*, 2019, **1**(2), 14.

34. P. H. McClelland, P. H McClelland, C. T. Kenney, F. Palacardo, N. L. S. Roberts, N. Luhende, J. Chua, J. Huang, P. Patel, L. Albertini Sanchez, W. J. Kim, J. Kwon, P. J. Christos and M. L. Finkel, *Int. J. Environ. Res. Public Health*, 2022, **19**(7), 4218.

35. W-P. Schmidt, I. Haider, M. Hussain, M. Safdar, F. Mustafa, T. Massey, G. Angelo, M. Williams, R. Gower, Z. Hasan, H. S. Waddington, N. Anjum and A. Biran, *Tropical Medicine & International Health*, 2022, **27**:7, 606.

36. S. V. Ajay, P. S. Kirankumar, A. Varghese and K. P. Prathish, *Expo Health,* 2022, **14**, 763.
37. Climate Bonds, *Waste Management Criteria: The Climate Bonds Standard & Certification Scheme's Waste Management Criteria*, Background Paper, August 2022.
38. L. Höglund-Isaksson, A. Gómez-Sanabria, Z. Klimont, P. Rafaj and W. Schöpp, *Environ. Res. Commun.*, 2020, **2**, 025004.
39. "Ghazipur Landfill Burns for 3rd Time in a Month", *Hindustan Times*, 21 April 2022.
40. D. Chavan, P. Lakshmikanthan, G. S. Majunatha, D. Singh, S. Khati, S. Arya, J. Tardio, N. Eshtiaghi, P. Mandal, S. Kumar and R. Kumar, *Waste Management*, 2022, **154**, 272.
41. Tropomi.eu, *Observing Our Future*, http://www.tropomi.eu/ (accessed May 2023).
42. GHGSat.com, *Global Emissions Monitoring*, http://www.tropomi.eu/ (accessed May 2023).
43. G. O. Badmus, O. S. Ogungbemi, O. V. Enuiyin, J. A. Adeyeye and A. T. Ogunyemi, *Scientific African*, 2022, **17**, e01308.
44. W. N. Igboama, O. S.Hammed, J. O Fatoba, M. T Aroyehun and J. C. Ehiabhili, *Appl Water Sci,* 2022, **12**, 130.
45. A. Ishaq, M. I. M. Said, S. Azman, M. F. Abdulwahab and M. I. Alfa, *Nigerian Journal of Technological Development*, 2022, **19**(3), 181.
46. S. Ma, C. Zhou, J. Pan, G. Yang, C. Sun, Y. Liu, X. Chen and Z. Zhao, *J. Clean. Prod.*, 2022, **333**, 130234.
47. M. Javahershenas, R. Nabizadeh, M. Alimohammadi and A. H. Mahvi, *Int. J. Environ. Anal. Chem.*, 2022, **102**, 558.
48. A. Kuleyin, Y. Ö and Y. Şişman, *Global NEST Journal*, 2020, **22**, 315.
49. M. B. Farhangi, Z. Ghasemzadeh, N. Ghorbanzadeh, M. Khalilirad and A. Unc, *J. Mater. Cycles Waste Manag.*, 2021, **23**, 1576.
50. A. F. P. Marinho, C. W. A. Nascimento and K. P. V. Cunha, *Environ. Monit. Assess.*, 2022, **194**, 459.
51. O. M. Olatunji, *Global Scientific Journals*, 2022, **10**, 1304.
52. S. Mor, P. Negi, and R. Khaiwal, *Environ Nanotechnol., Monit. Manag.*, 2018, **10**, 467.
53. F. Parvin and S. M. Tareq, *Appl. Water Sci.*, 2021, **11**, 100.
54. A. Ishaq, M. I. M. Said, S. Azman, M. F. Abdulwahab and M. I. Alfa, *Nigerian Journal of Technological Development*, 2022, **19**(3), 181.
55. World Resources Institute, *Food Loss and Waste Accounting and Reporting Standard*, WRI, Washington, DC, 2016, https://flwprotocol.org/wp-content/uploads/2017/05/FLW_Standard_final_2016.pdf (accessed May 2023).
56. Chilling Prospects, *Sustainable Energy for All*, Swiss Agency for Development and Cooperation, Geneva, 2022.
57. J. Sarr, J. L. Dupont and J. Guilpart, *The Carbon Footprint of the Cold Chain*, 7th Informatory Note on Refrigeration and Food. IIR, Paris, 2021.

58. T. Peters and L. Sayin, *Status of the Global Food Cold Chain: Summary Briefing*, Cool Coalition, UNEP, 2021.

59. World Resources Institute, *Reducing Food Loss and Waste*, WRI, Washington, DC, 2019, http://pdf.wri.org/reducing_food_loss_and_waste.pdf (accessed May 2023).

60. FAO, *Global Database of GHG Emissions Related to Feed Crops: Methodology*, Version 1, Livestock Environmental Assessment and Performance Partnership, FAO, Rome, 2017.

61. WWF (World Wildlife Fund), *No Food Left Behind, Part IV, Benefits & Trade-offs of Food Waste-to-Feed Pathways*, WWF, Gland, 2021.

62. C. H. Kim, J. Ryu, J. Lee, K. Ko, J.Y. Lee, K.Y. Park and H. Chung, *Processes*, 2021; **9**, 161.

63. I. Kinasih, Y. Suryani, E. Paujiah, R.A. Ulfa, S. Afiyati, Y.R. Adawiyah and R.E. Putra, *IOP Conf. Ser. Earth Environ. Sci.*, 2020, **593**, 012040.

64. S. A. Siddiqui, B Ristow, T. Rahayu, N. S. Putra, N. W. Yuwono, K. Nisa, B. Mategeko, S. Smetana, M. Saki, A. Nawaz and A. Nagdalian, *Waste Management*, 2022, **140**, 1.

65. I. Hölzle, M. Somani, G. V. Ramana and M. Datta, *J. Clean. Prod.*, 2022, **369**, 133136.

66. C. Scopetani, D. Chelazzi, A. Cincinelli, T. Martellini, V. Leiniö and J. Pellinen, *Chemosphere*, 2022, **293**, 133645.

67. E. Brassea-Pérez, C. J. Hernández-Camacho, V. Labrada-Martagón, J. P. Vázquez-Medina, R. Gaxiola-Roblesa and T. Zenteno-Savína, *Environmental Research*, 2022, **206**, 112636.

68. FAO and UNEP, *Global Assessment of Soil Pollution – Summary for Policy Makers*, FAO, Rome, 2021.

69. J. O'Connor, S. A. Hoang, L. Bradney, S. Dutta, X. Xiong, D. C. W. Tsang, K. Ramadass, A. Vinu, M. B. Kirkham and N. S. Bolan, *Environmental Pollution*, 2021, **272**, 115985.

70. R. Stehouwer, L. Cooperband, R. Rynk, J. Biala, J. Bonhotal, S. Antler, T. Lewandowski and H. Nichols, in *The Composting Handbook*, ed. R. Rynk, Academic Press, Cambridge, 2022, ch. 15.6, p. 765.

71. Ministerio del Medio Ambiente, *Estrategia Nacional de Residuos Orgánicos Chile 2040*, Gobierno de Chile, 2019.

72. Programa Reciclo Orgánicos, *Memoria Programa Reciclo Orgánicos (2017–2022)*, Ministerio del Medio Ambiente de Chile, Chile, Marzo 2022.

73. R. S. Mylavarapu, V. D. Nair and K. T. Morgan, *An Introduction to Biochars and Their Uses In Agriculture*, SL383, University of Florida, 2013.

74. DEFRA (Department for Environment, Food and Rural Affairs), *Review of the Potential Benefits, Costs and Issues Surrounding the Addition of Biochar to Soil: An Expert Elicitation Approach*, SP0676, London, 2009.

75. V. J. Bruckman and J. Pumpanen, *Developments in Soil Science*, 2019, **36**, 427.

76. Z. Elkhlifi, J. Iftikhar, M. Sarraf, B. Ali, M. H. Saleem, I. Ibranshahib, M. D. Bispo, L. Meili, A. Ercisli, E. Torun Kayabasi, N. Alemzadeh Ansari, A. Hegedűsova and Z. Chen, *Sustainability*, 2023, **15**(3), 2527.

77. D. D. Warnock, J. Lehmann, T. W. Kuyper and M. C. Rillig, *Plant Soil*, 2007, **300**, 9.

78. Z. Luo, B. Yao, X. Yang, L. Wang, Z. Xu, X. Yan, L. Tian, H. Zhou and Y. Zhou, *Chemosphere*, 2022, **287**, 132113.

79. UK Biochar Research Centre, What is biochar?, University of Edinburgh, 2009, https://www.biochar.ac.uk/what_is_biochar.php (accessed May 2023).

80. Biochar International, Biochar production and by-products, https://biochar.international/the-biochar-opportunity/biochar-production-and-by-products/ (accessed May 2023).

81. SUEZ, *Contributing to Net-zero Emissions: SUEZ Group and Airex Energy will Industrialize an Innovative Carbon Capture Solution to Restore Soil Quality and Act for the Climate*, Paris, 7 June 2021.

82. Microsoft, *Microsoft Carbon Removal: An Update with Lessons Learned in our Second Year*, 2022, https://query.prod.cms.rt.microsoft.com/cms/api/am/binary/RE4QO0D (accessed June 2023).

83. Microsoft, What is the role for biochar in Europe's quest for net zero?, Microsoft Corporate Blog, 22 November 2022, https://blogs.microsoft.com/eupolicy/2022/11/22/biochar-net-zero/#:~:text=Biochar%20has%20an%20important%20role,will%20enable%20it%20to%20scale (accessed May 2023).

84. T. A. Kurniawan, M. H. D. Othman, X. Liang, H. H. Goh, P. Gikas, K-K. Chong and K. W. Chew, *J. Environ. Manag.*, 2023, **332**, 117429.

85. L. Restuccia, G. A. Ferro, D. Suarez-Riera, A. Sirico, P. Bernardi, B. Belletti and A. Malcevschi A, *Procedia Struct. Integr.*, 2020, **25**, 226.

86. Z. Asadi Zeidabadi, S. Bakhtiari, H. Abbaslou and A. R. Ghanizadeh, *Constr. Build. Mater.*, 2018, **181**, 301.

87. I. Cosentino, L. Restuccia, G. A. Ferro and J-M. Tulliani, *Theoret. Appl. Fract. Mech.*, 2019, **103**, 102261.

88. L. Wang, L. Chen, D. C. W. Tsang, B. Guo, J. Yang, Z. Shen, D. Hou, Y. S. Ok and C. S. Poon, *J. Clean. Prod.*, 2020, **258**, 120678.

89. K. Kuoppamäki, M. Hagner, S. Lehvävirta and H. Setälä, *Ecol. Eng.*, 2016, **88**, 1.

90. C. T. N. Cao, C. Farrell, P. E. Kristiansen and J. P. Rayner, *Ecol. Eng.*, 2014, **71**, 368.

91. L. Chen, L. Huang, J. Hua, Z. Chen, L. Wei, A. I. Osman, S. Fawzy, D. W. Rooney, L. Dong and P-S. Yap, *Env. Chem. Lett.*, 2023, **21**, 1627–1657.

92. L. Chen, G. Msigwa, M. Yang, A. I. Osman, A. Fawzy, D. W. Rooney and P-S. Yap, *Environ. Chem. Lett.*, 2022, **20**, 2277.

93. L. Gan, A. Garg, H. Wang, G. Mei and J. Liu, *Acta Geophys.*, 2021, **69**, 2417.

94. K. Bernat, T. C. T. Le, M. Zaborowska and D. Kulikowska, *Energies*, 2023, **16**, 1264.

95. European Biogas Association, *EBA's Biomethane Fact Sheet*, Brussels, 2013, https://www.europeanbiogas.eu/wp-content/uploads/files/2013/10/eba_biomethane_factsheet.pdf (accessed May 2023).

96. World Biogas Association, *Global Potential of Biogas*, London, 2019, https://www.worldbiogasassociation.org/wp-content/uploads/2019/09/WBA-globalreport-56ppa4_digital-Sept-2019.pdf (accessed May 2023).

97. M. Spitoni, M. Pierantozzi, G. Comodi, F. Polonara and A. Arteconi, *J. Nat. Gas Sci. Eng.*, 2019, **62**, 32.

98. MNRE, *Bioenergy Overview, Current Status*, Ministry of New and Renewable Energy, Government of India, Delhi, 2020.

99. "Tata Power subsidiary inaugurates 5 kilowatt biogas plant in Bihar", *Business Standard*, 28 December 2020, https://www.business-standard.com/article/companies/tata-power-subsidiary-inaugurates-5-kilowatt-biogas-plant-in-bihar-120122800766_1.html (accessed May 2023).

100. PIB, "Rs. 2 lakh crores to be invested for setting up 5000 Compressed biogas in the country, says Petroleum Minister at MoU signing event for setting up 900 CBG plants", Press Information Bureau, Delhi, 20 November 2022.

101. S. A. Mueller, J. E. Anderson and T. J. Wallington, *Biomass and Bioenergy*, 2011, **35**, 1623.

102. OECD (Organisation for Economic Co-operation and Development), *Meeting Policy Challenges for a Sustainable Bioeconomy*, OECD Publishing, Paris, 2018.

103. Bio-based and Biodegradable Industries Association, *What is the bioeconomy?*, BBIA, Hertford, 2023.

104. International Advisory Council of the Global Bioeconomy Summit 2018, *Communiqué: Innovation in the Global Bioeconomy for Sustainable and Inclusive Transformation and Wellbeing*, 19–20 April, 2018, Berlin, https://www.biooekonomierat.de/media/pdf/archiv/international-gbs-2018-communique.pdf?m=1637836879& (accessed May 2023).

105. V. C. Igbokwe, F. N. Ezugworie, C. O. Onwosi, G. O. Aliyu and C. J. Obi, *J Environ. Manag.*, 2022, **305**, 114333.

106. M. B. Kannan and K. Ronan, *Waste Management*, 2017, **67**, 67.

107. A. Tarafdar, V. K. Gaur, N. Rawat, P. R. Wankhade, G. K. Gaur, M. K. Awasthi, N. A. Sagar and R. Sirohi, *Bioengineered*, 2021, **12**, 8247.

108. D. H. Zamil, R. M. Khan, T. L. Braun and Z. Y. Nawas, *J. Cosmet. Dermatol.*, 2022, **21**, 6056.

109. J. C. Solarte-Toro, J. A. Arrieta-Escobar, B. Marche and C. A. Cardona Alzate, *Computer Aided Chemical Engineering*, 2021, **50**, 1871.

110. INTERREG Europe, *The Biowaste Management Challenge*, Policy Brief, Brussels, 2021.

111. J-G. Rosenboom, R. Langer and G. Traverso, *Nature Reviews Materials*, 2022, 7, 117.

112. United States Department of Agriculture (USDA), *The BioPreferred Program and the Use of Biobased Products*, National Institute of Building Sciences, Washington, DC, 2023, https://www.wbdg.org/resources/bio-preferred-program-use-biobased-products (accessed May 2023).

CHAPTER 11

Sustainable Development Goals, Circularity and the Data Centre Industry: A Review of Real-world Challenges in a Rapidly Expanding Sector

DOMINIKA IZABELA PTACH*, DEBORAH ANDREWS AND SIMON P. PHILBIN

School of Engineering, London South Bank University, 103 Borough Road, London SE1 0AA, UK
*E-mail: ptachd2@lsbu.ac.uk

11.1 Introduction

11.1.1 Significance of the Data Centre Industry

Data centres play a crucial role in today's global society and digital economy. Whether we are searching for news on our smartphones or connecting to a virtual meeting at work, data centres are in the background of all online activities. Organisations across various sectors that process, store and disseminate extensive amounts of data rely heavily on data centres. Indeed, the world recognised their growing importance during the COVID-19 pandemic when many services moved to online provision and

Issues in Environmental Science and Technology No. 51
The Circular Economy: Meeting Sustainable Development Goals
Edited by Sadhan Kumar Ghosh and Gev Eduljee
© The Royal Society of Chemistry 2024
Published by the Royal Society of Chemistry, www.rsc.org

the volume of internet users increased remarkably. It appears to be problematic to grasp the true impact of digital infrastructure as the term "cloud" seems intangible; however, data centres are the engines of all digital products and services, and an estimated 7.2 million data centres with 70 million servers are continuously running to provide digital services across the globe.[1] The energy-intensive operations and infrastructure-embodied footprint of the Data Centre Industry (DCI) substantially impact the economy, society and the environment. On that account, data centres, as the backbone of all Information and Communications Technology (ICT), significantly impact several of the United Nations (UN) Sustainable Development Goals (SDGs), for instance Goal 7, Affordable and clean energy; Goal 9, Industry, innovation and infrastructure; Goal 11, Sustainable cities and communities; Goal 12, Sustainable consumption and production; and Goal 13, Climate action.

The DCI is one of the fastest-growing sectors. According to the report by DataReportal titled *Digital 2022: Global Overview*,[2] the world's population stood at 7.91 billion in January 2022, with 4.95 billion internet users, representing 62.5% of the world's total population. Furthermore, due to ever-growing digital adoption, internet users have doubled in the last decade, global internet traffic grew 15-fold between 2010 and 2020, and a further 40% growth occurred in 2020. Data processing and producing data is the primary driver of the industry. The last decade has seen 550% growth in data centre computing operations,[3] and predictions state a further fivefold increase in global demand and reliance on data centres by 2030.[4] This trend is driven by the growing popularity of services and technologies, such as video streaming, bitcoin mining and the Internet of Things (IoT).

Some sources claim that approximately 7.2 million data centres are working 24/7 worldwide to meet the constant and growing demand.[1,5] Different sources report various levels of activity in the sector. Intellect/TechUK[6] explains that the total number of data centres remains unknown due to inconsistent definitions of what a data centre is. Most data centres are congregated in four main countries across the world, namely the USA, Germany, the UK and China. More than half of Europe's data centres are concentrated in the north-west of the continent, with the highest number – 456 – in the UK.[7,8]

11.1.2 About Data Centres

Data centres originated from large computer rooms that were first built in the 1940s. In the 1980s, the introduction of microcomputers allowed the transfer of networking equipment to designated rooms, naming those spaces "data centres" for the first time. The industry experienced tremendous growth on the verge of the new millennium. Due to the "dot-com bubble" and the rising number of internet users, companies started to build Internet Data Centres (IDCs) rapidly. Since then, data centres have generally remained the same except for consecutive technological advances.

Data centres are the physical facilities housing the computer systems and supporting equipment. The Green Grid describes data centres as:

> ... *a building or portion of a building whose primary function is to house a computer room and its support areas. Data centres typically contain high-end servers and storage products with mission-critical functions.*[9]

Depending on the definition, data centres can vary from small cabinets through to designated rooms or entire floors to a complex of large warehouses – the largest data centre campus in China takes up to 1 million square metres.[10] Intellect/TechUK[6] reports that infrastructure qualifies as a data centre if operated with specific environmental and safety controls; therefore, a standalone rack of servers does not qualify as a data centre. Furthermore, data centres comprise the following components: a facility, computing infrastructure (including servers, storage and networking equipment, firewalls, cabling, and racks), support infrastructure (such as uninterruptable power supply; Heating, Ventilation And Cooling (HVAC); and security) and operational staff.[11]

11.1.3   Data Centre Business Models and Subsectors

There are various types of data centre, which organisations choose to engage with according to their requirements. The main types of data centre include the following:[11]

- **enterprise** (in-house data centre managed within the organisation)
- **cloud** (providers lease infrastructure as a service)
- **colocation** (rental spaces in a data centre, where organisations can install their own hardware, but the infrastructure is provided)
- **edge** (small data centres at a required location)
- **hyperscale** (large-scale providers such as Amazon, Google and Meta).

There are two primary business models: wholesale and colocation. In the first model, multiple customers store servers in rented racks in a shared facility; in the second model, one customer rents a third-party facility exclusively.[12]

The data centre industry is extensive and involves various actors and 11 subsectors, as consolidated by the Circular Economy for the Data Centre Industry (CEDaCI) project[13] (see Figure 11.1). The stakeholders include the following:

- suppliers (this includes suppliers of rack and cabling or energy)
- designers and manufacturers of computing hardware and ICT infrastructure
- building construction and maintenance companies
- data centre operations
- decommissioning and data sanitisation services
- recyclers, refurbishment services and remanufacturers.

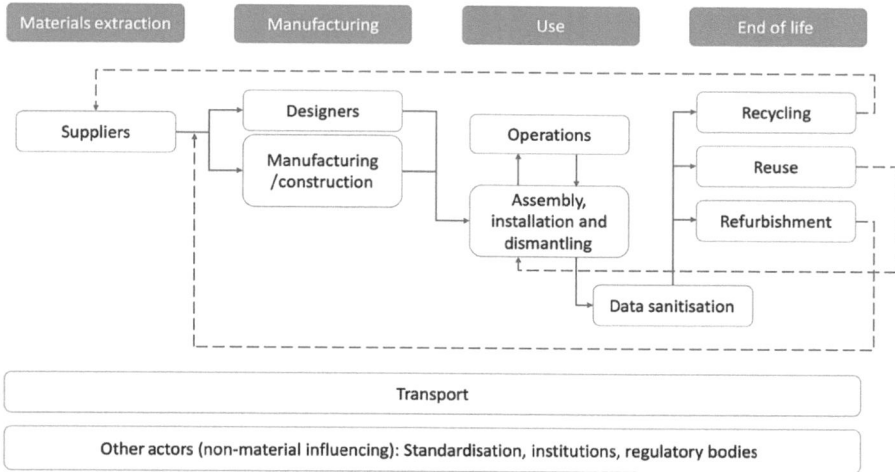

Figure 11.1 Data centre industry sub-sectors as identified by the CEDaCI Project. Adapted from ref. 13 with permission from CEDaCI Project.

11.2 Data Centre Industry Impacts Across the Triple Bottom Line

The structure of this section follows the three dimensions of sustainability as defined by the triple bottom line definition,[14] and explained in Chapter 2. Examples of positive and negative impacts are presented subsequently across the pillars of the economy, the environment and society. The environmental impacts are further divided into subsections concerning energy consumption, renewable energy, greenhouse gas emissions, water consumption and resource consumption.

11.2.1 Economic Impacts

Data centres are the backbone of the digital economy, which researchers, governments and businesses recognise as key to economic growth. The statistics are evidence of the economic significance of data centres globally. For instance, the data centre sector contributes over 5% of gross value added[15] ("value generated by any unit engaged in the production of goods and services"). Moreover, in the 2019 Digital Economy Report of the UN Conference on Trade and Development, it was estimated that the digital economy size ranges from 4.5% to 15.5% of Gross Domestic Product (GDP), depending on the scope of the definition.[16] According to Cloudscene,[8] the market for data centre providers in the UK is thriving, and the digital economy contributes 8% to the total GDP. The market for UK data centres is the second largest in the world and the largest in Europe, with a vital cluster in London. As identified by the TechUK[17] report, "Each new data centre contributes between £397 million and

£436 million gross value added per year to the UK economy while that of each existing data centre is estimated to lie between £291 million and £320 million per annum". The data centre sector, therefore, is and will remain of great value to local and global economies.

Data centres are essential to the public because of the services they deliver and the employment they provide. According to Uptime Institute[18] statistics, the data centre sector employs many highly skilled professionals – approximately 2 million full-time employees worldwide in 2019. Moreover, the resourcing requirement is projected to rise to 2.3 million people by 2025. The highest employee demand is likely to come from colocation, cloud and consumer internet companies, especially in the developing markets of the Asia-Pacific (APAC) region, and less so in North America and in Europe, the Middle East and Africa (EMEA). More mature markets in the USA and Western Europe will create opportunities for senior specialists. Often located in sites away from cities, data centres offer valuable and well-paid career opportunities for local communities, although others argue that data centres create few job openings due to innovations in automation, such as using Artificial Intelligence (AI) monitoring technologies. Some automated data centres have been built in the Arctic Circle, where no staff are required, for instance the Facebook data centre in Lulea, Sweden. However, Uptime Institute[18] predicts that the automation and use of AI are considered to have an insignificant impact on staffing demand.

The above trends significantly impact the economy and wider society, thereby creating educational paths and career opportunities. The DCI workforce requires relevant further or higher education as well as role-specific training. Therefore, the industry also offers many apprenticeship prospects. As Intellect UK outlines in the report titled *So What Have Data Centres Ever Done For Us?*,[15] the DCI sector involves various technical and additional roles. Furthermore, direct and indirect jobs can be differentiated to help understand the employment landscape. Direct employment comprises roles in the design and construction of the facility and management of data centre infrastructure. Indirect employment is much broader and involves people employed in supply chains and those in colocation services. Some examples of indirect roles, which demonstrate a wide range of careers in the sector, include the following:

- location consultants and location finders
- planning consultants and planning advocacy services
- real estate companies that sell data centre capacity and negotiate deals
- data centre search and selection companies
- lawyers and contract negotiators
- specialist accountants, finance consultants and providers
- energy managers and energy consultants
- public relations and media consultants
- industry associations, professional bodies and standards bodies
- conference organisers and specialist publishers.

11.2.2 Environmental Impacts

Energy consumption and associated greenhouse gas (GHG) emissions are receiving significant attention in the data centre sector, which is primarily due to the recently introduced decarbonisation policies as part of the Paris Agreement Under the United Nations Framework Convention on Climate Change, also called the Paris Agreement. There is limited data available on GHG emissions of the DCI sector. However, the International Energy Agency[19] (IEA) reported that data centres and the transmissions network produced approximately $300\,Mt\,CO_2$-eq in 2020 (including embodied emissions). Although the GHG emissions from data centres and network operations are significant, it is vital to underline the non-energy impacts of the sector. The sector needs to recognise environmental impacts associated with resource pressures of hardware and infrastructure manufacturing (such as raw materials' extraction, including critical raw materials and rare earth elements), water and energy consumption during operations and associated e-waste and end-of-life scenarios. Environmental and social impacts need more attention and action from industry, governments and researchers. Scoping and measuring both emissions and broader environmental impacts is complex, but assessment is crucial to understand the true level of sustainability, to reduce footprints and to reach climate agreements.

Researchers[20-23] agree that a life-cycle approach is necessary to assess the sustainability impacts of the DCI fully. The Life-cycle Assessment (LCA) method is systematised by the International Organization for Standardization (ISO) ISO 14044 standard and further adopted by the Green Grid,[24] specifically for the data centre sector. It considers all life-cycle stages, from materials' extraction and manufacturing of IT hardware and construction of a building to operations, decommissioning and end-of-life scenarios, thereby uncovering both embodied and operational environmental impacts, although the level of sophistication and a large number of data centre components often make LCAs problematic and time-consuming. Furthermore, there is a need for open-source, primary inventory data to conduct accurate assessments. There are a small number of research studies on the LCA of data centres, including studies by Letteri,[25] Shah[26] and Whitehead.[22,27] All these studies identified that operations are directly responsible for environmental impacts. However, the embodied carbon impact cannot be overlooked, and the significance of embodied impacts is likely to increase as a result of sectoral growth and improved operational energy efficiency.

Other evidence of possible non-energy impacts of digitalisation on the environment was presented in the report by the Öko-Institut, titled *Impacts of the Digital Transformation on the Environment and Sustainability*,[28] for the European Commission. As in other LCA studies, the authors pointed out that there are broader implications beyond GHG emissions and the high energy consumption of data centres. In this regard, they collected literature on resource depletion, water consumption, land use and land use change and biodiversity and presented a comprehensive review of broad systemic impacts.

11.2.2.1   Energy Consumption

The data centre industry is promoted as a high-energy consumer. However, industry practitioners argue that the sector is not the most significant offender when looking at the numbers and trends, and instead point at the user demand and energy-intensive technologies, such as blockchain ("a system in which a record of transactions made in bitcoin, or another cryptocurrency, is maintained across several computers that are linked in a peer-to-peer network"). The electricity use of a data centre was first researched by Koomey[29] in 2008, followed by many other researchers.[3,30,31] According to Greenpeace,[32] 21% of electricity consumption in the IT sector was used by data centres, 29% by networks, 16% by manufacturing and 34% by electronic devices. The most recent data from the IEA[19] reveals that data centres used between 220 and 320 terawatt hours (TWh) of energy in 2021. This is 10–60% more than in 2015, when the approximate usage accounted for 200 TWh. The energy utilised by data centres is equivalent to that of some nations, for instance the UK (294.4 TWh in 2021). In 2021, data centres accounted for 0.9–1.3% of global electricity demand. These numbers exclude the data centre transmission networks, which consumed an additional 220 TWh in 2015 and 260–340 TWh in 2021.

The increase in energy consumption in the last decade is limited due to the energy efficiency improvements in hardware and cooling. Another reason is clustering and integrating smaller, ineffective data centres into the bigger and more efficient facilities, which counterbalance the rise in energy consumption. Nevertheless, data centres need to be wary of *the energy efficiency rebound effect* as the savings from the above practices are constantly challenged by the ever-rising demand for digital services. Still, some forthcoming energy demand models are controversial and troubling. A projection created by Andrae[33] in 2015 assumed an increase in electricity consumption by a factor of 3–8% a year between 2019 and 2030.

11.2.2.2   GHG Emissions

Eurostat[34] defines *greenhouse gases* as a group of gases contributing to global warming and climate change. The Kyoto Protocol (1997) defines the seven gases, comprising carbon dioxide, methane, nitrous oxide and four fluorinated gases. For ease of measuring and comparison, GHGs are measured in carbon dioxide equivalents (CO_2-eq). Hence carbon is often the centre of focus; another term used interchangeably with GHG emissions is carbon footprint.

For 2020, IEA[19] reported an estimated 300 million tonnes of CO_2 embodied and operational emissions for data centres and transmission networks combined, approximately 0.6% of global emissions. For reference, this number is higher than the total CO_2 emissions of Poland in 2021. This is 16% higher than the anticipated emissions reported by The Climate Group on behalf of the Global eSustainability Initiative (GeSI) in 2008, which estimated the total carbon footprint to reach 259 Mt CO_2-eq in 2020.[35] Another source, The Shift Project, calculated that global emissions from cloud data centres range from

2.5% to 3.7% of all global GHG emissions. Therefore, we can ask the following question: What do these carbon emission numbers actually mean to the data centre industry?

The GHG Protocol Corporate Standard[36] describes a multi-level international accounting tool in which GHG emissions are categorised into three groups or "scopes". Scope 1 includes direct emissions from owned or controlled sources, for example running vehicles. Scope 2 includes indirect emissions from the generation of purchased electricity. Scope 3 includes all other indirect emissions both upstream and downstream of the company value chain. Indeed, it is particularly challenging to assess the GHG emissions for data centre services due to the complexity of the infrastructure and the often shared services. Therefore, the GHG Protocol provides comprehensive guidance for ICT businesses, and follows a life-cycle approach. The framework[37] considers three components for data centre services, namely emissions of the data centre, the network and the end-user devices, although accounting for the energy of the connected devices may escalate the numbers dramatically. For instance, the Microsoft Sustainability Report 2021[38] shows nearly 124 000 and 164 000 tonnes of CO_2 for scopes 1 and 2, respectively. At the same time, scope 3 emissions reached 13 785 000 tonnes of CO_2-eq. Scope 3 outweighed scopes 1 and 2 by nearly 48 times, regardless of the savings made in direct and indirect emissions. This example demonstrates the significance of supply chain emissions, which is the most carbon-intensive segment, but is often ignored, misunderstood and unreported, and accounts for up to 80% of total company impacts. The GHG Protocol emphasises that focusing on GHG emissions is a good point of reference for businesses to assess their environmental performance, and scopes 1 and 2 are good places to start due to data availability. Meanwhile, the Uptime Institute Data Centre Survey[39] from 2022 reports that only 37% of companies collect carbon footprint data, thereby indicating a substantial opportunity for improvement.

11.2.2.3   Renewable Energy

It can be observed that all the energy consumed by the sector would lead to enormous GHG emissions if the electricity were produced only by combusting fossil fuels. This is because burning oil, gas and coal leads to increased carbon in the atmosphere, which, in a broader context, causes climate change and has a negative impact on the environment overall and on human health. Consequently, under pressure from the media, legislators and clients, the data centre sector heavily invests in renewable energy to improve its reputation and environmental profile and avoid price volatility. Hyperscale cloud operators have the advantage in renewable energy procurement. Due to the nature of the cloud business model, such operators control their direct energy procurement and locate sites close to renewable resources. Hence hyperscalers are the primary purchasers of all corporate Power Purchase Agreements (PPAs), with Amazon Web Services, Microsoft, Meta and Google leading the way in this area.[19,40]

There are different sustainable strategies for procuring renewable energy. The most common are Renewable Energy Certificates (RECs) and PPAs. RECs are defined as "a market-based instrument that certifies the bearer owns one megawatt-hour (MWh) of electricity generated from a renewable energy resource".[41] However, this approach does not guarantee that the energy used is "green". For instance, the energy mix in the region of data centre operations may be fully supplied by fossil fuels, and the purchased renewable energy will be generated and used elsewhere. PPAs are contracts between suppliers and buyers to purchase energy at a specific cost and volume, and within a specific time frame. There are two types of PPAs, which are direct/physical or financial/virtual. Only the first type allows for the physical delivery of renewables to the buyer; the second type sells the power to the wholesale market.[42]

The aforementioned strategies can be regarded as offsetting mechanisms, which aim to alleviate previously generated emissions. However, those offsets do not reduce or remove emissions completely. PPAs and RECs still face problems in matching renewable energy purchased and the energy consumed by the facility. Moreover, they are often used to greenwash companies' sustainability portfolios.[42] Nevertheless, these procurement strategies enable renewable energy investments globally and help to decarbonise the energy systems.

11.2.2.4   Water Consumption

Water scarcity is becoming a global problem, and the data centre industry contributes to water stress and competition for water sources within local communities. Some controversial projects and recent environmental disasters have brought more attention to the water issue. For instance, Google's Luxembourg data centre proposal in 2019,[43] or recent news headlines questioning data centres' water use during the extreme drought in August 2022 in the UK.[44]

Data centres use water directly for cooling and indirectly through electricity generation (thermoelectric power). Limited studies are available on direct and indirect water uses; examples include the work of Mytton[45] and Ristic.[46] The total amount of sectoral water consumption remains unknown due to the lack of data. Nevertheless, the researchers were able to present water use estimations. Mytton calculated that US data centres consume 1.7 billion litres of water daily. Therefore, the US data centre yearly water footprint stands at 620.5 billion litres. This number is very close to the 2014 average of 626 billion litres reported by Shehabi,[30] who created a direct and an indirect water use forecast for 2014 to 2020. Although the forecast did not oversee the industry improvements in technologies and strategies, electricity generation is responsible for most of the water consumed. Furthermore, Rustic documented that data centres use between 4 and 544 litres of water per kWh, and outbound data traffic uses 1 to 205 litres of water per gigabyte. The Öko-Institut report[28] used these numbers to estimate the total water consumption for a 198 TWh data centre in 2018 to be 740–106 822 million m^3 annually. The range of calculations reflects the uncertainty in the exact number of data centres and the lack of data about water usage according to different variables. Moreover, water footprint studies have tended to explore the operations only. Nevertheless, when examining a data centre's whole life cycle, hardware and building manufacturing, decommissioning and end-of-life scenarios need to be considered. Moreover, Rustic emphasises the urgent need for more research regarding "the water intensity of different HVAC technologies in different climates and the water intensity of electricity generation" and recommends a systemic approach and implementation of standardised metrics. All the above studies suggest that water use in the data centre sector requires attention and transparent reporting, and further study of the sustainability impacts of water use is also needed.

Another area that requires more attention is alternative cooling methods and the free water cooling or waterside cooling. Some projects using water from natural reservoirs have been implemented thus far, for example Google's 2011 data centre in Hamina, Finland.[47] This facility is located near the Baltic Sea gulf, which supplies the data centre with seawater delivered through the granite tunnels built initially for a local paper mill. Cold sea water enters the system and takes the heat away from the site; before releasing it back into the sea, warm and cold water is mixed to reduce the temperature, so its temperature is similar to that of inlet seawater. It is claimed that this practice reduces marine life impacts, but data and independent studies that analyse the long-term effects of such actions are not available. In the IEEE Spectrum,[48] it was reported that Microsoft's experiment with submerging a data centre under the sea should have no negative results on marine life as "any heat generated by a Natick pod would rapidly be mixed with cool water and carried away by the currents. The water just meters downstream of a Natick vessel would get a few thousandths of a degree warmer at most." Nevertheless, scientists warn that ocean thermal pollution can adversely affect aquatic life, causing oxygen-level reductions, disruption to animal reproductive cycles and other physiological impacts.[49] Moreover, there is no legal control over utilising the ambient environment as coolants, which raises the question whether using seawater for cooling is truly environmentally friendly and green, as companies claim.

11.2.2.5   Physical Resource Consumption

A significant majority of concern focuses on climate change and the environmental impacts of data centre operations (*i.e.*, energy related), whereas the embodied impacts of a data centre and those at the end of life are often overlooked.[21] While manufacturing uses more than a quarter of electricity in the IT sector, electronic and electrical equipment cause 1–4% of environmental impacts in Europe alone.[50] Each piece of data centre hardware contains more than 50 materials: ferrous and non-ferrous metals, Platinum Group Metals (PGMs), Precious Metals (PMs), Rare Earth Elements (REEs), plastics and, in some cases, ceramics.[13] The European Commission has created a Critical Raw Materials (CRMs) list of 30 elements[51] that can be considered strategically significant resources for the growth of the European economy, and at high risk of limitations and disruptions to their supply chains. Of 30 CRMs, 20 are present in IT software. Raw materials are not finite; therefore, endless extraction and wasteful linear consumption threaten resource availability for future generations. Moreover, researchers have identified environmental impacts associated with material extraction and mining waste from linear consumption, including land use change, and air, water and soil pollution with heavy metals and radioactive particles. For instance, the largest cobalt and tantalum deposits are concentrated in the Democratic Republic of Congo, a politically unstable region associated with unethical artisanal mining, where extraction causes land degradation and soil pollution and threatens the agriculture sector.[52]

Researchers[4,23] emphasise the urgent need for greater material resource efficiency in data centre equipment to secure the future supply chain of critical materials. Furthermore, CEDaCI research identified Printed Circuit Boards (PCBs) as components with the highest environmental impact, advising recycling and refurbishment take-back schemes for economic value preservation and environmental impact reduction. The research also advocates responsible data centre equipment consumption and eco-design improvements to accommodate more efficient material recovery. Such practices and new business models, particularly the circular economy, need to be promoted in the data centre sector since most efforts concentrate on energy consumption only. The circular economy is key to the long-term benefits across all three pillars of sustainability, and therefore ensures a safe and prosperous future for many industries. The 2022 Uptime Institute Global Data Centre Survey further presents trends in reporting environmental data. For instance, only 28% of companies currently measure their e-waste generation or equipment life cycle.[39] Such poor practice might be a significant barrier to achieving circularity in the sector. One of the strategies is designing for a longer lifespan, which should be applied not only to consumer devices, but also to IT assets. Nevertheless, due to supply chain pressures, the industry's attention is switching from operational energy to resource efficiency, meaning that more and more companies are turning to circular thinking. With the popularity of cloud and infrastructure-as-a-service models, the sustainability choices and equipment reuse are in the providers' hands, as they have hardware ownership. Predominantly, hyperscalers and Original Equipment Manufacturers (OEMs) often introduce new circular plans. For example, in 2020, Microsoft launched Circular Centers, committed to reusing and repurposing equipment in its data centres; Google similarly refurbishes its IT assets and sells them in the secondary market (23% of data centre hardware components were refurbished in 2020) and worked closely with the Ellen MacArthur Foundation to analyse the ongoing circularity projects. OEMs, for instance Dell, HPE and Lenovo, offer a hardware-as-a-service model allowing clients to lease equipment for a fixed time.[53,54] Hopefully, more businesses will see the circular economy as an excellent opportunity for environmental footprint reduction and financial benefits, and the number of companies reporting e-waste or equipment life cycles will grow significantly in the near future.

11.2.3 Social Impacts

The data centre industry, and specifically the digital infrastructure, strongly influences the local and global economies and environments, and has numerous associated societal and political implications. Some reported spillover effects include unethical labour practices, gender inequalities, hazardous working conditions, water scarcity and pollution.[28] These potential systemic impacts are not direct effects of the industry operations and are primarily associated with the highly complex and long supply chains of electronics and electrical equipment. Nevertheless, there are also positive contributions of the industry to the digitalisation of developing countries, such as providing access to the internet and education for many remote areas in low-income countries. The sector needs mechanisms to pinpoint and address systemic social impacts and often overlooks this pillar of sustainability, which is paramount to the holistic perspective.

As a result of numerous corporate scandals in the 1990s, growing distrust in business activities, and rising public environmental awareness and concern, companies were pressured to act transparently and ethically. Ever since, businesses carefully monitor activities to maintain their reputation, as associations with unethical practices can lead to customer boycotts and damaging financial costs. Subsequently, a notion of Corporate Social Responsibility (CSR) was developed, and in 2008 the ISO formulated voluntary, universal guidance (ISO 260000) for implementing it in business practice. CSR states that an organisation has responsibility for the environmental and social impacts of its decisions, and advocates for transparency and ethical conduct. ISO 260000 is strongly encouraged in business practice, yet it is still voluntary in many countries. In the European Union, CSR reporting is outlined by Directive 2014/95/EU, commonly referred to as the Non-Financial Reporting Directive (NFRD). The directive covers non-financial information disclosure on subjects concerning the environment, social conduct towards employees, human rights, anti-corruption and bribery, and diversity of the company board. It is soon to be replaced by a broader scope covering all businesses regardless of size, the Corporate Sustainability Reporting Directive (CSRD). Concerning the UK, CSR is encouraged but not mandatory.

As mentioned previously, CSR coverage limits social sustainability to the borders of direct influence, such as immediate employees and headquarters. Furthermore, social sustainability can be interpreted in many ways, and it is hard to formulate a single definition.[55] Nonetheless, an understanding of indirect implications in a systemic context of business actions is needed.

One of the assessment methodologies currently recommended by the UN Environment Programme (UNEP) is Social Life-cycle Assessment (s-LCA) developed by the social life-cycle initiative.[56] The method is used for assessing the social and socio-economic impacts, as well as the potential impacts, of products and services. This method analyses good and adverse implications from a life-cycle perspective, that is from extraction and processing of raw materials, manufacturing, logistics, operations, reuse, maintenance, recycling and end-of-life scenarios. s-LCA is becoming of increasing interest to businesses, thereby allowing for transparency and mapping hotspots for improvement and overall sustainability, which is becoming not only an essential requirement for clients, but also an opportunity for unlocking new potential for businesses.

The most significant issue with s-LCA is the need for more open-source data. Compared with the popularity of life-cycle/environmental impact assessments (the most extensive database, Ecoinvent, has more than 18 000 datasets), s-LCA databases account for only 1000 to 5000 publicly available studies. Currently, limited or no known literature sources explicitly discuss the data centres' social impact. However, the social effects can be explored by looking at electronic and electrical equipment; data centre buildings, specifically materials used for manufacturing; and end-of-life scenarios. There are numerous resources exploring social and political issues associated with materials' extraction and e-waste, for instance, Global E-Waste Monitor[57] or the Ellen MacArthur Foundation e-waste reports.[58]

Data centre equipment has a significant embodied impact; Whitehead[27] estimated that, during the assumed 60 year life of a data centre, 15% of its impact derives from the building and facilities, while the remaining 85% derives from IT equipment. Two primary materials of data centre building are concrete and steel, while manufacturing digital infrastructure equipment requires three primary materials: "common" metals (including steel, copper, aluminium, zinc and brass), polymers (such as acrylonitrile butadiene styrene (ABS), high-density polyethylene (HDPE), polyurethane (PUR), polyvinyl chloride (PVC), general-purpose polystyrene (GPPS), polybutylene terephthalate (PBT) and ethylene-vinyl acetate (EVA)) and CRMs (including antimony, beryllium, chromium, cobalt, lithium, magnesium, palladium, silicon, dysprosium, neodymium, praseodymium and terbium). Although the CRMs account for 0.2% of the equipment, Andrews[4] argues that they are crucial to the electronics' functioning and, therefore, of high economic value. Moreover, some elements – gold and the 3Ts (tantalum, tin and tungsten) – are classified as Conflict Minerals (CMs), that is, minerals extracted in politically unstable regions. Mines of these conflict resources are located in African countries, particularly the eastern, remote regions of the Democratic Republic of Congo controlled by armed rebel groups. Consequently, extracting these precious elements results in local conflicts and even war and cruel exploitation of local communities, including children and women. A 2018 study by the UNEP[59] estimated that around 15 million people work in the Artisanal and Small-scale Gold Mining (ASGM) sector globally, including over 600 000 child workers and 4.5 million female workers. For many local communities, however, it is the only opportunity for income and a livelihood.

The data centre industry is one of the sectors contributing to Waste Electrical and Electronic Equipment (WEEE), and contributes to growing regional and global e-waste streams, reached 54 million tonnes per year in 2020.[57] Approximately 83% of e-waste end-of-life management is undocumented, and often the end of life of WEEE remains unknown. Possible scenarios include dumping, burning, trading or recycling in illegal or uncompliant conditions. The remaining 17.4% of recorded e-waste is still likely to be mishandled by either ending up in a landfill or exported to low-to-middle-income countries to avoid recycling costs. For instance, Nigeria and China both struggle with the infamous illegal e-waste industry, which has a damaging impact on local communities and the environment and is the primary hotspot for illegal workers. It is impossible to pinpoint a specific number of people employed in the entire illegal e-waste industry. However, a World Economic Forum report[57] speculates that, in Nigeria, the number of workers stands at approximately 100 000, while in China, the number possibly reaches 690 000. Together it accounts for more than the population of Leeds in the UK. Moreover, the ratio of female and child workers in Africa and Asia working in e-waste processing plants accounts for 30% of the workers. People engaged in uncontrolled and illegal recycling or disposal activities often lack protective equipment, suitable tools and safe working conditions, and unknowingly mishandle toxic substances. Some processing techniques include burning or melting the electronic parts in acid to collect valuable metals. The results of such practices and exposure to many highly toxic fumes and substances can have long-lasting effects on human health and well-being. Some of the adverse effects include carcinogenic diseases caused by exposure to heavy metals and radioactive uranium. Other consequences affect specifically women's sexual and prenatal health, causing miscarriages, premature births and stillbirths. Newborn babies are also at risk of low birth weights, defects and infant mortality. Furthermore, in a broader context, non-direct impacts of mishandling e-waste include water, air and soil pollution, and contaminated food chains and drinking water, which can have a catastrophic impact on human health (see also Chapter 3).

On the contrary, digital infrastructure plays an important role in providing services to the developing countries of the Global South (see Chapter 3). The digital divide between developed and developing countries significantly impacts many people. Furthermore, the COVID-19 pandemic has even further increased the inequalities in access to digital services. Connectivity is mostly provided thanks to mobile networks and the underpinning network towers. Interestingly, rapidly developing African countries are becoming attractive regions for new data centre projects. Currently, there are 86 colocation data centres in 15 countries in Africa.[60]

11.3   Current State and Opportunities for the Data Centre Sector to Support the 2030 Agenda

This section explores sustainability in the context of business and outlines the reasons why the private sector, specifically the data centre industry, should support sustainable transformation, circular economy business models and the pursuit of a trajectory towards realising the UN SDGs. It also discusses the sustainability maturity of the sector and reviews standards and initiatives in the sector, as well as current reporting requirements. Finally, the section presents the findings on reported metrics and SDGs from recent sustainability reports and other documents, such as CSR or Environmental, Social and Governance (ESG) reports, which are available in the public domain.

11.3.1   Business *vs*. Sustainability

Research often portrays business and sustainability as subjects that contradict each other due to conflicting prerogatives. Although the Brundtland Commission developed a standardised definition of sustainability, it is frequently used interchangeably with CSR or ESG definitions. Moreover, the intersection of business and sustainability linkages still needs further exploration in research. The concept of the three sustainability dimensions (*i.e.*, environmental, social and economic) has been in place for many years now. Still, many industries do not fully understand the holistic approach, as described by Dyllick,[61] who demonstrates the great divide between business sustainability and sustainable development in their work.

As explained in Chapter 2, the Triple Bottom Line (TBL) concept coined in 1994 was created with businesses in mind, initially as an accounting framework to measure business performance other than financial performance. Subsequently, the ESG framework for responsible investing gained popularity. Later, in 1997, the Global Reporting Initiative (GRI) was founded, which developed standards widely used by businesses worldwide and commonly used in the industrial sector. Nevertheless, with the private sector's immediate attention on financial growth, it is unlikely that the social or environmental values of Elkington's triple bottom line will be leading business concerns.[62,63] For decades, sustainability appeared to be only a CSR strategy. However, according to Azapagic and Perdan,[64] companies' performance in environmental and social contexts started to matter to the public in the new millennium, which is inspiring debates on how the private sector can support sustainable development.

11.3.2   The Private Sector – a Key Agent in the 2030 Agenda

The 2030 Agenda for Sustainable Development,[65] consolidated by the United Nations, outlines the scale, ambitions and means of implementing the proposed goals and targets. As the recent 2022 progress update suggests, initial progress on the SDGs is insufficient, and it is estimated that the targets will not be achieved in the given time frame. Moreover, the COVID-19 pandemic has further reversed the progress towards the 2030 Agenda (see Chapter 2), pushing many people back into poverty. Therefore, researchers and leaders agree that strengthened action and an efficient strategy are required to progress towards sustainable development.

To achieve the 2030 Agenda, the authors call for global partnership. Although the concept is proposed on the global level, primarily addressing national and regional governments of all countries, the UN recognises the importance of engaging other agents: the private sector, society, the UN structures and all other available resources. In the section on the means of implementation, the authors emphasise the crucial role of business:

> *We acknowledge the role of the diverse private sector, ranging from micro-enterprises to cooperatives to multinationals, and that of civil society organisations and philanthropic organisations in the implementation of the new Agenda.*[65]

Sachs[66] and Hák[67] also recognised the significance of the private sector in reaching the SDGs. With both multinational and Small and Medium-sized Enterprises (SMEs), private companies can accelerate all areas of sustainable development locally and globally. The private sector is the main constructive of the global economy and directly influences wider society. Therefore, it is necessary for the private sector to take leadership and responsibility for its operations by introducing measuring and reporting strategies, shaping policies and collaborating with stakeholders.[66]

There are significant opportunities for the private sector to advocate and lead sustainable development. However, the main driver behind the emerging transformation must be critical changes in social and economic models and the introduction of sustainable practices.[63,68] The current way of conducting business, that is, business as usual (BAU), concentrates solely on the financial benefit, and relies on the exploitation of the economic system and resources.[61] The BAU approach poses a question, as formulated by Scheyvens:[63]

> *Can, for example, a profit-motivated business really make a meaningful contribution to the achievement of the SDGs or are we likely to see "business-as-usual", which results in greater profits for some, and lost opportunities for many?*

The circular economy is one of the possibilities for securing a prosperous future without abusing the ecosystem; however, it requires changing current linear economic models and establishing a new paradigm for businesses.

11.3.3 Potential of the ICT Sector and Data Centre Infrastructure at its Core

Data centres are the backbone of all digital technologies. Furthermore, technology and the digital economy were recognised by the 2030 Agenda as enabling factors in achieving sustainable development globally. Likewise, the joint research of Earth Institute and Ericsson[69] identified the ICT sector as having a "great potential to accelerate human progress, to bridge the digital divide and to develop knowledge societies, as does scientific and technological innovation across areas as diverse as medicine and energy". Another report from Huawei[70] also presents favourable arguments for ICT to support the SDGs. Huawei's researchers conducted benchmarking exercises to showcase that digitally advanced countries reported higher progress towards the SDGs. Therefore, they suggest that innovation and investment in ICT and infrastructure can positively stimulate sustainable development through wealth creation and expanding a highly skilled and digitally literate society.

11.3.4 What the Data Centre Industry is Currently Doing Towards Sustainability and the SDGs

11.3.4.1 *Sustainability Maturity and Awareness*

Sustainability efforts in the data centre industry primarily focus on energy efficiency due to energy-intensive operations, meaning that other non-energy impacts remain overlooked, as explained in Section 11.2.2. Moreover, the term "sustainability" has become so fashionable and overused that it has lost its true meaning. Due to limited evidence, there is a need for research on sustainability and circularity awareness and the sustainability and circularity maturity of the sector. The available sources are reports from publishing bodies such as Uptime Institute or research executed by private companies, for example Supermicro.

Recent articles and reports from the Uptime Institute[39] warn that the sector's awareness is still low, and attitudes to green strategies differ between the regions. Europeans appear to see sustainability as an opportunity, but are critical of the progress made to date, whereas North America and Asia-Pacific countries are more sceptical. The Uptime Institute Global Data Centre Survey 2021 of 400 data centre owners and operators revealed that 38% of respondents genuinely agreed that data centre actions towards reducing energy use, water use and GHG emissions are meaningful. However, 45% still believe these actions are not delivering any change. Moreover, the survey declares that Europeans are critical of the progress, as 35% of operators think environmental advances are effective and call for more regulations for the sector. On the contrary, in North America, 45% of respondents believe current sustainability commitments are successful, while more than half (55%) disagree with this statement. Interestingly, the findings from the previous year revealed a tendency to deny climate change[71] and nearly a third of managers in data centres located in North America did not believe that human activity contributes to climate change or did not believe in climate change at all at the time of the survey.

In another report from Supermicro,[72] only 28% of industry decision-makers consider environmental issues when selecting data centre technology. The report explains further that the reasons behind such a small percentage are cost savings (29%), lack of knowledge and resources (27%), and environmental impacts being outside companies' main concerns (14%). Furthermore, the study shows that 58% of companies surveyed already had an environmental policy or were considering one at the time of the survey. Moreover, SMEs may lack the capacity to have such a policy due to a smaller budget to spend on a more comprehensive and meaningful strategy. Large companies are more likely to employ external experts or create departments dedicated to sustainability to create such a programme. An additional reason for not having one is that sustainability is a nice-to-have, and a voluntary addition to a company's picture.

11.3.4.2   Data Centre Industry Standards and Initiatives

The 2016 Paris Agreement on climate change was a wake-up call and laid the groundwork for new policies and regulations. Consequently, various initiatives concerning environmental impacts were initiated across governments, businesses and academia. In the data centre industry, however, evidence shows that a limited number of such proposals exist.

Presently, there are a few data centre sector regulations and certifications.[73] These mainly focus on the energy efficiency of the building or data centre components (*e.g.*, HVAC, servers, data storage), are voluntary or mandatory, and vary between countries or regions, which causes a lack of consensus in the industry. The internationally recognised credentials are, for instance:

- Eco-label Electronic Product Environmental Assessment Tool (EPEAT) for servers NSF/ANSI 426–2019.
- certification scheme by the Certified Energy Efficient Data Center Award (CEEDA)
- certification scheme by the Data Centre Alliance (DCA)
- building certification by the Leadership in Energy and Environmental Design (LEED) for building design and construction, and operations and maintenance, of data centres.

In the European Union, the following have been established:

- Regulation on eco-design requirements for servers and data storage products EU 2019/424
- Energy Efficiency Directive (under Fit for 55 legislative package)
- EU Code of Conduct on Data Centre Energy Efficiency.

EU 2019/424 regulation (also known as Lot 9) sets out the sustainable design practice requirements for data centre electronics, but mostly focuses on server design. The Energy Efficiency Directive by the European Commission, proposed as part of the Fit for 55 legislative package (including policies that aim to lower net GHG emissions by at least 55% by 2030), outlines energy-related sustainability reporting requirements for data centre operators. Furthermore, the directive incentivises data centres to explore waste heat reuse potentials by requiring a cost–benefit assessment of waste heat utilisation in data centres above the 100 kW capacity threshold. In February 2022, the forenamed directive was strengthened to ensure data collection transparency and quality. The EU Code of Conduct on Data Centre Energy Efficiency is a very comprehensive, voluntary guideline for various DCI stakeholders, outlining best practices and setting out the targets and requirements for energy consumption reduction in a "cost-effective manner without hampering the mission-critical function of data centres".[74] Furthermore, it is the only independent initiative that monitors energy consumption and has mobilised 290 participating data centres. As concluded by Avgerinou,[75] the EU Code of Conduct on Data Centre Energy Efficiency is an incredibly effective non-regulatory act.

In addition, a sustainability certification, Building Research Establishment Environmental Assessment Methods (BREEAM), for data centres functions explicitly in the UK.

To allow for monitoring of the progress of any of the above measures, there is a need for standard practices and metrics. Different bodies, official, commercial and non-profit, proposed an overwhelming number of metrics, making the scope difficult to navigate and causing inconsistencies in calculating practices. The primary triad of standardisation establishments in Europe are the European Committee for Standardization (CEN), the European Committee for Electrotechnical Standardization (CENELEC) and the European Telecommunications Standards Institute (ETSI). ISO and the International Electrotechnical Commission (IEC) are working jointly to develop international standards and guidelines. Previously, many industry-developed metrics were subsequently standardised and officially approved, for instance Power Usage Effectiveness (PUE), Water Usage Effectiveness (WUE) or Carbon Usage Effectiveness (CUE), which were standardised by Green Grid – a non-profit industry consortium. The recent effort to create industry-agreed global standards is the *EN 5060051 standard series*, also known as *ISO/IEC 30134: Information technology – Data Centres – Key Performance Indicators*.[76] This series focuses on the energy and sustainability of data centres and consists of the following metrics:

- Power Usage Effectiveness (PUE)
- Renewable Energy Factor (REF)
- IT Equipment Energy Efficiency for servers (ITEEsv)
- IT Equipment Utilisation for Servers (ITEUsv)
- Energy Reuse Factor (ERF)
- Cooling Efficiency Ratio (CER)
- Carbon Usage Effectiveness (CUE)
- Water Usage Effectiveness (WUE).

The industry itself is leading innovation in many practices and standards. Further significant initiatives in the sector concerning energy efficiency, resource efficiency and sustainability are the Climate Neutral Data Centre Pact, the Circular Economy for the Data Centre Industry (CEDaCI) project, the Open Compute Project (OCP), the Circular Electronics Partnership (CEP) and Free ICT Europe (FIE).

11.3.4.2.1   Climate Neutral Data Centre Pact

As data centre operators and trade associations committed to the European Green Deal with ambitions to become climate neutral by 2050, an initiative – the Climate Neutral Data Centre Pact[77] – was formed in 2018. The initiative is a set of conformed actions across areas such as energy efficiency and conservation, clean (renewable) energy procurement, water conservation, circular economy and governance. This self-regulatory initiative claims to be a one-step-ahead response to the upcoming 2024 EU regulations for the sector.

11.3.4.2.2   Circular Economy for the Data Centre Industry Project

Another initiative, the Circular Economy for the Data Centre Industry (CEDaCI) project,[78] pioneers building a circular economy for the sector in north-west Europe. CEDaCI proceedings aim to raise awareness and increase equipment reuse and remanufacture while reducing e-waste rich in precious virgin materials and CRMs, and securing a resilient and viable supply chain. Furthermore, the project advocates for improved eco-design guidelines and product life extension and educates SMEs from various data centre subsectors about circularity-friendly decision-making and conscious hardware consumption.

11.3.4.2.3   Open Compute Project

The OCP[79] is a collaborative, transnational project based in Prineville, Oregon, in the USA. Its mission is to advance the efficiency, sustainability and scalability of data centre hardware and infrastructure. The unique selling point of this initiative is an innovative approach to reducing the complexity of the designs and manufacturing and strongly promoting the open-source approach over ownership.

11.3.4.2.4   Circular Electronics Partnership

CEP[80] is a global partnership of leading tech, consumer goods and waste management organisations with a common mission to transform electronics and electrical equipment industries towards the circular economy. This initiative will set a sectoral roadmap, and design solutions and interventions to achieve circular economy principles by 2030.

11.3.4.2.5   Free ICT Europe (FIE)

FIE[81] is a not-for-profit foundation of independent bodies representing and supporting fair and open IT hardware and software markets. Foundation supports the ICT secondary market in Europe and advocates for standardised practices for the reuse and resale of equipment and software. FIE collaborates with decision-makers and businesses to actively shape EU legislation for sustainable hardware transformation.

11.3.4.3 Sustainability Reporting

Sustainability reporting aims to disseminate companies' non-financial information to increase transparency about business operations and showcase positive contributions to society. It can be voluntary or mandatory, depending on the nature of the business and region. As mentioned in Section 11.2.3, in the European Union, sustainability reporting is mandatory for specific companies and is outlined by the Non-Financial Reporting Directive (NFRD) and the Corporate Sustainability Reporting Directive (CSRD), whereas, in the UK, it is not required but encouraged. Nevertheless, more mandatory reporting requirements are on the way. New policies are most likely to incentivise companies to report data by imposing penalties or similar forms of reprimand. In the EU, firms need to disclose specific environmental and social information; however, there is no standard format. Companies can use European, international or national guidance, depending on their requirements. There are various initiatives including the Organisation for Economic Cooperation and Development (OECD) guidelines, United Nations Global Compact's Communication on Progress (COP), the International Integrated Reporting Council (IIRC) and leading guidelines from the Global Reporting Initiative (GRI) to name a few.

The GRI is an independent, international body with the most extensive and widely relevant benchmark for sustainability reporting. The GRI has also acknowledged the private sector's significance in achieving SDGs and collaborated with the UN to encourage governments and businesses to measure their environmental, social and economic contributions. As a result, the GRI, together with the UN Global Compact and the World Business Council for Sustainable Development, designed the SDG Compass[82] to guide on "how they can align their strategies as well as measure and manage their contribution to the realisation of the SDGs". It is a directory of relevant business tools that might help build SDG reporting following GRI standards.

Sustainability reporting practices across all sectors, including the data centre industry, require an agreed, consistent approach. Sector-specific guidance is required to prevent misinterpretation, resulting in ineffective and inaccurate disclosures. For instance, more clarity on the span of companies' environmental and societal responsibilities is needed, especially in light of scope 3 emission reporting, from which companies currently refrain, possibly because it accounts for the highest GHG emissions.

Sustainability reports vary between data centre providers; hyperscale cloud suppliers and large colocation companies tend to have a more meticulous approach and publish extensive environmental data, which they have collected over longer periods of time. To explore differences and similarities in disclosed metrics, several recent (mostly from 2021, due to the industry reporting period) sustainability reports, or equivalents available publicly on various companies' websites, were reviewed (see Table 11.1).[38,83–88]

For instance, the Environmental Report 2022[87] by Google, the 2021 Sustainability Report[88] by Meta and the 2021 Environmental Sustainability Report[38]

Table 11.1 The review of the chosen metrics disclosed in chosen sustainability reports.

	2021 Equinix Sustainability Report[83]	2021 Annual Sustainability Report Colt Group[84]	Environmental, Social and Governance Report 2021 by Digital Realty[85]	Sustainability Report 2020 by Virtus Data Centres[86]	Environmental Report 2022 by Google[87]	2021 Sustainability Report by Meta[88]	2021 Environmental Sustainability Report by Microsoft[38]
Energy consumption	✓	✓	✓		✓	✓	✓
Renewable energy consumption	✓	✓	✓		✓	✓	✓
Renewable energy [%]	✓	✓	✓	✓	✓	✓	✓
Power Usage Effectiveness (PUE)	✓				✓	✓	
Water Usage Effectiveness (WUE)						✓	
Scope 1	✓	✓	✓	✓	✓	✓	✓
Scope 2	✓	✓	✓	✓	✓	✓	✓
Scope 3		✓	✓	✓	✓	✓	✓

by Microsoft record detailed disaggregated data on scopes 1, 2 and 3 emissions; general energy data (including consumption, efficiency and renewable); and detailed water-related figures (covering consumption, withdrawal and discharge). In addition, the Environmental Report 2022 by Google and the 2021 Environmental Sustainability Report by Microsoft include carbon intensity and full waste-related statistics. Google and Meta reports follow GRI reporting practice standards, whereas the report from Microsoft follows the Sustainability Accounting Standards Board (SASB) guidelines. Sustainability Report 2020[86] by Virtus Data Centres includes less information and significantly fewer statistics than the aforementioned reports. All reviewed reports disclosed data on renewable energy mix ratios and covered scopes 1 and 2 of GHG emissions. Amongst those documents, reports by Equinix,[83] Google and Meta feature energy efficiency data by presenting the PUE metric. Lastly, the 2021 Sustainability Report by Meta stated the annual WUE of their facilities.

The reason for differences in the amount of data and level of detail is possibly due to business size. The report by Virtus Data Centres is less exhaustive, possibly because it is published by a much smaller business than other reports mentioned, which originate from companies that operate hundreds of facilities.

11.3.4.4   SDGs in Sustainability Reporting

Business plays a crucial role in sustainable development, and SDGs are setting the direction in which companies can aim their strategies and activities. Matters across climate change, resource consumption, production, biodiversity preservation, human health and well-being, and education are equally important to transnational governments and the private sector. In addition, SDGs can bring new business opportunities in sustainable innovation and transformation, building reliable and worthwhile supply chains or engaging and shaping regional and national environmental policies.

However, companies should refrain from using sustainability reporting as a marketing opportunity or competitive advantage. A common practice is to disclose only positive and attractive details while ignoring unfavourable information, consequently altering the perception of the company's performance, called greenwashing. Furthermore, "SDGs-washing" and even "blue-washing" have been recognised, by declaring positive contributions to SDGs while ignoring the negative impacts and using the UN logo to proclaim sustainability without genuine action, respectively.

There are numerous reports on SDGs in the private sector, for instance *SDG Compass*[82] (by the GRI, the World Business Council for Sustainable Development and the UN Global Compact), *SDG Industry Matrix*[89] and *Making the SDGs Relevant to Business*[90] (by PRé Sustainability and 2.-0 LCA consultants). All those white papers aim to translate the SDGs to business needs and opportunities, provide sectoral guidance and educate for a sustainability-orientated business journey. Moreover, the PRé Sustainability report encourages connecting SDGs to life-cycle assessment, and KPMG's *How to Report on SDGs*[91] proposes how to measure progress towards global goals. Moreover, SDGs opened a market for a myriad of tools and aids for building SDG-based business strategies, for instance *SDG Action Manager*,[92] *SDG Monitor*[93] and *SDG Impact Assessment Tool*.[94] Different tools require a mixed level of previous sustainability and SDG knowledge from the user and provide various options. Most often, these tools provide a platform to organise ideas for acting on SDGs. Although there is an abundance of private sector guidance and platforms available, resources specific to the data centre sector are missing. The literature search identified only one current study that discusses the sector's impact on the SDGs, by Hoosain *et al.*[95] Unfortunately, the arguments given by the authors lack correct references. Furthermore, this investigation into SDGs in the DCI seems oversimple and incomplete, leaving an unfulfilled research opportunity.

Studying the latest available sustainability reports and equivalents,[83–86,88,96,97] SDGs mentioned in the reports were identified and organised (see Table 11.2).

Most of the reviewed documents mentioned SDGs in their strategies, apart from the following: the Sustainability Report 2020 by Virtus Data Centres, the 2021 Environmental Sustainability Report by Microsoft (a special report on SDGs was issued separately) and the Environmental Report 2022 by Google (SDGs are not disclosed in the report, but Google takes part in the Business for 2030 Initiative). Although the approach and total number of SDGs mentioned varies, the 2021 Environmental Sustainability Report by Microsoft mentioned all 17 goals in its extensive document, which primarily focuses on four goals – SDGs 4 (Quality education), 8 (Decent work and economic growth), 13 (Climate action) and 16 (Peace, justice and strong institutions) – and showcases positive contributions to all other SDGs. Another report with comprehensive SDG coverage is the 2021 Annual Sustainability Report Colt Group, which records not only goals but also associated targets. The other documents include seven or fewer goals. The most commonly declared SDGs are as follows:

- SDG 7, Affordable and clean energy
- SDG 8, Decent work and economic growth
- SDG 9, Industry, innovation and infrastructure
- SDG 13, Climate change (environmental dimension)
- SDG 17, Partnership for the goals.

Only SDG 13 links to the environmental dimension, and only SDG 7 links to society. The remaining SDGs, 8, 9 and 17, refer to the economy.

Commonly, the declared SDGs tend to record positive actions and most often refer only to business operations and direct employees while overlooking trade-offs or indirect implications. The selected SDGs appear to frequently fit into ongoing business community initiatives, such as education programmes, donations to charities or similar projects, and it is difficult to ascertain whether the companies have plans to increase SDG coverage in the future.

Table 11.2   The review of the SDGs mentioned in chosen sustainability reports.

Sustainability, CSR, ESG or equivalent report	Goal 1	Goal 2	Goal 3	Goal 4	Goal 5	Goal 6	Goal 7	Goal 8	Goal 9	Goal 10	Goal 11	Goal 12	Goal 13	Goal 14	Goal 15	Goal 16	Goal 17	Total number of SDGs
2021 Equinix Sustainability Report[83]					✓		✓	✓	✓	✓			✓					6
2021 Annual Sustainability Report Colt Group[84]			✓	✓	✓		✓	✓	✓	✓	✓	✓	✓			✓	✓	12
Environmental, Social and Governance Report 2021 by Digital Realty[85]									✓		✓	✓	✓					4
Sustainability Report 2020 by Virtus Data Centres[86]																		0

[a]Business for 2030 initiative – Google[96]			✓			✓			✓							✓	4
2021 Sustainability Report by Meta[88]		✓	✓			✓	✓		✓			✓			✓	✓	7
[b]Microsoft and the United Nations Sustainable Development Goals[97]	✓	✓	✓	✓	✓	✓	✓	✓	✓	✓	✓	✓	✓	✓	✓	✓	17

[a] SDGs are not disclosed in the sustainability report, but Google takes part in the Business for 2030 Initiative.
[b] A special SDG report by Microsoft focuses on 4 goals, but states contribution to all 17.

11.4   The DCI Sector, SDGs and Opportunities for Further Research

The DCI sector has the potential to, directly and indirectly, contribute to the 2030 Agenda for Sustainable Development, which emphasises the digital economy and technology as integral enabling elements of the future world. Currently, a limited number of SDG goals and targets are mentioned in business sustainability/CSR/ESG reports, and documents frequently focus on SDGs that correspond to direct employees, GHG emissions and community service projects. Moreover, SDGs are often used to demonstrate positive contributions while overlooking associated trade-offs. The data centre industry has the opportunity to directly influence several of the SDGs (see Table 11.3).

Furthermore, it is likely that the DCI can have a further impact on the SDGs shown in Table 11.4.

Table 11.3   Examples of SDGs goals and targets that the DCI can directly influence. Reproduced from ref. 65 with permission from United Nations.

SDG	Target
Goal 7. Ensure access to affordable, reliable, sustainable, and modern energy for all	7.2 By 2030, increase substantially the share of renewable energy in the global energy mix
Goal 8. Promote sustained, inclusive and sustainable economic growth, full and productive employment, and decent work for all	
Goal 9. Build resilient infrastructure, promote inclusive and sustainable industrialisation and foster innovation	9.c Significantly increase access to information and communications technology and strive to provide universal and affordable access to the internet in the least developed countries by 2020
Goal 12. Ensure sustainable consumption and production patterns	12.2 By 2030, achieve the sustainable management and efficient use of natural resources
	12.5 By 2030, substantially reduce waste generation through prevention, reduction, recycling and reuse
	12.6 Encourage companies, especially large and transnational companies, to adopt sustainable practices and to integrate sustainability information into their reporting cycle
Goal 17. Strengthen the means of implementation and revitalise the Global Partnership for Sustainable Development	17.8 Fully operationalise the technology bank and science, technology, and innovation capacity-building mechanism for the least developed countries by 2017 and enhance the use of enabling technology, in particular information and communications technology

Table 11.4   Examples of SDGs that the DCI can indirectly influence. Reproduced from ref. 65 with permission from United Nations.

SDG
Goal 1. End poverty in all its forms everywhere
Goal 3. Ensure healthy lives and promote well-being for all at all ages
Goal 4. Ensure inclusive and equitable quality education and promote lifelong learning opportunities for all
Goal 5. Achieve gender equality and empower all women and girls
Goal 6. Ensure availability and sustainable management of water and sanitation for all
Goal 8. Promote sustained, inclusive and sustainable economic growth, full and productive employment and decent work for all
Goal 11. Make cities and human settlements inclusive, safe, resilient and sustainable
Goal 16. Promote peaceful and inclusive societies for sustainable development, provide access to justice for all and build effective, accountable and inclusive institutions at all levels

Nonetheless, due to the complexity and vastness of this unique sector, this subject requires further attention and investigation from researchers inside and outside the industry. In order to determine and translate relevant goals and targets for the DCI, there is a need for multidimensional research into all SDGs in the sector's value chain and analysis of the industry landscape for opportunities and barriers. Future research should address the following research questions:

- Which UN SDGs are relevant to the data centre industry?
  - How can the goals and targets in the context of the DCI be systematised?
  - What are the synergies and trade-offs?
- How can DCI companies monitor and report on their progress towards the Sustainable Development Goals?

This highlights the gaps in current practice and the need to consolidate relevant, cross-sectoral metrics and indicators, tools or frameworks, and to further explore challenges that stakeholders may face when reflecting on SDGs. Insights from the data centre business stakeholders would be beneficial to identify the current approach to addressing and monitoring SDGs, the knowledge and awareness of the 2030 Agenda framework and the prospects of advancing the SDG progress within the DCI. Such research would be of value to researchers and industry stakeholders working towards implementing the UN SDGs in this extraordinary industry and could inspire other sectors.

11.5   Conclusions

This chapter described examples of the favourable and damaging impacts aligned with the three pillars of sustainability – social, economic and environmental – outlined by the *triple bottom line* definition. The scope of the demonstrated implications covered fundamental hotspots in the data centre life cycle, from materials' extraction, through operations to electronic waste at the end of life, and sectoral and systemic challenges for the data centre sector.

The presence of data centre infrastructure is paramount to the digital economy. It transformed the traditional economy paradigms and enabled business innovation, automation and instantaneous access to services as diverse as healthcare, leisure and entertainment, and e-payments. During the recent coronavirus pandemic, the digital economy has allowed flexibility for working from home and delivering many traditional services digitally, significantly increasing the number of internet users and the volume of data traffic once more. There are countless examples of the economic and social benefits of the digital economy, but the environment remains a substantial trade-off. The data centre infrastructure is hidden away in windowless warehouses, invisible even if located at the heart of a city. The 24/7 operations consume considerable amounts of energy, similar to the demands of some regions or countries. Moreover, it is expected that there will be a further increase in energy usage in the future due to the ever-rising demand for digital services. Furthermore, the thousands or millions of servers, the other networking equipment, the extensive supporting infrastructure and the facility structure all include embodied carbon footprints.

It is essential to utilise the whole-system approach (making sense of the complexity and interlinkages) and life-cycle thinking (quantifying environmental impacts throughout a product or service lifespan) to fully understand the influence of the DCI on overall sustainability, equally positive and negative.

References

1. P. Thibodeau, Data centers decline as users turn to rented servers, Computerworld, 2017, https://www.computerworld.com/article/3188885/data-centers-decline-as-users-turn-to-rented-servers.html (accessed December 2022).
2. S. Kemp, *Digital 2022: Global Overview Report*, Global Digital Insights, DataReportal, 2022, https://datareportal.com/reports/digital-2022-global-overview-report (accessed December 2022).
3. E. Masanet, A. Shehabi, N. Lei, S. Smith and J. Koomey, *Science*, 2020, **367**, 984.
4. D. Andrews and B. Whitehead, in *Sustainable Innovation 2019: 22nd International Conference Road to 2030: Sustainability, Business Models, Innovation and Design,* Epsom, Surrey, 2019.
5. P. Taylor, Global number of data centers worldwide 2015–2021, Statista, 2022, https://www.statista.com/statistics/500458/worldwide-datacenter-and-it-sites/ (accessed December 2022).
6. E. Fryer, *Er, What IS a Data Centre?*, Intellect/TechUK, London, 2013.

7. P. Taylor, Data centers worldwide by country 2022, Statista, 2022, https:// www.statista.com/statistics/1228433/data-centers-worldwide-by-country/ (accessed December 2022).

8. Cloudscene, Cloudscene Data Center Market Overview – United Kingdom, Cloudscene, https://cloudscene.com/market/data-centers-in-united-kingdom/all (accessed December 2022).

9. The Green Grid, The Green Grid glossary, 2022, https://www.thegreengrid .org/en/resources/glossary (accessed 14 December 2022).

10. S. Moss, In search of the world's largest data center, DCD, 2022, https:// www.datacenterdynamics.com/en/analysis/in-search-of-the-worlds-largest-data-center/ (accessed December 2022).

11. K. Yasar, P. Loshin and B. Lutkevich, What is a data center?, TechTarget. com, 2022, https://www.techtarget.com/searchdatacenter/definition/ data-center (accessed December 2022).

12. C. Longbottom, *Retail colocation vs. wholesale data centers: How to choose*, TechTarget.com, 2022, https://www.techtarget.com/searchdatacenter/tip/ How-does-a-wholesale-data-center-differ-from-retail (accessed December 2022).

13. B. WeLOOP, *A Situational Analysis of a Circular Economy in the Data Centre Industry*, WeLOOP, Loos-En-Gohelle, 2020, https://www.weloop.org/ wp-content/uploads/2021/09/2020_04_16_CEDaCI_situation_analysis_ circular_economy_report_VF.pdf (accessed May 2023).

14. J. Elkington and I. H. Rowlands, *Alternatives Journal (Waterloo)*, 1999, **25**, 42.

15. E. Fryer, *So What Have Data Centres Ever Done For Us?* Intellect/TechUK, London, 2013.

16. United Nations Conference on Trade and Development, *Digital Economy Report 2019. Value Creation and Capture: Implications for Developing Countries*, New York, 2019.

17. TechUK, *The UK Data Centre Sector. The Most Important Industry You've Never Heard Of*, TechUK, London, 2020.

18. R. Ascierto, H. Dooley and N. Novak, *The People Challenge: Global Data Center Staffing Forecast 2021–2025*, Uptime Institute, New York, 2021, https://uptimeinstitute.com/global-data-center-staffing-forecast-2021-2025 (accessed June 2023).

19. G. Kamiya, *Data Centres and Data Transmission Networks*, IEA, Paris, 2022.

20. S. Flucker, R. Tozer and B. Whitehead, *Building Services Engineering Research and Technology*, 2018, **39**, 173.

21. A. Laurent and M. Dal Maso, *Environmental Sustainability of Data Centres: A Need for a Multi-impact and Life Cycle Approach*, Copenhagen Centre on Energy Efficiency, Copenhagen, 2020.

22. B. Whitehead, D. Andrews, A. Shah and G. Maidment, *Build Environ*, 2014, **82**, 151.

23. D. Andrews, E. J. Newton, N. Adibi, J. Chenadec and K. Bienge, *Sustainability*, 2021, **13**(11), 6319.

24. The Green Grid, *Data Centre Life Cycle Assessment Guidelines*, White Paper no 45, v2, The Green Grid, 2012, https://www.on365.co.uk/files/3714/3203/1600/ Data-Centre-Life-Cycle-Assessment-Guidelines.pdf (accessed May 2023).

25. D. J. Lettieri, *Expeditious Data Center Sustainability, Flow, and Temperature Modeling: Life-Cycle Exergy Consumption Combined with a Potential Flow Based, Rankine Vortex Superposed, Predictive Method*, University of California, Berkeley, 2012, https://escholarship.org/uc/item/9pf8k8wk (accessed May 2023).

26. A. J. Shah, Y. Chen and C. E. Bash, in *IEEE International Symposium on Sustainable Systems and Technology*, Boston, 2012.

27. B. Whitehead, D. Andrews, A. Shah and G. Maidment, *Build. Environ.*, 2015, **93**, 395.

28. R. Liu, P. Gailhofer, C.-O. Gensch, A. Köhler and F. Wolff, *Impacts of the Digital Transformation on the Environment and Sustainability*, Öko-Institut, Berlin, 2019.

29. J. G. Koomey, *Growth in Data Center Electricity Use 2005 to 2010*, Analytics Press, Oakland, 2011.

30. A. Shehabi, S. Smith, D. Sartor, R. Brown, M. Herrlin, J. Koomey, E. Masanet, N. Horner, I. Azevedo and W. Lintner, *United States Data Center Energy Usage Report*, LBNL-1005775, Ernest Orlando Lawrence Berkeley National Laboratory, Berkeley, 2016.

31. M. Avgerinou, P. Bertoldi and L. Castellazzi, *Energies*, 2017, **10**, 1470.

32. G. Cook, J. Lee, T. Tsai, A. Kong, J. Deans, B. Johnson and E. Jardim, *Clicking Clean: Who is Winning a Race to Build a Green Internet?* Greenpeace, Beijing, 2017.

33. A. S. G. Andrae and T. Edler, *Challenges*, 2015, **6**, 117.

34. Eurostat, Glossary: Greenhouse gas (GHG), Eurostat, Luxembourg, 2016, https://ec.europa.eu/eurostat/statistics-explained/index.php?title=Glossary:Greenhouse_gas_(GHG) (accessed December 2022).

35. The Climate Group on behalf of the Global eSustainability Initiative (GeSI). *SMART 2020: Enabling the Low Carbon Economy in the Information Age*, 2008, https://gesi.org/research/download/7 (accessed June 2023).

36. World Business Council for Sustainable Development (WBCSD) and World Resources Institute (WRI), *GHG Protocol Initiative: A Corporate Accounting and Reporting Standard*, Geneva and Washington, DC, 2004, https://ghgprotocol.org/corporate-standard (accessed May 2023).

37. Global e-Sustainability Initiative and Carbon Trust, *ICT Sector Guidance built on the GHG Protocol Product Life Cycle Accounting and Reporting Standard Chapter 1: Introduction and General Principles*, The Carbon Trust, London, 2017.

38. Microsoft, *2021 Environmental Sustainability Report. From Pledges to Progress*, Microsoft, Redmond, 2021.

39. J. Davis, D. Bizo, A. Lawrence, O. Rogers and M. Smolaks, *Uptime Institute Global Data Center Survey 2022*, Uptime Institute, New York, 2022, https://uptimeinstitute.com/resources/research-and-reports/uptime-institute-global-data-center-survey-results-2022 (accessed May 2023).

40. G. Kamiya and O. Kvarnström, Data centres and energy – from global headlines to local headaches? – Analysis, IEA, Paris, 2019, https://www.iea.org/commentaries/data-centres-and-energy-from-global-headlines-to-local-headaches (accessed December 2022).

41. J. Chen, Renewable Energy Certificate (REC): Definition, types, example, Investopedia, 2022, https://www.investopedia.com/terms/r/rec.asp (accessed December 2022).

42. D. Mytton, *Renewable Energy for Data Centers: Renewable Energy Certificates, Power Purchase Agreements and Beyond*, Uptime Institute, New York, 2021.

43. S. Moss, Luxembourg's economy minister calls on Google to make decision about delayed $1bn data center, DCD, https://www.datacenterdynamics.com/en/news/luxembourgs-economy-minister-calls-on-google-to-make-decision-about-delayed-1bn-data-center/ (accessed December 2022).

44. C. Hancock, UK utility probes data centers' water usage during UK drought, Bloomberg UK, https://www.bloomberg.com/news/articles/2022-08-23/uk-utility-probes-data-centers-water-usage-during-drought (accessed December 2022).

45. D. Mytton, *npj Clean Water*, 2021, **4**(1), 1.

46. B. Ristic, K. Madani and Z. Makuch, *Sustainability*, 2015, **7**, 11260.

47. Google, *Google Data Centers: Hamina, Finland*, Google, 2022, https://www.google.com/about/datacenters/locations/hamina/ (accessed December 2022).

48. B. Cutler, S. Fowers, Kramer Jeffrey and E. Peterson, *Want an Energy-Efficient Data Center? Build It Underwater,* IEEE Spectrum, New York, 2017, https://spectrum.ieee.org/want-an-energyefficient-data-center-build-it-underwater (accessed December 2022).

49. J. G. Speight, *Natural Water Remediation*, Butterworth-Heinemann, Elsevier, Oxford, 2020.

50. E. Labouze, V. Monier and J.-B. Puyou, *Study on External Environmental Effects Related to the Life Cycle of Products and Services – Final Report*, BIO Intelligence Service, Paris, 2003.

51. K. Doyle, What are Critical Raw Materials?, Techbuyer, 2020, https://www.techbuyer.com/uk/blog/what-are-critical-raw-materials-crm (accessed December 2022).

52. C. Banza Lubaba Nkulu, L. Casas, V. Haufroid, T. de Putter, N. D. Saenen, T. Kayembe-Kitenge, P. Musa Obadia, D. Kyanika Wa Mukoma, J. M. Lunda Ilunga, T. S. Nawrot, O. Luboya Numbi, E. Smolders and B. Nemery, *Nat. Sustain.*, 2018, **1**, 495.

53. D. Swinhoe, Data centre dynamics – Re-use, refurb, recycle: Circular economy thinking and data center IT assets, DCD, https://www.datacenterdynamics.com/en/analysis/re-use-refurb-recycle-circular-economy-thinking-and-data-center-it-assets/ (accessed January 2023).

54. Google, Once is never enough, Google, 2018, https://sustainability.google/progress/projects/circular-economy/ (accessed January 2023).

55. S. Vallance, H. C. Perkins and J. E. Dixon, *Geoforum*, 2011, **42**, 342.

56. UN Environment Programme (UNEP), *Guidelines for Social Life Cycle Assessment of Products and Organisations 2020*, UNEP, Paris, 2020.

57. World Economic Forum, *A New Circular Vision for Electronics Time for a Global Reboot*, World Economic Forum, Geneva, 2019.

58. V. Odumuyiwa, A. Adelopo and E. A. Nubi, in *Circular Economy in Africa: Examples and Opportunities*, Ellen MacArthur Foundation, Cowes, 2021.
59. UN Environment Programme (UNEP), *Global Mercury Assessment 2018*, UNEP, Geneva, 2018.
60. Data Center Map, *Colocation Africa – Data Centers*, https://www.datacentermap.com/africa/ (accessed December 2022).
61. T. Dyllick and K. Muff, *Organization & Environment*, 2015, **29**(2), 156.
62. S. C. Berning, *European Journal of Sustainable Development*, 2019, **8**, 194.
63. R. Scheyvens, G. Banks and E. Hughes, *Sustainable Development*, 2016, **24**, 371.
64. A. Azapagic and S. Perdan, *Process Safety and Environmental Protection*, 2000, **78**, 243.
65. United Nations, *Transforming Our World: The 2030 Agenda for Sustainable Development*, United Nations, New York, 2015.
66. J. D. Sachs, *The Lancet*, 2012, **379**, 2206.
67. T. Hák, S. Janousková and B. Moldan, *Ecol. Indic.*, 2020, **60**, 565.
68. B. Adams and G. Luchsinger, Are FFD3 and Post 2015 striking the right public–private balance?, Civil Society Financing for Development Group, 2015, https://csoforffd.org/third-ffd-conference/cso-contributions-to-ffd3/other-cso-inputs-to-ffd/csos-papers-other-docs/429-2/ (accessed May 2023).
69. The Earth Institute and Ericsson, *How Information and Communications Technology can Accelerate Action on the Sustainable Development Goals*, ICT & SDGs Final Report, The Earth Institute, Columbia University, New York, 2017.
70. Huawei, *Accelerating SDGs through ICT Digital Solutions Next Steps Recommendations for Accelerating Digitally Enabled Sustainable Development*, Huawei, Shenzhen, 2018.
71. R. Ascierto and A. Lawrence, *Global Data Center Survey*, Uptime Institute, New York, 2020.
72. Super Micro Computer Inc., *Data Center s & The Environment: The State of Global Environmental Sustainability in Data Center Design*, Supermicro Computers, San Jose, 2019.
73. Fiona Brocklehurst, *International Review of Energy Efficiency in Data Centres Acknowledgements*, Ballarat Consulting, 2021.
74. European Commission, The European code of conduct for energy efficiency in data centres, Brussels, 2008, https://joint-research-centre.ec.europa.eu/energy-efficiency/energy-efficiency-products/code-conduct-ict/code-conduct-energy-efficiency-data-centres_en (accessed December 2022).
75. M. Avgerinou, P. Bertoldi and L. Castellazzi, *Energies*, 2017, **10**, 1470.
76. N. Cappella, The EN 50600: How to meet the European standard for data centres, Techerati, 2021, https://www.techerati.com/features-hub/opinions/the-en-50600-how-to-meet-the-european-standard-for-data-centres/ (accessed December 2022).

77. Climate Neutral Data Centre, *Climate Neutral Data Centre Pact*, 2021, https://www.climateneutraldatacentre.net/ (accessed December 2022).

78. CEDaCI, *Circular Economy for Data Centre Industry project*, CEDaCI.org, 2019, https://www.cedaci.org/ (accessed December 2022).

79. Open Compute Project, https://www.opencompute.org/ (accessed December 2022).

80. Circular Electronics Partnership, *CEP2030*, Geneva, 2022, https://cep2030.org/ (accessed January 2023).

81. Free ICT Europe Foundation, *Free ICT Europe – Freedom To Support, Repair & Resell*, Gouda, Netherlands, https://www.freeict.eu/ (accessed January 2023).

82. SDG Compass, *The Guide for Business Action on the SDGs*, SDG Compass, 2015, https://sdgcompass.org/ (accessed December 2022).

83. Equinix, *Sustainability Report FY2021*, Equinix, Redwood City, 2022.

84. Colt Group, *2021 Annual Sustainability Report Colt Group*, Colt Data Centre Services, London, 2022.

85. Digital Realty, *Environmental, Social and Governance Report 2021*, Digital Realty, London, 2022.

86. Virtus Data Centres, *Sustainability Report 2020*, Virtus, London, 2021.

87. Google, *Google Environmental Report 2022*, Google, Mountain View, 2022.

88. Meta, *2021 Sustainability Report*, Meta, Menlo Park, 2022.

89. United Nations Global Compact (UNGC) and KPMG, *SDG Industry Matrix*, UNGC, New York, 2015.

90. B. Weidema, M. Goedkoop and E. Mieras, *Making the SDGs Relevant to Business: Existing Knowledge on the Linking of SDGs to Business Needs and the Role of LCA in Meeting the Needs and Filling the Gaps*, PRé Sustainability & 2.-0 LCA consultants, Amersfoort and Aalborg, 2018.

91. KPMG, *How to Report on the SDGs. What Good Looks Like and Why it Matters*, KPMG, Amstelveen, 2018.

92. United Nations Global Compact (UNGC), *SDG Action Manager*, UNGC, New York, 2020, https://app.bimpactassessment.net/get-started/partner/ungc (accessed December 2022).

93. SDG Monitor, Sustainability Measurement Tool, https://www.sdgmonitor.co/ (accessed December 2022).

94. Gothenburg Centre for Sustainable Development, SDG Impact Assessment Tool, https://sdgimpactassessmenttool.org/en-gb (accessed December 2022).

95. M. S. Hoosain, B. S. Paul, S. Kass and S. Ramakrishna, *Circular Economy and Sustainability*, 2023, **3**, 173–197.

96. Google, *Business for 2030*, Google Mountain View, http://www.businessfor2030.org/google (accessed December 2022).

97. Microsoft, *Microsoft and the United Nations Sustainable Development Goals 2021 Report*, Microsoft, Redmond, 2019.

Subject Index

www.ingramcontent.com/pod-product-compliance
Lightning Source LLC
Chambersburg PA
CBHW041836190326
41458CB00027BA/6468